Modern
Elementary
Statistics

JOHN E. FREUND

Arizona State University

PRENTICE-HALL, INC., Englewood Cliffs, New Jersey 07632

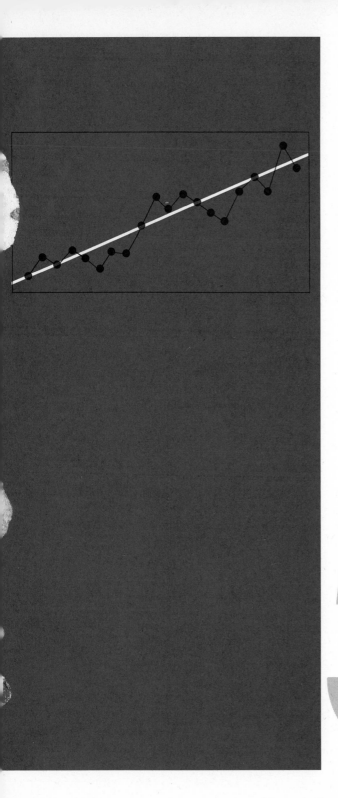

Modern
Elementary
Statistics

FIFTH EDITION

Library of Congress Cataloging in Publication Data

FREUND, JOHN E.
    Modern elementary statistics.

    Includes bibliographical references and index.
    1. Statistics.   I. Title.
HA29.F685 1979   519.5   78-18178
ISBN   0-13-593491-5

Modern Elementary Statistics   Fifth Edition *by John E. Freund*

10 9 8 7 6 5 4 3

Editorial/production supervision by Nancy Milnamow
Interior design and cover design by Walter A. Behnke
Manufacturing buyer: Phil Galea

PRENTICE-HALL INTERNATIONAL, INC., *London*
PRENTICE-HALL OF AUSTRALIA PTY. LIMITED, *Sydney*
PRENTICE-HALL OF CANADA, LTD., *Toronto*
PRENTICE-HALL OF INDIA PRIVATE LIMITED, *New Delhi*
PRENTICE-HALL OF JAPAN, INC., *Tokyo*
PRENTICE-HALL OF SOUTHEAST ASIA PTE. LTD., *Singapore*
WHITEHALL BOOKS LIMITED, *Wellington, New Zealand*

As is explained in the Preface, all sections marked ★
may be regarded as optional

Contents

Preface

1

2

3

4

**Possibilities
and
Probabilities**

Page 69

Feb. 1

5

**Some Rules
of
Probability**

Page 93

Feb. 1

6

Feb. 1

**Expectations
and
Decisions**

Page 131

7

Probability Distributions

Page 152

Feb. 8

Jan. 11

8

The Normal Distribution

Page 186   *Feb.8*

9

Sampling and Sampling Distributions

Page 210

Jan. 11

Further Exercises for Chapters 7, 8, and 9, 239

10

Inferences About Means

Page 246

11

Inferences About Standard Deviations

Page 285

12

Inferences About Proportions

Page 296

Jan. 25

also march 22

13

Nonparametric Methods

March 1

Page 332

March 8

14

Regression

Page 359

Jan. 25

March 15

X

15

Correlation

Page 391

Jan.25 *March 15* *Jan.25*

16

Analysis of Variance

Page 412

March 8

xi

Preface

The foremost objective of this edition, like that of the previous editions, is to acquaint beginning students in the various sciences with the fundamentals of modern statistics. Although the basic organization has remained the same, there are substantial changes; most apparent, perhaps, are those in the format, which should make the book easier to read and easier to teach.

Most of the examples and exercises are new, and, as in previous editions, they are distributed over a wide variety of applications. New also are the five sets of further exercises, or review exercises, strategically spaced throughout the book. Altogether, there are over 750 numbered exercises. Most of them are drawn from actual problems, but many have been modified and scaled down somewhat to simplify the computational load.

There are numerous changes in the text, based largely on the suggestions of colleagues and students, who generously took the time to share their experiences and thoughts. For instance, the discussion of descriptive statistics has been rearranged so that the material on grouped data can easily be omitted, methods of survey sampling are taken up in the chapter on sampling and sampling distributions, there is more material on simulation (looked upon as a teaching aid and also as a tool in research), the section on probability and odds has been rewritten, the continuity correction is used more extensively in tests based on count data and also in some of the nonparametric tests, the chapter on analysis of variance has been reworked completely to include an introduction to some of the concepts of experimental design, and there is an optional section on the nature of measurements (or scaling) in Chapter 1. Also, some topics (for instance, the weighted mean and the median test) have been moved from the text proper to the exercises, and others (for instance, the exercises dealing with index numbers) have been deleted altogether.

As in the previous editions, the author shall resist the temptation to tell anyone specifically what chapters, sections, or subjects to study or teach, but it should be observed that there are topics which may be omitted without loss of continuity. Upon the urging of many colleagues, and our publishers in particular, some sections are thus marked *, meaning that they are optional. This posed all sorts of difficulties, for what one person may consider optional another person may well consider essential. For instance, many instructors will not cover the material on the description of grouped data, yet a person who has to work with published government data may well consider it essential. Although quite a few colleagues indicated that they do not cover the material on subjective probability, to mark it optional would be an affront to those who hold the subjectivist point of view. However, risking their wrath, the sections dealing with Bayes' theorem and Bayesian estimation are marked optional.

As indicated in previous editions, the study of statistics may not only be directed toward applications in various specialized fields of inquiry, but it may also be presented at various levels of mathematical difficulty and in almost any balance between theory and application. As it is more important, in the author's opinion, to understand the meaning and implications of basic ideas than it is to memorize an impressive list of formulas, some of the details that are sometimes included in introductory courses in statistics have been sacrificed. This may be unfortunate in some respects, but it should prevent the reader from getting lost in an excessive amount of detail which could easily obscure the more important issues. It is hoped that this will avoid some of the unfortunate consequences which often result from the indiscriminate application of so-called standard techniques without a thorough understanding of the basic ideas that are involved.

It cannot be denied that a limited amount of mathematics is a prerequisite for any course in statistics, and that a thorough study of the theoretical principles of statistics would require a knowledge of mathematical subjects taught ordinarily only on the graduate level. Since this book is designed for students with relatively little background in mathematics, the aims, and, therefore, also the prerequisites needed here are considerably more modest. Actually, the mathematical background needed for this study of statistics is amply covered in a course of college algebra; in fact, even a good knowledge of high school algebra provides a sufficient foundation.

The author would like to express his appreciation to the many colleagues and students whose helpful suggestions and criticisms contributed greatly to previous editions of this text as well as to this fifth edition. In particular, the author would like to thank his nephew Ray and his son Doug for reading various drafts of the manuscript, checking the calculations in the text, and working on the answers to the exercises, and his son John for helping with the proofreading. The author would also like to express his appreciation to the staff of Prentice-Hall, Inc., particularly to Harry Gaines, for their courteous cooperation in the preparation of the manuscript and in the production of this book. Thanks also go to the reviewers of this manuscript: Professor Arnold Levine, Tulane University, Professor Constance Elson, Ithaca College, Professor Lowell Doerder, Black Hawk College, Professor Larry Roi, University of Michigan, Professor Maita Levine, University of Cincinnati, and Professor H. Z. Solomon, Los Angeles City College.

Finally, the author is indebted to Professor E. S. Pearson and the *Biometrika* trustees for permission to reproduce parts of Tables 8, 12, 18, and 41 from their *Biometrika Tables for Statisticians*; and to the Rand Corporation for permission to reproduce the sample pages of random normal numbers shown in Table X.

JOHN E. FREUND

Scottsdale, Arizona

XV

1

Everything dealing even remotely with the collection, processing, analysis, and interpretation of numerical data belongs to the domain of statistics. This includes such diversified tasks as calculating a baseball player's batting average, collecting and presenting data on marriages and divorces, evaluating the merits of commercial products, or studying the vibrations of airplanes or guided missiles. The word "statistics" itself is used in several ways. It denotes a collection of data such as those found in the financial pages of newspapers or, say, the *Statistical Abstract of the United States*, published annually by the Department of Commerce. A second meaning of "statistics" is that of the totality of methods employed in the collection, processing, analyzing, or inter-preting of any kind of data; in this sense, statistics is a branch of applied mathematics, and it is this field of mathematics which is the subject matter of this book.

Introduction

1.1

The Study of Statistics

There are essentially two reasons why the scope of statistics and the need to study statistics have grown enormously in the last few decades. One reason is that the amount of data that is collected, processed, and disseminated to the public for one reason or another has increased almost beyond comprehension. To act as watchdogs, more and more persons with some knowledge of statistics are needed to take an active part in the collection of the data, in the analysis of the data, and, what is equally important, in all of the preliminary planning. Without the latter, it is frightening to think of all the things that can go wrong in the compilation of statistical data. As we shall see in Chapter 9, the results of costly surveys can be useless if questions are ambiguous or asked in the wrong way, if they are asked of the wrong persons, in the wrong place, or at the wrong time.

The second, and even more important, reason why the scope of statistics and the need to study statistics have grown so tremendously in recent years is the increasingly quantitative approach employed in all the sciences, as well as in business and many other activities which directly affect our lives. This includes the use of mathematical techniques in the evaluation of anti-pollution controls, in inventory planning, in the analysis of traffic patterns, in the study of the effects of various kinds of medications, in the evaluation of teaching techniques, in the analysis of competitive behavior of business-men and governments, in the study of diet and longevity, and so forth.

Numerous textbooks have been written on business statistics, educational statistics, medical statistics, psychological statistics, . . . , and even on statistics for historians. Although problems arising in these various disciplines will sometimes require special statistical techniques, none of the basic methods discussed in this text is restricted to any particular field of application. To illustrate this point, consider Exercise 12 on page 351, made up by the author.

12 In a random sample of 200 retired persons, 137 stated that they prefer living in an apartment to living in a one-family home. At the 0.05 level of significance, does this refute the claim that at most 60 percent of all retired persons prefer living in an apartment to living in a one-family home?

Except for the reference to the level of significance, the question asked here should be clear, and it should also be clear that the answer would be of interest mainly to social scientists or to persons in the construction industry. However, if we wanted to cater to the special interests of students of biology, engineering, education, or ecology, we might rephrase the exercise as follows:

12 In a random sample of 200 citrus trees exposed to a 20° frost, 137 showed some damage to their fruit. At the 0.05 level of significance, does this refute the claim that at most 60 percent of all citrus trees exposed to a 20° frost will show some damage to their fruit?

12 In a random sample of 200 transistors made by a given manufacturer, 137 passed an accelerated performance test. At the 0.05 level of significance, does this refute the claim that at most 60 percent of all the transistors made by the manufacturer will pass the test?

12 In a random sample of 200 high school seniors in a large city, 137 said that they will go on to college. At the 0.05 level of significance, does this refute the claim that at most 60 percent of all the high school seniors in this city will go on to college?

12 In a random sample of 200 cars tested for the emission of pollutants, 137 failed to meet a state's legal standards. At the 0.05 level of significance, does this refute the claim that at most 60 percent of all cars tested in this state will fail to meet its legal emission standards?

So far as the work in this book is concerned, the statistical treatment of all these versions of Exercise 12 is the same, and with some imagination the reader should be able to rephrase it for almost any field of specialization. We could present, and so designate, special problems for readers with special interests, but this would defeat our main goal of impressing upon the reader the importance of statistics in all of science, business, and everyday life. To attain this goal, the examples and the exercises in this text cover a wide spectrum of interests.

1.2

Descriptive Statistics and Statistical Inference

The origin of modern statistics may be traced to two areas of interest which, on the surface, have very little in common: games of chance and what we now call political science. Mid-eighteenth-century studies in probability (motivated largely by interest in games of chance) led to the mathematical treatment of errors of measurement and the theory which now forms the foundation of statistics. In the same century, interest in the numerical description of political units (cities, provinces, counties, etc.) led to the development of methods which nowadays come under the heading of **descriptive statistics.** This includes any kind of data processing which is designed to summarize, or describe, important features of the data without going any further; that is, without attempting to infer anything that goes beyond the data themselves.

EXAMPLE Tests performed in 1976 on eight small imported cars showed that they were able to accelerate from 0 to 60 mph in 18.7, 19.2, 16.2, 19.0, 13.5, 12.3, 17.6, and 13.9 seconds. If we report the fact that half of these cars were able to accelerate from 0 to 60 mph in less than 17.0 seconds, while a fourth of them took 19.0 seconds or more, our work belongs to the domain of descriptive statistics. This would also be the case if we claimed that these eight cars averaged

$$\frac{18.7 + 19.2 + 16.2 + 19.0 + 13.5 + 12.3 + 17.6 + 13.9}{8}$$

$$= 16.3 \text{ seconds}$$

On the other hand, we would do more than merely describe, if we concluded that half of all small cars imported that year were able to accelerate from 0 to 60 mph in less than 17.0 seconds, or if we applied our results to comparable cars imported, say, in 1978.

Although descriptive statistics is an important branch of statistics and it continues to be widely used, statistical information usually arises from samples (from observations made on only part of a large set of items), and this means that its analysis will require generalizations which go beyond the data. As a result, the most important feature of the recent growth of statistics has been a shift in emphasis from methods which merely describe to methods which serve to make generalizations; that is, a shift in emphasis from descriptive statistics to the methods of **statistical inference.**

EXAMPLE Such methods are required, for instance, to predict the operating life span of a hand-held calculator (on the basis of the performance of several such calculators); to estimate the 1985 assessed value of all property in Pima County, Arizona (on the basis of business trends, population projections, and so forth); to compare the effectiveness of two reducing diets (on the basis of the weight losses of persons who have been on the diets); to determine the most effective dose of a new medication (on the basis of tests performed with volunteer patients from selected hospitals); or to predict the flow of traffic on a freeway which has not yet been built (on the basis of past traffic counts on alternative routes).

In each of the situations described in this example, there are uncertainties because there is only partial, incomplete, or indirect information, and it is the job of statistics to judge the merits of the various alternatives and, perhaps, suggest a "most profitable" choice, a "most promising" prediction, or a "most reasonable" course of action.

1.3

The Nature of Statistical Data ★[†]

Statistical data are the raw material of any statistical investigation, and they arise whenever measurements are made or observations are classified. They may be weights of animals, measurements of personality traits, or earthquake intensities, and they may be simple "Yes or No" answers or descriptions of persons' marital status as single, married, widowed, or divorced. Since we said on page 1 that statistics deals with numerical data, this requires some explanation, because "Yes or No" answers and descriptions of marital status would hardly seem to qualify as being numerical. Observe, however, that we can record "Yes or No" answers to a question as 0 and 1 (or as 1 and 2, or perhaps as 29 and 30 if we are referring to the 15th "Yes or No" question of a questionnaire), and that we can record a person's marital status as 1, 2, 3, or 4, depending on whether the person is single, married, widowed, or divorced. In this artificial, or nomial way, categorical (qualitative, or descriptive) data can be made into numerical data, and if we thus code the various categories, we refer to the numbers we record as **nominal data.**

Nominal data are numerical in name only, because they do not share any of the properties of the numbers we deal with in ordinary arithmetic. For instance, if we record marital status as 1, 2, 3, or 4 as suggested above, we cannot write $3 > 1$ or $2 < 4$, and we cannot write $2 - 1 = 4 - 3$, $1 + 3 = 4$, or $4 \div 2 = 2$. It is important, therefore, always to check whether mathematical calculations performed in a statistical analysis are really legitimate.

EXAMPLE   In mineralogy, the hardness of solids is sometimes determined by observing "what scratches what." If one mineral can scratch another, it receives a higher hardness number, and on Mohs' scale the numbers from 1 to 10 are assigned, respectively, to talc, gypsum, calcite, fluorite, apatite, feldspar, quartz, topaz, sapphire, and diamond. With these numbers we can write $6 > 3$, for example, or $7 < 9$, since feldspar is harder than calcite and quartz is softer than sapphire. On the other hand, we cannot write $10 - 9 = 2 - 1$, for example, because the difference in hardness between diamond and sapphire is actually much greater than that between gypsum and talc. Also, it would be meaningless to say that topaz is twice as hard

---

[†] As is explained in the Preface, all sections marked ★ may be regarded as optional. Although the material in this section is meant to serve as a general warning against the indiscriminant mathematical manipulation of statistical data, it is most relevant to students of the behavioral and social sciences, where artificial scales serve to measure such things as neurotic tendencies, happiness, or conformity to social standards.

as fluorite simply because their respective hardness numbers on Mohs' scale are 8 and 4.

EXAMPLE   Suppose we are given the following temperature readings (in degrees Fahrenheit): 63°, 68°, 91°, 107°, 126°, and 131°. In this case we can write 107° > 68°, for example, or 91° < 131°, which simply means that 107° is warmer than 68° and that 91° is cooler than 131°. Also, we can write 68° − 63° = 131° − 126°, for example, since equal temperature differences are equal in the sense that the same amount of heat is required to raise the temperature of an object from 63° to 68° or from 126° to 131°. On the other hand, it would not mean much if we said that 126° is twice as hot as 63°, even though 126 ÷ 63 = 2. To show why, we have only to change to the centigrade scale, where the first temperature becomes $\frac{5}{9}(126 - 32) = 52°$, the second temperature becomes $\frac{5}{9}(63 - 32) = 17°$, and the first figure is now more than three times the second. This difficulty arises from the fact that the Fahrenheit and centigrade scales both have artificial origins (zeros); that is, the number 0 of neither scale is indicative of the absence of whatever quantity we are trying to measure. See, however, part (e) of Exercise 5 on page 8.

EXAMPLE   Data on which we can perform all the customary operations of arithmetic are not difficult to find. They include all the usual measurements (or determinations) of length, height, money amounts, weight, volume, area, pressure, time (though not calendar time), sound intensity, density, brightness, velocity, and so forth.

The examples we have given here illustrate three general types of data. If we cannot do anything except set up inequalities (as was the case in our first example), we refer to the data as **ordinal data.** In mathematics, the symbol > means "greater than," but in connection with ordinal data it may be used to designate "happier than," "preferred to," "more difficult than," "tastier than," and so forth. If we can also form differences (as in the second example), we refer to the data as **interval data**; and if we can also form quotients (as is the case, for example, when we measure height or weight), we refer to the data as **ratio data.** The distinction we have made here between nominal, ordinal, interval, and ratio data is important, for as we shall see, the nature of a set of data may suggest the use of particular statistical techniques.

Questions concerning scales of measurement, or types of measurements, are of general interest, of course, but they are most pertinent to students of the behavioral and social sciences. Clearly, measuring or quantitatively describing such things as social conformity, intelligence, or marital adjustment is much less obvious (and requires much closer attention) than measuring physical weight, biological age, or a person's financial assets.

1.4

Figures Don't Lie, But . . .

The amount of statistical information that is disseminated to the public for one reason or another is phenomenal, and what part of it is "good" statistics and what part is "bad" statistics is anybody's guess. Certainly, all of it cannot be accepted uncritically. Sometimes entirely erroneous conclusions are based on sound data, as is illustrated by the following:

EXAMPLE   A city once claimed to be the "nation's healthiest city," since its death rate was the lowest in the country. Even if we go along with their definition that "healthy" means "not dead," there is another factor that was not taken into account. Since the city had no hospital, its citizens had to be hospitalized elsewhere, and their deaths were recorded in the cities in which death actually occurred. The following are two other *non sequiturs* based on otherwise sound statistical data: "Statistics show that there were fewer airplane accidents in 1920 than in 1975; hence, flying was safer in 1920 than in 1975." "Since there are more automobile accidents in the daytime than there are at night, it is safer to drive at night."

Sometimes identical data are made the basis for directly opposite conclusions, as in collective-bargaining disputes when the same data are used by one side to show that employees are getting rich and by the other side to show that they are on the verge of starvation.

In view of examples like these, it is understandable that some persons are inclined to feel that figures can be made to show pretty much what one wants them to show. This may be uncomfortably close to the truth, but it would not be, if enough persons had some knowledge of statistics. Hopefully, the reader will learn how to distinguish between "good statistics" and "bad statistics," between statistical methods properly applied and statistical methods shamefully misapplied, and between statistical information correctly analyzed and interpreted and statistical information intentionally or unintentionally perverted.

EXERCISES   1 Rephrase the exercise referred to on page 2 so that it would be of special interest to
  (a) a cosmetics salesman;
  (b) a musician;
  (c) an archaeologist.

2 In five chemistry tests, a student received grades of 47, 71, 75, 79, and 88. Which of the following conclusions can be obtained from these figures by purely descriptive methods and which require generalizations? Explain your answers.
  (a) Only one of the grades exceeds 80.
  (b) Probably, the student did so poorly on the first test because he was sick.
  (c) The student's grades increased from each test to the next.
  (d) The student must have studied harder for each successive test.
  (e) Two of the grades are below 75 and two are above 75.
  (f) The difference between the highest and lowest grades is 41.

3 On three consecutive days, a traffic policeman issued 9, 14, and 10 speeding tickets, and 5, 10, and 12 tickets for going through red lights. Which of the following conclusions can be obtained from these data by purely descriptive methods and which require generalizations? Explain your answers.
  (a) Altogether on these three days, the policeman issued more speeding tickets than tickets for going through red lights.
  (b) On two of the three days, the policeman issued more speeding tickets than tickets for going through red lights.
  (c) The policeman issued the smallest number of tickets on the first day because he was new on the job.
  (d) This policeman will seldom give more than 15 speeding tickets on any one day.
  (e) For the three days, the number of tickets which the policeman issued for going through red lights increased from day to day.

4 Will we get nominal data or ordinal data if
  (a) mechanics have to say whether changing the spark plugs on a new model car is very difficult, difficult, no worse or better than most other cars, easy, or very easy;
  (b) the religion of persons attempting suicide is coded 1, 2, 3, 4, or 5, representing Protestant, Catholic, Jewish, other, and none;
  (c) consumers must say whether they prefer brand B to brand A, like them equally, prefer brand A to brand B, or have no opinion?

5 Are the following nominal, ordinal, interval, or ratio data? Explain your answers.
  (a) Social Security numbers.
  (b) The number of passengers on buses from Los Angeles to San Diego.
  (c) Vocational interest scores consisting of the total number of "Yes" answers given to a set of questions, if it can be assumed that each "Yes" answer represents the same increment of vocational interest.
  (d) Military ranks.
  (e) Temperatures measured on the Kelvin scale.

6 IQ scores are sometimes looked upon as interval data. What assumption would this entail about the differences in intelligence of three persons with IQ's of 95, 105, and 135? Is this assumption reasonable?

7 On page 6 we indicated that data pertaining to calendar time (for instance, the years in which Army beat Navy in football) are not ratio data. Explain why. What kind of time measurements do constitute ratio data?

BIBLIOGRAPHY   A brief and informal discussion of what statistics is and what statisticians do may be found in a pamphlet titled *Careers in Statistics*, which is published by the American Statistical Association. It may be obtained by writing to this organization at 806 15th Street, N.W., Washington, D.C., 20005. Among the few books on the history of statistics, there is on the elementary level

WALKER, H. M., *Studies in the History of Statistical Method.* Baltimore: Williams & Wilkins Co., 1929.

and on the more advanced level

PEARSON, E. S., and KENDALL, M. G., eds., *Studies in the History of Statistics and Probability.* Darien, Conn.: Hafner Publishing Co., 1970.

KENDALL, M. G., and PLANCKETT, R. L., eds., *Studies in the History of Statistics and Probability, Vol. II.* New York: Macmillan Publishing Co., 1977.

A more detailed discussion of the nature of statistical data and the general problem of **scaling** (namely, the problem of constructing scales of measurement) may be found in

SIEGEL, S., *Nonparametric Statistics for the Behavioral Sciences.* New York: McGraw-Hill Book Co., 1956.

2

Summarizing Data: Frequency Distributions

In recent years the collection of statistical data has grown at such a rate that it would be impossible to keep up with even a small part of the things which directly affect our lives unless this information is disseminated in "predigested" or summarized form. Of course, we do not always deal with very large sets of data (in many cases, they are prohibitively costly and hard to collect), but the problem of putting mass data into a suitable form is so important that it requires special attention.

The most common method of summarizing data is to present them in condensed form in tables or charts, and this used to take up the better part of elementary courses in statistics. Nowadays, the scope of statistics has expanded to such an extent that much less time is devoted to this kind of work—in fact, we shall talk about it only in this very brief introductory chapter.

2.1
Frequency Distributions

When we deal with large sets of data, a good over-all picture and sufficient information can often be conveyed by grouping the data into a number of classes.

EXAMPLE To present information about the total billings of 8,644 selected law firms in a recent year, a financial analyst summarized the data as follows:

Total billings	Number of firms
Less than $100,000	1,406
$100,000 to $249,999	4,352
$250,000 to $499,999	1,833
$500,000 to $749,999	489
$750,000 to $999,999	163
$1,000,000 or more	401
Total	8,644

EXAMPLE To summarize 2,439 complaints about comfort-related characteristics of its airplanes, an airline's customer service department issued the following table:

Nature of complaint	Number of complaints
Inadequate leg room	719
Uncomfortable seats	914
Narrow aisles	146
Insufficient carry-on facilities	218
Insufficient restrooms	58
Miscellaneous other complaints	384
Total	2,439

Tables like these are called **frequency distributions** (or simply **distributions**). The one in the first example shows how the law firms' billings are distributed among the chosen classes, and the one in the second example shows how the complaints are distributed among the different kinds. If data are grouped according to numerical size, as in the first example, the resulting table is called a **numerical** or **quantitative distribution.** In contrast, if data are grouped into nonnumerical categories, as in the second example, the resulting table is called a **categorical** or **qualitative distribution.**

Frequency distributions present data in a relatively compact form, give a good over-all picture, and contain information that is adequate for many purposes, but some things which can be determined from the original data cannot be determined from a distribution. For instance, given only the distribution of our first example, we can find neither the exact size of the lowest and highest of the law firms' total billings, nor the exact total or average of the billings of the 8,644 firms. Similarly, from the table of our second example we cannot tell how many of the complaints pertaining to uncomfortable seats were about the width of the seats, whether the complaints about insufficient carry-on facilities applied to particular size luggage, and so forth. Nevertheless, frequency distributions present **raw** (unprocessed) data in a more readily usable form, and the price we pay for this—the loss of certain information—is usually a fair exchange.

The construction of a numerical distribution consists essentially of three steps: (1) choosing the classes, (2) sorting (or tallying) the data into these classes, and (3) counting the number of items in each class. Since the last two steps are purely mechanical, we shall concentrate here on the first step; namely, that of choosing suitable classifications.

The two things that we must consider in choosing a classification are the number of classes we should use and the range of values each class should cover; that is, from where to where each class should go. Both choices are arbitrary to some extent, but they depend on the nature of the data and on the purpose the distribution is to serve. The following are some rules that are usually observed:

We seldom use fewer than 6 or more than 15 classes; the exact number we use in a given situation will depend mainly on the number of measurements or observations we have to group. Clearly, we would lose more than we gain if we group 6 observations into 12 classes, and we would probably give away too much information if we group 10,000 measurements into 3 classes.

We always make sure that each item (measurement or observation) will go into one and only one class. To this end we must make sure that the smallest and largest values fall within the classification, that none of the values can fall into possible gaps between successive classes, and that the classes do not overlap (namely, that successive classes have no values in common).

Whenever possible, we make the classes cover equal ranges of values. If we can, we also make these ranges multiples of numbers that are easy to work with, such as 5, 10, or 100, for this will facilitate constructing, reading, and using the distribution.

EXAMPLE If we assume that the law firm billings were all rounded to the nearest dollar, only the last of these rules was violated in the construction of the distribution on page 11. (Had the billings been given to the nearest cent, however, a billing of, say, $249,999.53 would have fallen between the second class and the third class, and we would also have violated the second rule.) Actually, the last rule was violated in two ways: The interval from $100,000 to $249,999 and the interval from $250,000 to $499,999 cover unequal ranges of values, and the first class and the last class have, respectively, no specific lower and upper limits.

In general, we refer to any class of the "less than," "or less," "more than," or "or more" type as an **open class.** If a set of data contains a few values which are much greater than or much smaller than the rest, open classes are quite useful in reducing the number of classes required to accommodate the data. However, they are avoided when possible because they make it difficult, or even impossible, to calculate values of interest, such as an average or a total.

As we have suggested, the appropriateness of a classification may depend on whether the data are rounded to the nearest dollar or to the nearest cent. Similarly, it may depend on whether the data are given to the nearest inch or to the nearest hundredth of an inch, whether they are given to the nearest second or to the nearest microsecond, whether they are rounded to the nearest percent or to the nearest tenth of a percent, and so on.

EXAMPLE To group the heights of children, we could use the first of the following three classifications if the heights are given to the nearest inch, the second if the heights are given to the nearest tenth of an inch, and the third if the heights are given to the nearest hundredth of an inch:

Height (inches)	Height (inches)	Height (inches)
25–29	25.0–29.9	25.00–29.99
30–34	30.0–34.9	30.00–34.99
35–39	35.0–39.9	35.00–39.99
40–44	40.0–44.9	40.00–44.99
45–49	45.0–49.9	45.00–49.99
etc.	etc.	etc.

To illustrate what we have been discussing in this section, let us go through the actual steps of grouping a given set of data into a frequency distribution.

EXAMPLE The following data, obtained in an air pollution study, are 80 determinations of the daily emission of sulfur oxides (in tons) of an industrial plant:

15.8	26.4	17.3	11.2	23.9	24.8	18.7	13.9	9.0	13.2
22.7	9.8	6.2	14.7	17.5	26.1	12.8	28.6	17.6	23.7
26.8	22.7	18.0	20.5	11.0	20.9	15.5	19.4	16.7	10.7
19.1	15.2	22.9	26.6	20.4	21.4	19.2	21.6	16.9	19.0
18.5	23.0	24.6	20.1	16.2	18.0	7.7	13.5	23.5	14.5
14.4	29.6	19.4	17.0	20.8	24.3	22.5	24.6	18.4	18.1
8.3	21.9	12.3	22.3	13.3	11.8	19.3	20.0	25.7	31.8
25.9	10.5	15.9	27.5	18.1	17.9	9.4	24.1	20.1	28.5

Since the smallest observation is 6.2 and the largest is 31.8, we might choose the six classes 5.0–9.9, 10.0–14.9, . . . , and 30.0–34.9, we might choose the seven classes 5.0–8.9, 9.0–12.9, . . . , and 29.0–32.9, or we might choose the nine classes 5.0–7.9, 8.0–10.9, . . . , and 29.0–31.9. Note that in each case the classes accommodate all of the data, they do not overlap, and they are all of the same size.

Deciding upon the second of these classifications, we now tally the 80 observations and obtain the results shown in the following table:

Tons of sulfur oxides	Tally	Frequency
5.0– 8.9	///	3
9.0–12.9	///// /////	10
13.0–16.9	///// ///// ////	14
17.0–20.9	///// ///// ///// ///// /////	25
21.0–24.9	///// ///// //// //	17
25.0–28.9	///// ////	9
29.0–32.9	//	2
	Total	80

The numbers given in the right-hand column of this table, which show how many items fall into each class, are called the **class frequencies.** The smallest and largest values that can go into any given class are referred to as its **class limits,** and for the distribution of the emission data they are 5.0 and 8.9, 9.0 and 12.9, 13.0 and 16.9, . . . , and 29.0 and 32.9. More specifically, 5.0, 9.0, 13.0, . . . , and 29.0 are called the **lower class limits,** and 8.9, 12.9, 16.9, . . . , and 32.9 are called the **upper class limits.**

Using the terminology introduced in this example, we can now say that the choice of the class limits depends on the extent to which the numbers that are to be grouped are rounded. If our data are weights to the nearest pound, the class 150–159 actually contains all weights between 149.5 and 159.5; and if our data are lengths rounded to the nearest tenth of an inch, the class 5.0–7.4 actually contains all lengths between 4.95 and 7.45. These pairs of values are usually referred to as **class boundaries** or **real class limits.**

EXAMPLE   For the distribution of the emission data, the class boundaries are 4.95, 8.95, 12.95, . . . , and 32.95; namely, the midpoints between the respective class limits. Of course, these values cannot occur, and it is important to remember that class boundaries must be, by their very nature, "impossible" values; that is, values which cannot occur among the data being grouped. To make sure of this, we have only to observe the extent to which the data are rounded. For instance, for the distribution of the law firm billings on page 11 the class boundaries, where they exist, are the impossible values $99,999.50, $249,999.50, $499,999.50, $749,999.50, and $999,999.50. They are impossible for figures rounded to the nearest dollar.

Numerical distributions also have what we call **class marks** and **class intervals.** Class marks are simply the midpoints of the classes, and they are obtained by adding the lower and upper limits of a class (or the lower and upper boundaries of a class) and dividing by 2. A class interval is merely the length of a class, or the range of values it can contain, and it is given by the difference between its class boundaries. If the classes of a distribution are all equal in length, their common class interval, which we refer to as the **class interval of the distribution,** is also given by the difference between any two successive class marks.

EXAMPLE   For the distribution of the emission data, the class marks are $\frac{5.0 + 8.9}{2} = 6.95$, $\frac{9.0 + 12.9}{2} = 10.95, \ldots$, and $\frac{29.0 + 32.9}{2} = 30.95$, and the class interval of the distribution is $10.95 - 6.95 = 4$. Note that the class intervals are *not* given by the differences between the respective class limits, which in our example would all equal 3.9 instead of 4.

There are essentially two ways in which frequency distributions can be modified to suit particular needs. One way is to convert a distribution into a **percentage distribution** by dividing each class frequency by the total number of items grouped, and then multiplying by 100.

EXAMPLE   For the distribution of the emission data, we would show instead of the class frequencies that the first class contains $\frac{3}{80} \cdot 100 = 3.75$ percent of the data, the second class contains $\frac{10}{80} \cdot 100 = 12.50$ percent of the data, . . . , and the seventh class contains $\frac{2}{80} \cdot 100 = 2.50$ percent of the data.

Percentage distributions are often used when it is desired to compare two or more distributions; for instance, if we wanted to compare the emission of sulfur oxides of the plant of our example with that of a plant at a different location.

The other way of modifying a frequency distribution is to convert it into a "less than," "or less," "more than," or "or more" **cumulative distribution.** To this end we simply add the class frequencies, starting either at the top or at the bottom of the distribution.

EXAMPLE   For the emission data, we thus obtain the following "less than" cumulative distribution:

Tons of sulfur oxides	Cumulative frequency
Less than 5.0	0
Less than 9.0	3
Less than 13.0	13
Less than 17.0	27
Less than 21.0	52
Less than 25.0	69
Less than 29.0	78
Less than 33.0	80

Note that instead of "less than 5.0" we could have written "4.9 or less" or "less than 4.95," instead of "less than 9.0" we could have written "8.9 or less" or "less than 8.95," and so on.

So far we have discussed only the construction of numerical distributions, but the general problem of constructing categorical (or qualitative) distributions is about the same. Here again we must decide how many categories (classes) to use and what kind of items each category is to contain, making sure that all the items are accommodated and that there are no ambiguities. Since the categories must often be chosen before any data are actually collected, it is prudent to include a category labeled "others" or "miscellaneous."

For categorial distributions, we do not have to worry about such mathematical details as class limits, class boundaries, and class marks. On the other hand, there is often a serious problem with ambiguities and we must be very careful and explicit in defining what each category is to contain. For instance, if we had to classify items sold at a supermarket into "meats," "frozen foods," "baked goods," and so forth, it would be difficult to decide, for example, where to put frozen beef pies. Similarly, if we had to classify occupations, it would be difficult to decide where to put a farm manager, if our table contained (without qualification) the two categories "farmers" and "managers." For this reason, it is advisable, where possible, to use standard categories developed by the Bureau of the Census and other government agencies. References to lists of such categories may be found in the book by P. M. Hauser and W. R. Leonard given on page 26.

2.2
Graphical Presentations

When frequency distributions are constructed primarily to condense large sets of data and display them in an "easy to digest" form, it is usually advisable to present them graphically. The most common form of graphical presentation of statistical data is the **histogram,** an example of which is shown in Figure 2.1. A histogram is constructed by representing the measurements or observations that are grouped (in Figure 2.1, the sulfur oxides emission data) on a horizontal scale, the class frequencies on a vertical scale, and drawing rectangles whose bases equal the class interval and whose heights are determined by the corresponding class frequencies. The markings on the horizontal scale can be the class limits as in Figure 2.1, the class boundaries, the class marks, or arbitrary key values. For easy readability, it is usually better to indicate

FIGURE 2.1
Histogram of the distribution of the emission data.

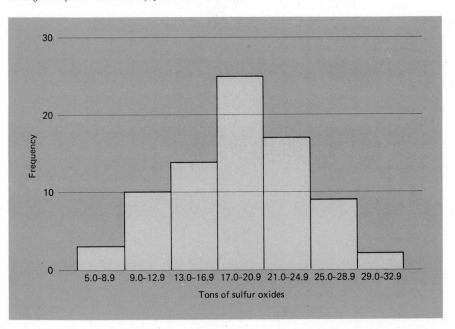

the class limits, although the rectangles actually go from one class boundary to the next. Histograms cannot be used in connection with frequency distributions having open classes, and they must be used with extreme care if the class intervals are not all equal (see the discussion on page 20).

Another, less widely used form of graphical presentation is the **frequency polygon** (see Figure 2.2). In these the class frequencies are plotted at the class marks and the successive points are connected by means of straight lines. Note that we added classes with zero frequencies at both ends of the distribution to "tie down" the graph to the horizontal scale. If we apply the same technique to a cumulative distribution, we obtain what is called an **ogive.** However, the cumulative frequencies are plotted at the class boundaries instead of the class marks—it stands to reason that the cumulative frequency corresponding, say, to "less than 13.0" should be plotted at 12.95, the class boundary, since "less than 13.0" actually includes everything up to 12.95. Figure 2.3 shows an ogive corresponding to the "less than" distribution of the emission data.

Although the visual appeal of histograms, frequency polygons, and ogives exceeds that of frequency tables, there are ways in which distributions can be presented even more dramatically and often more effectively. Two kinds of such pictorial presentations (often seen in newspapers, magazines, and reports of various sorts) are illustrated by the **pictogram** of Figure 2.4 and the **pie chart** of Figure 2.5.

FIGURE 2.2

Frequency polygon of the distribution of the emission data.

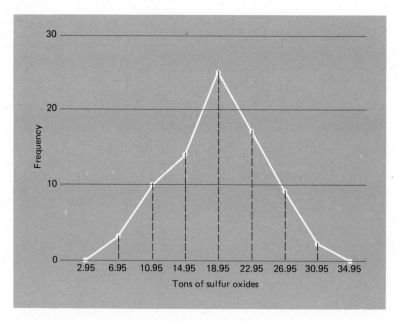

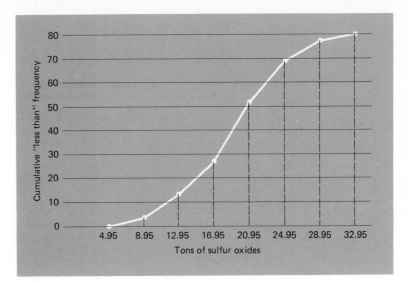

FIGURE 2.3
Ogive of the distribution of the emission data.

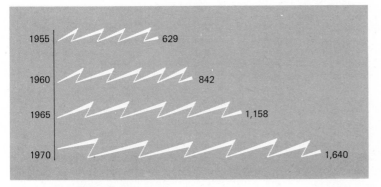

FIGURE 2.4
Electric energy production in the U.S. (billions of kilowatt-hours).

FIGURE 2.5
State and the local government revenue and expenditure dollars: 1972. (Source: U.S. Bureau of the Census.)

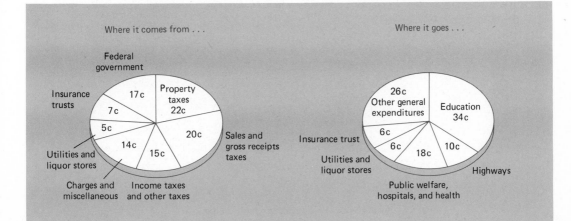

2.3
A Word of Caution

Intentionally or unintentionally, frequency tables, histograms, and other pictorial presentations are sometimes very misleading. Suppose, for instance, that when we grouped the emission data we combined the two classes 17.0–20.9 and 21.0–24.9 into one class, the class 17.0–24.9. This new class has a frequency of $25 + 17 = 42$, but in Figure 2.6, where we still use the heights of the rectangles to represent the class frequencies, we get the erroneous impression that this class contains about two thirds of the data (instead of slightly more than one half). This is due to the fact that when we compare the sizes of rectangles, triangles, or other plane figures, we instinctively compare their areas and not their sides. This does not matter when the class intervals are all equal, but in Figure 2.6 the class 17.0–24.9 is twice as wide as the others, and we should compensate for this by dividing the height of the

FIGURE 2.6

Incorrectly modified histogram of the distribution of the emission data.

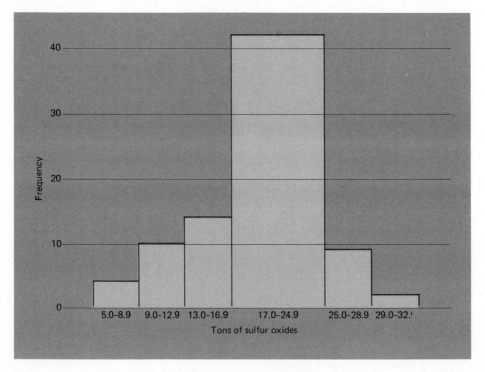

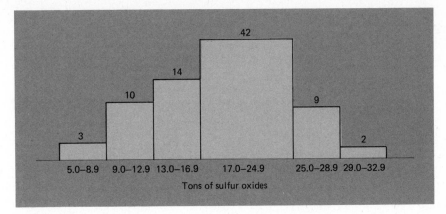

FIGURE 2.7

Correctly modified histogram of the distribution of the emission data.

rectangle by 2. Figure 2.7 (where the vertical scale, which has lost its significance, has been omitted) shows the result of this adjustment. Now we get the correct impression that the class 17.0–24.9 contains just about half the data; 42 out of 80 to be exact.

EXERCISES

1 The number of empty seats on flights from Atlanta to St. Louis are grouped into a table having the classes 0–4, 5–9, 10–14, 15–19, 20–24, 25–29, and 30 or more. Is it possible to determine from this distribution the number of flights on which there were
    (a) at least 15 empty seats?
    (b) more than 15 empty seats?
    (c) more than 14 empty seats?
    (d) exactly 9 empty seats?
    (e) at most 24 empty seats?
    (f) anywhere from 10 through 19 empty seats?
    (g) anywhere from 25 through 34 empty seats?
    (h) at most 40 empty seats?

2 Is it possible to determine from the distribution of law firm billings on page 11 how many of the firms had billings (to the nearest dollar) of
    (a) at least $500,000?
    (b) more than $750,000?
    (c) more than $1,500,000?
    (d) exactly $1,000,000?
    (e) anywhere from $100,000 to $750,000?
    (f) less than $250,000?
    (g) at most $500,000?
    (h) $999,999 or less?

3 The following is the distribution of the weights of 125 mineral specimens collected on a field trip:

Weight (grams)	Number of specimens
0– 19.9	19
20.0– 39.9	38
40.0– 59.9	35
60.0– 79.9	17
80.0– 99.9	11
100.0–119.9	3
120.0–139.9	2
Total	125

If possible, find the number of specimens which weigh
  (a) at most 40.0 grams;
  (b) 40.0 grams or more;
  (c) more than 140.0 grams;
  (d) less than 79.9 grams;
  (e) less than 40.0 grams;
  (f) at least 119.9 grams;
  (g) anywhere from 59.9 to 99.9 grams;
  (h) exactly 20.0 grams.

4 The number of congressmen absent each day during a session of Congress are grouped into a distribution having the classes 5–19, 20–34, 35–49, 50–64, 65–79, 80–94, 95–109, and 110–124. Determine
  (a) the class boundaries;
  (b) the class marks;
  (c) the class interval of the distribution.

5 The weights of 297 applicants for positions in the fire department of a large city are measured to the nearest tenth of a pound with the lowest being 142.9 pounds and the highest being 226.1 pounds. Give the class limits of a distribution with nine classes into which these weights might be grouped.

6 Measurements of the boiling point of a silicon compound, given to the nearest degree centigrade, are grouped into a table having the class boundaries 129.5, 139.5, 149.5, 159.5, 169.5, 179.5, and 189.5 degrees centigrade. Find
  (a) the class limits;
  (b) the class marks;
  (c) the class interval of the distribution.

7 The class marks of a distribution of the daily number of surgeries that are performed at a hospital are 3, 8, 13, 18, 23, 28, and 33. What are the class limits of this distribution and what is its class interval?

8 The following are the burning times of certain solid-fuel rockets given to the nearest tenth of a second:

4.1	5.0	4.8	4.3	4.2	5.3	4.2	3.6	4.5	4.4
4.5	3.2	4.0	3.8	3.8	5.3	4.5	4.6	4.0	5.2
5.2	4.4	4.7	4.1	4.6	4.9	4.1	5.8	4.2	4.2
4.8	4.1	5.6	4.5	5.1	4.6	4.3	5.2	4.7	3.2
4.0	4.6	4.0	4.2	4.5	3.5	4.7	4.9	3.9	4.8
3.7	5.4	4.9	4.6	4.3	5.4	5.0	4.5	4.7	4.3

(a) Group these burning times into a distribution having the classes 3.0–3.4, 3.5–3.9, 4.0–4.4, . . . , and 5.5–5.9 seconds.

(b) Draw a histogram of the distribution obtained in part (a).

(c) Convert the distribution obtained in part (a) into a cumulative "less than" distribution.

(d) Draw an ogive of the cumulative distribution obtained in part (c).

9 The following are the body weights (in grams) of 80 rats used in a study of vitamin deficiencies:

132	125	117	124	108	112	110	127	96	129
130	122	118	114	103	119	106	125	114	100
125	128	106	111	116	123	119	114	117	143
136	92	115	118	121	137	132	120	104	125
119	115	101	129	87	108	110	133	135	126
127	103	110	126	118	82	104	137	120	95
146	126	119	119	105	132	126	118	100	113
106	125	117	102	146	129	124	113	95	148

(a) Group these figures into a table having the classes 80–89, 90–99, 100–109, . . . , 130–139, and 140–149.

(b) Convert the distribution obtained in part (a) into a percentage distribution.

(c) Draw a frequency polygon of the percentage distribution obtained in part (b).

(d) Convert the distribution obtained in part (a) into a cumulative "or less" distribution.

(e) Draw an ogive of the cumulative distribution obtained in part (d).

10 A week's records of a cab company show the following amounts (in dollars) spent on gasoline by each of its 48 cabs:

72.17	44.63	33.21	49.30	37.80	36.45	22.16	20.45
28.95	43.57	18.75	42.70	33.80	71.88	33.68	56.13
37.87	24.75	38.67	32.45	31.55	50.55	64.50	39.01
69.49	52.83	53.41	60.75	21.45	47.82	40.58	30.56
37.51	34.69	41.88	30.24	15.25	27.63	24.65	45.14
20.11	31.22	41.35	26.27	36.00	38.76	25.68	23.65

(a) Group these figures into a table having the classes $15.00–$24.99, $25.00–$34.99, $35.00–$44.99, . . . , and $65.00–$74.99.

(b) Draw a histogram of the distribution obtained in part (a).

(c) Convert the distribution obtained in part (a) into a cumulative "or more" percentage distribution.

(d) Draw an ogive of the cumulative distribution obtained in part (a).

11 The following is a distribution of the final examination grades which 180 students obtained in a freshman course in philosophy:

Grade	Frequency
0–19	18
20–39	51
40–59	66
60–79	32
80–99	13

(a) Find the class boundaries.
(b) Find the class marks.
(c) Draw a histogram of this distribution.
(d) Combine the grades from 40 to 79 into one class and draw a histogram of the resulting distribution.
(e) Draw a frequency polygon of the original distribution.
(f) Convert the original distribution into a cumulative "less than" distribution.
(g) Draw an ogive of the distribution obtained in part (f).

12 A **bar chart** is a form of graphical presentation that is very similar to a histogram. In this kind of chart (see Figure 2.8), the lengths of the bars are proportional to

FIGURE 2.8

Bar chart of the distribution of the emission data.

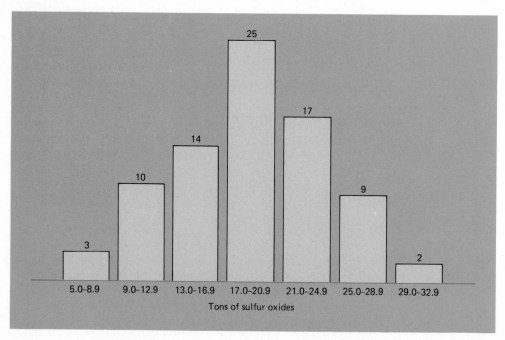

the class frequencies, but there is no pretense of having a continuous horizontal scale.

(a) Draw a bar chart of the distribution obtained in part (a) of Exercise 8.

(b) Draw a bar chart of the distribution obtained in part (a) of Exercise 9.

(c) Draw a bar chart of the distribution obtained in part (a) of Exercise 10.

(d) Draw a bar chart of the distribution given in Exercise 11.

13 Categorical distributions are often presented as **pie charts** such as those of Figure 2.5, where a circle is divided into sectors which are proportional in size to the frequencies (or percentages) of the categories they represent. In each sector, 1 percent is represented by a central angle of one hundredth of 360 degrees; namely, by a central angle of 3.6 degrees.

(a) Construct a pie chart of the following distribution, which shows how the dogs entered in a dog show are distributed according to A.K.C. classifications:

Group	Number
Sporting dogs	26
Hounds	34
Working dogs	76
Terriers	17
Toys	24
Nonsporting dogs	23
Total	200

(b) Draw a pie chart of the distribution on page 11 of the complaints received by the airline's customer service department.

(c) Draw a pie chart to display the information that a large hospital's expenses are as follows: 73 percent for salaries, professional medical fees, and employee benefits; 13 percent for medical and surgical supplies and equipment; 8 percent for maintenance, food, and power; and 6 percent for administrative services.

14 A survey made at a resort city showed that 50 tourists arrived by the following means of transportation: car, train, plane, plane, plane, bus, train, car, car, car, plane, car, plane, train, car, car, bus, car, plane, plane, train, train, plane, plane, car, car, train, car, car, plane, car, car, plane, bus, plane, bus, car, plane, car, car, train, train, car, plane, bus, plane, car, car, train, and bus. Construct a categorical distribution showing the frequencies corresponding to the different means of transportation, and present it pictorially in the form of a pie chart.

15 Asked to rate the maneuverability of a car as excellent, very good, good, fair, poor, or very poor, forty drivers responded as follows: very good, good, good, fair, excellent, good, good, good, very good, poor, good, good, good, good, very good, good, fair, good, good, very poor, very good, fair, good, good, excellent, very good, good, good, good, fair, fair, very good, good, very good, excellent, very good, fair, good, good, and very good.

(a) Construct a categorical distribution showing the frequencies corresponding to these ratings of the maneuverability of the car.

$7,000 in 1960

$14,000 in 1975

FIGURE 2.9

Pictogram for Exercise 16.

(b) When categorical data are of an ordinal nature (that is, when the nature of the categories implies a definite order), they are often presented graphically by means of bar charts (as defined in Exercise 12). Draw a bar chart of the distribution obtained in part (a).

16 The pictogram of Figure 2.9 is intended to illustrate that in a certain region average family income has doubled from $7,000 in 1960 to $14,000 in 1975. Explain why this pictogram does not convey a fair impression of the actual change, and suggest how it might be modified.

BIBLIOGRAPHY Discussions of what not to do in the presentation of statistical data may be found in

CAMPBELL, S. K., *Flaws and Fallacies in Statistical Thinking*. Englewood Cliffs, N.J.: Prentice-Hall, Inc., 1974.

HUFF, D., *How to Lie with Statistics*. New York: W. W. Norton & Company, Inc., 1954.

REICHMAN, W. J., *Use and Abuse of Statistics*. Baltimore: Penguin Books, Inc., 1971.

Useful references to lists of standard categories may be found in

HAUSER, F. M., and LEONARD, W. R., *Government Statistics for Business Use, 2nd ed*. New York: John Wiley & Sons, Inc., 1956.

3

When we describe sets of data we try to say neither
too little nor too much. So, depending on the purpose
they are to serve, statistical descriptions can be
very brief or very elaborate. Sometimes it may be
satisfactory to present data just as they are, in raw
form, and let them speak for themselves; on other
occasions it may be necessary only to group the data
and present their distribution in tabular or graphical
form. Usually, though, data have to be summarized
further, and in this chapter we shall discuss the two
most widely used kinds of statistical descriptions,
called **measures of location** and **measures of variation.** Other kinds of statistical descriptions will be
taken up later, when needed.

Summarizing
Data:
Statistical
Descriptions

3.1

Samples and Populations

Before we study specific statistical descriptions, let us make the following distinction: If a set of data consists of all conceivably possible (or hypothetically possible) observations of a certain phenomenon, we call it a **population**; if a set of data contains only a part of these observations, we call it a **sample.** We added the qualification "hypothetically possible" in this definition to take care of such clearly hypothetical situations as where we look at the outcomes (heads or tails) of 12 flips of a coin as a sample from the population of all possible flips of the coin, where we look at the weights of ten 30-day-old lambs as a sample of the weights of all (past, present, or future) 30-day-old lambs, or where we look at four determinations of the uranium content of an ore as a sample of all possible determinations of the uranium content of the ore. In fact, we often look at the results of an experiment as a sample of what we might obtain if the experiment were repeated over and over again.

On page 3 we said that statistics dealt originally with the description of human populations (political units), but as it grew in scope, the term "population" took on the much wider connotation given to it above. Whether or not it sounds strange to refer to the heights of all the trees in a forest or the speeds of all the cars passing a check point as populations is beside the point—in statistics, "population" is a technical term with a meaning of its own.

Although we are free to call any group of items a population, what we do in practice depends on the context in which they are to be viewed. Suppose, for instance, that we are offered a lot of 4,000 ceramic tiles, which we may or may not buy depending on their strength. If we measure the breaking strength of 20 of these tiles in order to estimate the average breaking strength of all the tiles, these 20 measurements are a sample from the population which consists of the breaking strengths of the 4,000 tiles. In another context, however, if we consider entering into a long-term contract calling for the delivery of tens of thousands of such tiles, we would look upon the breaking strengths of the original 4,000 tiles only as a sample. Similarly, the complete figures for a recent year, giving the elapsed times between the filing and disposition of divorce suits in San Diego County, can be looked upon as either a population or a sample. If we are interested only in San Diego County and that particular year, we would look upon the data as a population; on the other hand, if we want to generalize about the time that is required for the disposition of divorce suits in the entire United States, in some other county, or in some other year, we would look upon the data as a sample.

As we have used it here, the word "sample" has very much the same meaning as it has in everyday language. A newspaper considers the attitudes of 150 readers toward a proposed school bond to be a sample of the attitudes of all its readers toward the bond; and a consumer considers a box of Mrs. See's candy a sample of the firm's product. Later, we shall use the word "sample" only when referring to data which can reasonably serve as the basis for valid generalizations about the populations from which they came; in this more technical sense, many sets of data which are popularly called samples are not samples at all.

In this chapter we shall make descriptions without making generalizations, but it is important even here to distinguish between populations and samples. For future use, we shall use different symbols depending on whether we are describing populations or samples, and sometimes even different formulas.

3.2
Measures of Location

It is often necessary to represent a set of data by means of a single number which, in its way, is descriptive of the entire set. Exactly what sort of number we choose depends on the particular characteristic we want to describe. In one study we may be interested in the extreme (smallest and largest) values among the data; in another, in the value which is exceeded by only 10 percent of the data; and in still another, in the total of all the values. In the next section, we shall consider several measures which somehow describe the center or middle of a set of data—appropriately, they are called **measures of central location.**

3.3
The Mean, the Median, and the Mode

Among the different measures of central location, by far the best known and the most widely used is the **arithmetic mean,** or simply the **mean,** which we define as follows: **The mean of a set of values is the sum of the values divided by their number.** In everyday language, the mean is often called the "average," and on occasion we shall call it that ourselves. As we shall see, however, there are other "averages" in statistics, and we cannot afford to speak loosely when there is any danger of ambiguity.

EXAMPLE Given that the 89th through 93rd Congresses of the United States enacted 1,283, 1,002, 941, 768, and 295 measures (bills, acts, and resolutions), we find that the mean number of measures enacted was

$$\frac{1{,}283 + 1{,}002 + 941 + 768 + 295}{5} = \frac{4{,}289}{5} = 857.8$$

EXAMPLE Given that during the 12 months of 1976 a person charged 3, 1, 2, 5, 7, 6, 8, 3, 6, 2, 1, and 4 restaurant meals to BAC, we find that the monthly mean number of restaurant meals charged by this person to BAC was

$$\frac{3 + 1 + 2 + 5 + 7 + 6 + 8 + 3 + 6 + 2 + 1 + 4}{12} = \frac{48}{12} = 4$$

Since we shall have occasion to calculate the means of many different sets of data, it will be convenient to have a simple formula that is always applicable. This requires that we represent the figures to be averaged by some general symbol such as x, y, or z; the number of values in a sample, the **sample size**, is usually denoted by the letter n. Choosing the letter x, we can refer to the n values in a sample as x_1 (which is read "x sub-one"), $x_2, x_3, \ldots$, and x_n, and write

$$\text{sample mean} = \frac{x_1 + x_2 + x_3 + \ldots + x_n}{n}$$

This formula is perfectly general and it will take care of any set of sample data, but it can be made more compact by assigning the sample mean the symbol $\bar{x}$ (which is read "x bar") and using the $\sum$ **notation.** The symbol $\sum$ is capital *sigma*, the Greek letter for S. In this notation, we let $\sum x$ stand for "the sum of the x's" (that is, $x_1 + x_2 + x_3 + \cdots + x_n$), so that we can write

Sample mean

$$\bar{x} = \frac{\sum x}{n}$$

If we refer to the measurements as y's or z's, we write their mean as $\bar{y}$ or $\bar{z}$. In the formula for $\bar{x}$, the term $\sum x$ does not state explicitly which values of x are to be added; let it be understood, however, that $\sum x$ always refers to the sum of all the x's under consideration in a given situation. In Section 3.10 on page 60, the use of the sigma notation is discussed in more detail.

The mean of a population of N items is defined in the same way. It is the sum of the N items, $x_1 + x_2 + x_3 + \cdots + x_N$, or $\sum x$, divided by the **population size** N. Assigning the population mean the symbol μ (lowercase *mu*, the Greek letter for m), we write

Population mean

$$\mu = \frac{\sum x}{N}$$

with the reminder that $\sum x$ is now the sum of all N values of x which constitute the population.

To distinguish between descriptions of samples and descriptions of populations, statisticians not only use different symbols, but they refer to the former as **statistics** and to the latter as **parameters,** and usually denote parameters by Greek letters.

EXAMPLE To illustrate all the terminology and notation introduced in this section, suppose we want to know the mean lifetime of a production lot (considered to be a population) of $N = 40{,}000$ light bulbs. Obviously, we cannot test all of the light bulbs for there would be none left to use or sell, so we take a sample, calculate $\bar{x}$, and use this quantity to estimate μ. If $n = 5$ and the light bulbs in the sample last 967, 949, 940, 952, and 922 hours, we have

$$\bar{x} = \frac{967 + 949 + 940 + 952 + 922}{5} = 946 \text{ hours}$$

If these lifetimes constitute a sample in the technical sense (that is, a set of data from which valid generalizations can be made), we can estimate the mean lifetime μ of all the 40,000 light bulbs as 946 hours.

For non-negative data (that is, numbers which must be positive or zero), the mean not only describes the middle of the data, it also puts some limitations on their size. This follows from the fact that the total of the values, $\sum x$, must equal, and hence cannot exceed, $n \cdot \bar{x}$. To prove this, we have only to multiply by n both sides of the equation $\bar{x} = \dfrac{\sum x}{n}$.

EXAMPLE If the mean annual salary paid to the top three executives of a firm is \$64,000, can one of them receive an annual salary of \$200,000? Since $n = 3$ and $\bar{x} = 64{,}000$, we get $\sum x = 3 \cdot 64{,}000 = \$192{,}000$, and since the total of the three salaries is \$192,000, it is impossible for any one of the executives to receive more than that.

EXAMPLE If nine high school juniors averaged 41 on the verbal part of the PSAT/NMSQT test, at most how many of them could have scored 65 or more? Since $n = 9$ and $\bar{x} = 41$, we get $\sum x = 9 \cdot 41 = 369$, and since 65 goes into 369 at most 5 times ($369 = 5 \times 65 + 44$), it follows that at most 5 of these high school juniors could have received scores of 65 or more.

The popularity of the mean as a measure of the "middle" or "center" of a set of data is not accidental. Aside from the fact that it is a simple, familiar measure, it has the following desirable properties: **(1) It can be calculated for any set of numerical data, so it always exists; (2) a set of numerical data has one and only one mean, so it is always unique; (3) it lends itself to further statistical treatment (for instance, the means of several sets of data can be combined into the over-all mean of all the data);** and **(4) it is relatively reliable in the sense that means of many samples drawn from the same population generally do not fluctuate, or vary, as widely as other statistics used to estimate the population mean μ.** This last property is of fundamental importance in statistical inference.

There is another property of the mean which, on the surface, seems desirable: **(5) it takes into account every item of the data.** Sometimes, though, samples contain very small or very large values which are so far removed from the main body of the data that the appropriateness of including them in a sample is questionable. Such values may be due to chance, or they may be due to gross errors in recording the data, gross errors in calculations, malfunctioning of equipment, or other identifiable sources of contamination. In any case, when such values are averaged in with the other values, they can affect the mean to such an extent that it is debatable whether it really provides a useful description of the "middle" of the data.

EXAMPLE The ages of six students taking part in a geology field trip are 18, 19, 20, 17, 19, and 18, and the age of the instructor who conducts the field trip is 50. The mean age of these seven persons is

$$\frac{18 + 19 + 20 + 17 + 19 + 18 + 50}{7} = \frac{161}{7} = 23$$

but any statement to the effect that the average age of the group is 23 would probably be misinterpreted.

EXAMPLE Referring to the light-bulb example on page 31, suppose that the second value is incorrectly recorded as 499 instead of 949. This makes the mean of the sample

$$\bar{x} = \frac{967 + 499 + 940 + 952 + 922}{5} = 856 \text{ hours}$$

and it introduces an error of $946 - 856 = 90$ hours into our estimate of the mean lifetime μ of all the 40,000 light bulbs.

To avoid the possibility of being misled by very small or very large values, we sometimes describe the "middle" or "center" of a set of data with other kinds of statistical measures. One of these is the **median**, which is defined as **the value of the middle item, or the mean of the values of the two middle items, when the data are arranged in an increasing or decreasing order**

of magnitude. Like the mean of n sample values $x_1, x_2, x_3, \ldots$, and x_n, the sample median, which we denote $\tilde{x}$, can be used to estimate the population mean μ; in fact, this accounts for our main interest in this statistical measure.

If there is an odd number of items in a set of data, there is always a middle item whose value is the median.

EXAMPLE   Given that 42, 39, 31, 35, and 38 bighorn sheep were killed by hunters in Arizona in the years 1969 through 1973, the median is 38 (not 31, since the numbers must first be arrayed according to size).

EXAMPLE   Asked to how many charitable organizations they gave cash donations in 1977, eleven taxpayers responded that they gave money to 10, 15, 12, 16, 12, 11, 14, 13, 9, 10, 13 charities. Arranging these figures according to size, we get

$$9 \quad 10 \quad 10 \quad 11 \quad 12 \quad 12 \quad 13 \quad 13 \quad 14 \quad 15 \quad 16$$

and it can be seen that the median is 12. Note that there are two 12's among the data and that we do not refer to either of them as *the* median—the median is a number and not necessarily a particular measurement or observation.

Generally speaking, the median of a set of n items, where n is odd, is the value of the $\dfrac{n+1}{2}$ th largest item.

EXAMPLE   The median of 25 numbers is the value of the $\dfrac{25+1}{2} = $ 13th largest number, and the median of 71 numbers is the value of the $\dfrac{71+1}{2} = $ 36th largest number.

For a set of n items, where n is even, there is no single middle item and the median is the mean of the values of the two middle items. Yet, interpreted correctly, the formula $\dfrac{n+1}{2}$ still serves to locate the position of the median.

EXAMPLE   If $n = 6$, for instance, $\dfrac{6+1}{2} = 3.5$, and we interpret this as "halfway between the values of the third and fourth items." So, the median of 8, 6, 3, 15, 10, and 13 is the mean of 8 and 10, namely, $\dfrac{8+10}{2} = 9$.

Also, if $n = 100$, then $\dfrac{100+1}{2} = 50.5$, and the median is the mean of the values of the 50th and 51st largest items.

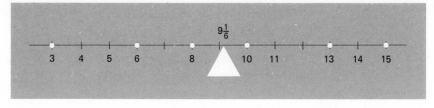

FIGURE 3.1

The mean as a center of gravity.

It is important to remember that the formula $\dfrac{n+1}{2}$ is not a formula for the median, itself; it merely tells us the position of the median, namely, how many of the ordered values we have to count until we reach the item whose value is the median (or the two items whose values have to be averaged to obtain the median).

In the last example we showed that the median of 8, 6, 3, 15, 10, and 13 is 9. If we calculate the mean of these numbers, we get $\dfrac{8 + 6 + 3 + 15 + 10 + 13}{6}$ $= 9\frac{1}{6}$, and it should not really come as a surprise that the two results are not the same. Each of these averages describes the "middle" or center" of the data in its own way. The median is average in the sense that it splits the data into two parts so that the values of half the items are less than or equal to the median, and the values of the other half are greater than or equal to the median. The mean, on the other hand, is typical in the sense of a center of gravity, as is illustrated by Figure 3.1. To put it differently, the mean is typical in the sense that if each value in a set of data is replaced by the same number but the total remains unchanged, this number will have to be the mean. This follows immediately from the formula $n \cdot \bar{x} = \sum x$, which we gave on page 31.

Like the mean, the median always exists and is unique for any set of data. Also like the mean, the median is simple enough to find once the data have been arranged according to size, but it should be kept in mind that ordering large sets of data manually can be a very tedious job. Unlike the mean, the median is not so easily affected by extreme values.

EXAMPLE On page 31 we showed that the mean of 967, 949, 940, 952, and 922 (the lifetimes of five light bulbs) is 946, and on page 32 we showed that if we misread 949 as 499, the mean would be 856. As we pointed out, this introduces an error of 90 hours into our estimate of the mean lifetime μ of all the 40,000 light bulbs from which the sample came. Now, if we use the median instead of the mean, we find that for the original data the median is 949, when 949 is misread as 499 the median is 940, so that the error introduced by this mistake is 9 hours. That is only one tenth of the error we made when we used the mean.

Also unlike the mean, the median can be used to define the middle of a number of objects, properties, or qualities; namely, the middle or the most typical of a set of nominal data.

EXAMPLE   It is possible, for instance, to rank a number of tasks according to their difficulty and then describe the middle (or median) one as being of "average" difficulty; also, we might rank samples of chicken gravy according to their consistency and then describe the middle (or median) one as having "average" consistency.

Perhaps the most important difference between the properties of the median and the mean is that in problems of inference (estimation, prediction, and so on) the mean is usually more reliable than the median. This is meant to say that the medians of many samples drawn from the same population usually vary more widely than the corresponding sample means (see Exercise 14 on page 38, Exercise 8 on page 48, and Exercise 12 on page 234).

Another measure which is sometimes used to describe the "middle" of a set of data is the **mode**. It is defined simply as the value which occurs with the highest frequency, and its two main advantages are that (1) it requires no calculations, and (2) it can be determined for quantitative as well as qualitative data.

EXAMPLE   If 20 golf professionals score 68, 70, 73, 71, 78, 70, 69, 70, 72, 74, 72, 69, 67, 70, 70, 71, 69, 74, 71, and 72 playing at the Indian Wells Country Club, then 70 (the value which occurs most often) is the **modal score.**

EXAMPLE   If more visitors to California want to see Disneyland than any other tourist attraction, we say that Disneyland is their **modal choice.**

Two definite disadvantages of the mode are that (1) it may not exist (which is the case when no two values in a set of data are alike), and (2) the mode may not be unique.

EXAMPLE   A sample from the records of a motor vehicle bureau show that ten drivers in a certain age group received 0, 1, 3, 0, 4, 3, 1, 2, 0, and 3 tickets in 1977. As can easily be seen, the numbers 2 and 4 each occur once, the number 1 occurs twice, and the numbers 0 and 3 each occur three times, so there are two modes: 0 and 3.

The presence of more than one mode is sometimes indicative of the fact that the data are not homogeneous; namely, that they can be looked upon as a combination of several sets of data. In the preceding example, we might infer that there are many very good drivers and many very poor (or reckless) drivers, while fewer drivers fall into the categories between these two extremes.

There are many other measures of central location besides the mean, the median, and the mode, and some of them will be introduced in the exercises that follow. The question of what particular "average" should be used in a given situation is not always easily answered, and the fact that there is a good deal of arbitrariness in the selection of statistical descriptions has led some persons to believe that the magic of statistics can be used to prove almost anything. A famous nineteenth-century British statesman said that there are three kinds of lies: lies, damned lies, and statistics, and Exercises 11 and 12 on page 37 describe a situation where this kind of criticism would be justified.

EXERCISES

1 Suppose that we are given complete information about the total sales tax collected by the 112 drugstores in a city during a given month. Give one illustration each of a problem in which these data would be looked upon as
   (a) a sample;
   (b) a population.

2 Suppose that the final election returns from a county show that the two candidates for a certain office received 14,283 and 12,695 votes. What office might these candidates be running for so that these figures would constitute
   (a) a sample?
   (b) a population?

3 The following are the amounts of time (in minutes) which 15 American students spent viewing a famous painting in a museum in Paris, France: 2, 10, 15, 8, 6, 17, 2, 10, 3, 9, 5, 9, 1, 10, and 13. Determine
   (a) the mean;
   (b) the median;
   (c) the mode.

4 The following are the weight losses (in pounds) of ten persons following a prescribed diet for two weeks: 4.3, 3.4, 3.8, 5.2, 4.4, 2.9, 3.7, 5.6, 4.1, and 3.6. Calculate
   (a) the mean;
   (b) the median.

5 At their first inaugurations, the first ten presidents of the United States were 57, 61, 57, 57, 58, 57, 61, 54, 68, and 51 years old. Find
   (a) the mean age of these presidents at their first inaugurations;
   (b) the modal age of these presidents at their first inaugurations.

6 The following are the IQ's of twenty persons empaneled for jury duty by a court: 108, 98, 103, 122, 87, 105, 101, 112, 95, 106, 100, 91, 99, 118, 103, 110, 107, 102, 93, and 101.
   (a) Find the mean IQ of the twenty persons on this panel.
   (b) Recalculate the mean of the twenty IQ's by first subtracting 100 from each value, finding the mean of the numbers thus obtained, and then adding

100 to the result. What general simplification does this suggest for the calculation of a mean?

7 By mistake, an instructor has erased the grade which one of the ten students in his class received in a final examination. However, he knows that the ten students averaged (had a mean grade of) 72 in the examination, and that the other students received grades of 48, 71, 79, 95, 45, 57, 75, 83, and 97. What must have been the erased grade?

8 An elevator in a department store is designated to carry a maximum load of 3,000 pounds. Is it overloaded if at one time it carries
   (a) 18 passengers whose mean weight is 135 pounds?
   (b) 12 women whose mean weight is 123 pounds and 9 men whose mean weight is 175 pounds?

9 At a recent auction, 800 ancient gold coins sold for an average (mean) price of $219. At most how many of these gold coins could have brought $20,000 or more?

10 Each of 15 persons soliciting funds for a charitable organization was assigned a quota (amount of money) he or she should raise, and the following are the percentages of their respective quotas which they actually attained: 92, 107, 353, 90, 78, 80, 74, 92, 102, 86, 106, 109, 95, 102, and 91. Calculate the mean and the median of these percentages, and indicate which of the two measures is a better indication of these person's "average" performance.

11 A consumer testing service obtained the following miles per gallon in five highway test runs performed with each of three imported sports cars:

Car A: 34.7, 34.4, 33.6, 33.4, 30.9 $\bar{x} = 33.4$
Car B: 34.1, 31.7, 34.3, 31.7, 34.2 $\bar{x} = 33.2$
Car C: 32.7, 35.1, 31.5, 32.1, 31.6 $\bar{x} = 32.6$

   (a) If the manufacturers of car A want to advertize that their car performed best in this test, which of the "averages" discussed in this text could they use to substantiate their claim? *mean*
   (b) If the manufacturers of car B want to advertize that their car performed best in this test, which of the "averages" discussed in this text could they use to substantiate their claim? *median*

12 Another measure of central location, the **midrange** is simply the mean of the smallest and largest values in a sample or population. With reference to Exercise 11, show that if the manufacturers of car C used the midrange to average the five values obtained for each car, this would substantiate the claim that their car performed best in the test.

13 Forty registered voters were asked whether they considered themselves Democrats, Republicans, or Independents. Use the following results to determine their modal choice: Democrat, Republican, Independent, Independent, Democrat, Independent, Republican, Republican, Independent, Democrat, Democrat, Independent, Democrat, Independent, Republican, Independent, Independent, Independent, Democrat, Democrat, Republican, Independent, Independent, Republican, Republican, Democrat, Republican, Democrat, Independent, Independent, Democrat, Democrat, Independent, Republican, Independent, Independent, Democrat, Independent, Republican, Democrat.

14 To verify the claim that the mean is generally more reliable than the median (namely, that it is subject to smaller chance fluctuations), a student conducted an experiment consisting of 12 tosses of three dice. The following are his results: 2, 4, and 6; 5, 3, and 5; 4, 5, and 3; 5, 2, and 3; 6, 1, and 5; 3, 2, and 1; 3, 1, and 4; 5, 5, and 2; 3, 3, and 4; 1, 6, and 2; 3, 3, and 3; 4, 5, and 3.

    (a) Calculate the twelve medians and the twelve means.

    (b) Group the medians and the means obtained in (a) into separate distributions having the classes 1.5–2.5, 2.5–3.5, 3.5–4.5, and 4.5–5.5. (Note that there will be no ambiguities since the medians of three whole numbers and the means of three whole numbers cannot equal 2.5, 3.5, or 4.5.)

    (c) Draw histograms of the two distributions obtained in (b) and explain how they illustrate the claim that the mean is generally more reliable than the median.

15 Repeat Exercise 14 with your own data by repeatedly rolling three dice (or one die three times) and construct corresponding distributions for the twelve medians and the twelve means. (If no dice are available, simulate the experiment mentally or by drawing numbered slips of paper out of a hat.)

16 In averaging quantities it is often necessary to account for the fact that not all of them are equally important in the phenomenon being described. For instance, in 1975 the average annual cost of heating and cooling a home by means of a heat pump was $885.89 in Philadelphia, $688.80 in Houston, and $1,044.91 in Concord, Mass., and the mean of these three figures is $873.20, but we cannot very well say that this is the average annual cost of operating a heat pump in one of these cities. The three figures do not carry equal weight because there are not equally many heat pumps used to heat and cool homes in the three cities. In order to give quantities being averaged their proper degree of importance, it is necessary to assign them (relative importance) **weights,** and then calculate a **weighted mean.** In general, the weighted mean $\bar{x}_w$ of a set of numbers x_1, x_2, $x_3, \ldots$, and x_n, whose relative importance is expressed numerically by a corresponding set of numbers $w_1, w_2, w_3, \ldots$, and w_n, is given by

Weighted mean

$$\bar{x}_w = \frac{w_1 x_1 + w_2 x_2 + \cdots + w_n x_n}{w_1 + w_2 + \cdots + w_n} = \frac{\sum w \cdot x}{\sum w}$$

If all the weights are equal, this formula reduces to that of the ordinary (arithmetic) mean.

    (a) If somebody invests $2,000 at 6 percent, $5,000 at 7 percent, and $25,000 at 8 percent, what is the average return on these investments?

    (b) If an instructor counts the final examination in a course four times as much as each one-hour examination, what is the weighted average grade of a student who received grades of 70, 54, 73, and 67 in four one-hour examinations and a final examination grade of 76?

    (c) In 1974, cod, flounder, tuna, and haddock brought commercial fishermen 19.6, 26.5, 41.0, and 36.9 cents per pound. Given that they caught 59 million pounds of cod, 156 million pounds of flounder, 386 million pounds of tuna, and 8 million pounds of haddock, what is the over-all average price they received per pound?

17 The following is a special case of the weighted mean of Exercise 16. Given k sets of data having the means $\bar{x}_1, \bar{x}_2, \bar{x}_3, \ldots,$ and $\bar{x}_k$, and consisting, respectively, of $n_1, n_2, n_3, \ldots,$ and n_k measurements or observations, the **grand mean** of all the data is

Grand mean of combined data

$$\bar{\bar{x}} = \frac{n_1\bar{x}_1 + n_2\bar{x}_2 + \cdots + n_k\bar{x}_k}{n_1 + n_2 + \cdots + n_k} = \frac{\sum n \cdot \bar{x}}{\sum n}$$

where the numerator represents the total of all the observations and the denominator represents the total number of observations.

(a) In an anthropology class there are 22 juniors, 18 seniors, and 10 graduate students. If the juniors averaged 71 in the midterm examination, the seniors averaged 78, and the graduate students averaged 89, what is the mean for the entire class?

(b) In 1976, a college paid its 42 instructors a mean salary of $9,200, its 86 assistant professors a mean salary of $11,800, its 67 associate professors a mean salary of $13,900, and its 55 full professors a mean salary of $17,500. What was the mean salary paid to the 250 members of this faculty?

(c) In 1977, a sample survey yielded the following data on the average number of times per year that 200 persons in various age groups visit a doctor:

Age group	Number of persons in the sample	Mean number of visits
Under 6	20	6.5
6–16	55	2.9
17–44	68	5.0
45 and over	57	6.2

What is the mean for all 200 persons in the sample?

18 The **geometric mean** of n positive numbers is the nth root of their product. For example, the geometric mean of 3 and 12 is $\sqrt{3 \cdot 12} = \sqrt{36} = 6$ and the geometric mean of $\frac{1}{3}$, 1, and 81 is $\sqrt[3]{\frac{1}{3} \cdot 1 \cdot 81} = \sqrt[3]{27} = 3$. The geometric mean is used mainly to average ratios, rates of change, economic indexes, and the like.

(a) Find the geometric mean of 8 and 32.

(b) Find the geometric mean of 1, 2, 8, and 16.

(c) During a recent flu epidemic, 12 cases were reported on the first day, 18 on the second day, and 48 on the third day. Thus, from the first day to the second day the number of cases reported was multiplied by $\frac{18}{12}$, and from the second day to the third day the number of cases was multiplied by $\frac{48}{18}$. Find the geometric mean of these two growth rates and (assuming that the growth pattern continues) predict the number of cases that will be reported on the fourth and fifth days.

In actual practice, geometric means are usually calculated by making use of the fact that the logarithm of the geometric mean of a set of positive numbers equals the arithmetic mean of their logarithms.

19 The **harmonic mean** of n numbers $x_1, x_2, x_3, \ldots,$ and x_n is n divided by the sum of the reciprocals of the n numbers; namely, $\dfrac{n}{\sum 1/x}$. For instance, if \$12 is spent on vitamin pills costing 40 cents a dozen and another \$12 is spent on vitamin pills costing 60 cents a dozen, the average price is not $\dfrac{40 + 60}{2} = 50$ cents a dozen.

Since a total of \$24 is spent on a total of $\dfrac{1,200}{40} + \dfrac{1,200}{60} = 50$ dozen vitamin pills, the average price is $\dfrac{2,400}{50} = 48$ cents a dozen.

   (a) Verify that the harmonic mean of 40 and 60 is 48, so that it gives the appropriate "average" in the above example.
   (b) If an investor buys \$9,000 worth of a company's stock at \$45 a share and \$9,000 worth at \$36 a share, calculate the average price which the investor pays per share, and verify that it is the harmonic mean of \$45 and \$36.
   (c) If a jet travels the first third of a trip at 300 miles per hour, the next third at 450 miles per hour, and the final third at 360 miles per hour, use the harmonic mean to determine its average speed for the whole trip.

In actual practice, the harmonic mean is used only when dictated by special circumstances, as in the above examples.

3.4
Measures of Variation

One of the most important characteristics of almost any set of data is that the values are not all alike; indeed, the extent to which they are unalike, or vary among themselves, is of basic importance in statistics. Measures of central location describe one important aspect of a set of data—their "middle" or their "average"—but they tell us nothing about this other basic characteristic. Consequently, we require ways of measuring the extent to which data are dispersed, or spread out, and the statistical measures which provide this information are called **measures of variation.** The following are some examples which will illustrate the importance of measuring the variability of statistical data.

EXAMPLE   Suppose that a none-too-honest land developer claims that the average temperature at his development is a "comfortable" 75 degrees. This figure may be correct, but it would certainly be misleading if part of the time the temperature is a frigid 35 degrees, while more than half the time it is close to 100 degrees. What is needed here besides the average is some indication of the actual fluctuations in temperature.

EXAMPLE   Suppose that in a hospital each patient's pulse rate is taken in the morning, at noon, and in the evening, and that on a certain day the records show

	Pulse rate (beats per minute)		
Patient *A*	72	76	74
Patient *B*	72	91	59

The mean pulse rates of the two patients are the same (74, as can easily be checked), but observe the difference in variability. Whereas patient A's pulse rate appears to be quite stable, that of patient B fluctuates widely, and this may well be reason for serious concern.

EXAMPLE The concept of variability is of special importance in statistical inference (estimation, tests of hypotheses, prediction, etc.). Suppose, for example, that we have a coin which is slightly bent and we wonder whether it is still balanced or "fair;" that is, whether it will still come up heads about 50 percent of the time. So, we toss the coin 100 times and get 28 heads and 72 tails. Is there anything out of the ordinary about this result? Specifically, does the shortage of heads—only 28 where we might have expected 50—imply that the coin is not fair?

To answer this question, we must have some idea about the magnitude of the fluctuations, or variations, brought about by chance in the number of times a coin comes up heads when it is tossed 100 times. Suppose, thus, that we take a coin in mint condition, repeatedly toss it 100 times, and that in 10 sets of 100 tosses each we get, respectively, 44, 59, 50, 53, 40, 46, 51, 48, 54, and 56 heads. Since the number of heads varies from 40 to 59, we might conclude that a shortage or excess of 10 heads from the expected 50 heads is not unusual, but that the shortage of 22 heads, which we got with the bent coin, is so much larger that we would be reluctant to attribute it to chance. In other words, it would seem reasonable to conclude that the original coin is not fair.

We have given these three examples to show that the concept of variability plays an important role in the analysis of statistical data. Next, we shall see how it can actually be measured.

3.5
The Range

To introduce one way of measuring variability, let us refer back to the second of the examples of the preceding section, and let us observe that the pulse rate of patient *A* varied from 72 to 76, while that of patient *B* varied from 59 to 91. These extreme (smallest and largest) values give us an indication of the variability of the respective sets of data, and more or less the same thing is accomplished by taking the differences between the respective extremes. For patient *A* we obtain a **range** of $76 - 72 = 4$, and for patient *B* we obtain a range of $91 - 59 = 32$.

EXAMPLE   For the air pollution data on page 14, the smallest value is 6.2, the largest value is 31.8, so the range of the data is $31.8 - 6.2 = 25.6$; for the data on the lifetimes of light bulbs on page 31, the smallest value is 922, the largest value is 967, so the range is $967 - 922 = 45$.

The range is easy to calculate and easy to understand, but despite these advantages it is generally not a very useful measure of variation. Its main shortcoming is that it tells us nothing about the dispersion of the values which fall between the two extremes.

EXAMPLE   Each of the following three sets of data

6	18	18	18	18	18	18	18	18	18
6	6	6	6	6	18	18	18	18	18
6	7	9	11	12	14	15	16	17	18

has a range of $18 - 6 = 12$, but the dispersions of the data are hardly the same.

Nevertheless, when the sample size is quite small, the range can be an adequate measure of variation. For instance, it is used widely in industrial quality control to keep a close check on the consistency of raw materials or products, or on the uniformity of a process, by taking small samples at regular intervals of time and observing the range of the data.

3.6

The Standard Deviation

To introduce the **standard deviation**, by far the most generally useful measure of variation, let us observe that the dispersion of a set of data is small if the values are closely bunched about their mean, and that it is large if the values are scattered widely about their mean. It would seem reasonable, therefore, to measure the variation of a set of data in terms of the amounts by which the values deviate from their mean. If a set of numbers $x_1, x_2, x_3, \ldots$, and x_n, constituting a sample, has the mean $\bar{x}$, the differences $x_1 - \bar{x}, x_2 - \bar{x}, x_3 - \bar{x}, \ldots$, and $x_n - \bar{x}$ are called the **deviations from the mean,** and it suggests itself that we might use their average (namely, their mean) as a measure of the variation of the sample. Unfortunately, this will not do. Unless the x's are all equal, some of the deviations will be positive, some will be negative, and as the reader will be asked to show in Exercise 9 on page 62, the sum of the deviations from the mean, $\sum (x - \bar{x})$, and consequently also their mean, is always zero.

Since we are really interested in the magnitude of the deviations, and not in whether they are positive or negative, we might simply ignore the signs and define a measure of variation in terms of the absolute values of the deviations from the mean. Indeed, if we add the deviations from the mean as if they were all positive or zero and divide by n, we obtain the statistical measure which is called the **mean deviation.** This measure has intuitive appeal, but because of the absolute values it leads to serious difficulties, theoretical ones, in problems of inference.

An alternative approach is to work with the squares of the deviations from the mean, as this will also eliminate the effect of the signs. Squares of real numbers cannot be negative; in fact, squares of the deviations from a mean are all positive unless a value happens to coincide with the mean. Then, if we average the squared deviations from the mean and take the square root of the result (to compensate for the fact that the deviations were squared), we get

$$\sqrt{\frac{\sum (x - \bar{x})^2}{n}}$$

and this is how, traditionally, the standard deviation used to be defined. Expressing literally what we have done here mathematically, it is also called the **root-mean-square deviation.**

Nowadays, it is customary to modify this formula by dividing the sum of the squared deviations from the mean by $n - 1$ instead of n. Following this practice, which will be explained below, let us define the **sample standard deviation,** denoted s, as

Sample standard deviation

$$s = \sqrt{\frac{\sum (x - \bar{x})^2}{n - 1}}$$

and its square, the **sample variance,** as

Sample variance

$$s^2 = \frac{\sum (x - \bar{x})^2}{n - 1}$$

3.6   The Standard Deviation

To facilitate the calculation of standard deviations, a table of square roots, Table XIII, is given at the end of the book; the use of this table is explained on page 479.

These formulas for the standard deviation and the variance apply to samples, but if we substitute μ for $\bar{x}$ and N for n, we obtain analogous formulas for the standard deviation and the variance of a population. It has become fairly general practice to denote the population standard deviation σ (lower-case *sigma*, the Greek letter for s) when dividing by N, and S when dividing by $N - 1$; symbolically

Population standard deviation

$$\sigma = \sqrt{\frac{\sum (x - \mu)^2}{N}} \quad \text{and} \quad S = \sqrt{\frac{\sum (x - \mu)^2}{N - 1}}$$

Ordinarily, the purpose of calculating a sample statistic (such as the mean, the standard deviation, or the variance) is to estimate the corresponding population parameter. If we actually took many samples from a population which has the mean μ, calculated the sample means $\bar{x}$, and then averaged all these estimates of μ, we should find that their average is very close to μ. However, if we calculated the variance of each sample by means of the formula $\dfrac{\sum (x - \bar{x})^2}{n}$, and then averaged all these supposed estimates of σ^2, we would probably find that their average is less than σ^2. Theoretically, it can be shown that we can compensate for this by dividing by $n - 1$ instead of n in the formula for s^2. Estimators which have the desirable property that their values will on the average equal the quantity they are supposed to estimate are said to be **unbiased**; otherwise, they are said to be **biased**. So, we say that $\bar{x}$ is an unbiased estimator of the population mean μ, and that s^2 is an unbiased estimator of the population variance σ^2. It does not follow from this, however, that s is also an unbiased estimator of σ; but when n is large the bias is small, so we can use s as an estimate of σ.

In calculating the sample standard deviation using the formula by which it is defined, we must (1) find $\bar{x}$, (2) determine the n deviations from the mean $x - \bar{x}$, (3) square these deviations, (4) add all the squared deviations, (5) divide by $n - 1$, and (6) take the square root of the result obtained in step 5.

EXAMPLE   Suppose that a bacteriologist found 6, 12, 9, 10, 6, and 8 micro-organisms of a certain kind in six cultures. First calculating $\bar{x}$, we get

$$\bar{x} = \frac{6 + 12 + 9 + 10 + 6 + 8}{6} = 8.5$$

and the work required to find $\sum (x - \bar{x})^2$ may be arranged as in the following table[†]:

x	$x - \bar{x}$	$(x - \bar{x})^2$
6	-2.5	6.25
12	3.5	12.25
9	0.5	0.25
10	1.5	2.25
6	-2.5	6.25
8	-0.5	0.25
	0.0	27.50

Then, dividing by $6 - 1 = 5$ and taking the square root, we get

$$s = \sqrt{\frac{27.50}{5}} = \sqrt{5.5} = 2.3$$

rounded to one decimal.

It was easy to calculate s in this example because the data were whole numbers and the mean was exact to one decimal. In general, the calculations required by the formulas defining s and s^2 can be quite tedious, and it may be advantageous to use instead the formula

Computing formula for the sample standard deviation

$$s = \sqrt{\frac{n(\sum x^2) - (\sum x)^2}{n(n - 1)}}$$

or the corresponding formula for s^2.

EXAMPLE   Referring again to the number of microorganisms observed in the six cultures, we first calculate the two sums

$$\sum x = 6 + 12 + 9 + 10 + 6 + 8 = 51$$

and

$$\sum x^2 = 6^2 + 12^2 + 9^2 + 10^2 + 6^2 + 8^2 = 461$$

[†] Note that the sum of the entries in the middle column is equal to zero. Since this must always be the case, it provides a check on the calculations.

Then, substitution into the formula yields

$$s = \sqrt{\frac{6(461) - (51)^2}{6 \cdot 5}} = \sqrt{\frac{165}{30}} = \sqrt{5.5} = 2.3$$

rounded to one decimal.

The result which we obtained here is the same as before. Indeed, the computing formula for s gives the exact value of s, not an approximation, and its main advantage is that we do not have to work with the deviations from the mean—all we need is the sum of the x's and the sum of their squares. Aside from its advantage in manual calculations, the computing formula for s (or a slight modification of it) is the one usually preprogrammed into electronic statistical calculators.

In Section 3.4 we showed that there are many situations in which knowledge of the variability of a set of data can be of importance. Another application arises in the comparison of numbers belonging to different sets of data.

EXAMPLE The final examination of a French course consists of two parts: vocabulary and grammar. If a certain student scored 66 points on the vocabulary part and 80 points on the grammar part, it would seem that she did much better on the second part than on the first. However, if all the students in the class averaged 51 points on the vocabulary part with a standard deviation of 12 points, and 72 points on the grammar part with a standard deviation of 16 points, we can argue that the given student's score on the vocabulary part was $\dfrac{66 - 51}{12} = 1.25$ standard deviations above the class average, while her score on the grammar part was only $\dfrac{80 - 72}{16} = 0.50$ standard deviation above the class average. Whereas the original scores cannot be meaningfully compared, these new scores, expressed in terms of standard deviations, can. Clearly, the given student rates much higher on her command of French vocabulary than on her knowledge of French grammar, compared to the rest of the class.

What we did in this example consisted of converting the raw scores into **standard units**, or **z-scores**. In general, if x is a measurement belonging to a set of data having the mean $\bar{x}$ (or μ) and the standard deviation s (or σ), then its value in standard units, denoted by the letter z, is

Formula for converting to standard units

$$z = \frac{x - \bar{x}}{s} \quad or \quad z = \frac{x - \mu}{\sigma}$$

depending on whether the data constitute a sample or a population. In these units, z tells how many standard deviations a value lies above or below the mean of the set of data to which it belongs. Standard units will be used frequently in later chapters.

3.7

Chebyshev's Theorem

In the argument which led to the definition of the standard deviation, we observed that the dispersion of a set of data is small if the values are bunched closely about their mean, and that it is large if the values are scattered widely about their mean. Correspondingly, we can now say that if the standard deviation of a set of data is small, the values are concentrated near the mean, and if the standard deviation is large, the values are scattered widely about the mean. To present this argument on a less intuitive basis (after all, what is small and what is large?), let us refer to a theorem named after the Russian mathematician P. L. Chebyshev (1821–1894). This theorem states that

Chebyshev's theorem

> *For any set of data (population or sample) and any constant k greater than 1, at least $1 - 1/k^2$ of the data must lie within k standard deviations on either side of the mean.*

Thus, we can be sure that at least $1 - \frac{1}{4} = \frac{3}{4}$, or 75 percent, of the values in any set of data must lie within 2 standard deviations on either side of the mean; at least $1 - \frac{1}{9} = \frac{8}{9}$, or about 88.9 percent, of the values in any set of data must lie within 3 standard deviations on either side of the mean; and at least $1 - \frac{1}{25} = \frac{24}{25}$, or 96 percent, of the values in any set of data must lie within 5 standard deviations of the mean.

EXAMPLE   If all the 1-pound cans of coffee filled by a food processor have a mean weight of 16.00 ounces with a standard deviation of 0.02 ounce, we can be sure that at least 75 percent of the cans contain between 15.96 and 16.04 ounces of coffee, and that at least 96 percent of the cans contain between 15.90 and 16.10 ounces of coffee.

EXAMPLE   If two populations have the same mean $\mu = 140$, what can we say about the percentages of the values that must lie between 125 and 155, given that their respective standard deviations are $\sigma = 10$ and $\sigma = 1.5$. For the first population the values must lie within $\frac{15}{10} = 1.5$ standard deviations of the mean, and for the second population the values must lie within $\frac{15}{1.5} = 10$ standard deviations of the mean.

Therefore, at least $1 - \dfrac{1}{(1.5)^2} = \dfrac{5}{9}$, or about 55.6 percent, of the values of the first population must lie between 125 and 155, while the corresponding percentage for the second population is at least $1 - \dfrac{1}{10^2} = \dfrac{99}{100}$, or 99 percent. This illustrates how the magnitude of the standard deviation "controls" the concentration of a set of data about its mean.

EXERCISES

1   A recent study showed that the cost of buying a typical "shopping basket" of 50 food items for consumption at home was $183.40 in Stockholm, $175.10 in Geneva, $252.15 in Tokyo, $109.55 in London, $142.15 in Paris, $153.75 in Frankfurt, and $117.85 in New York. Determine the median and the range of these data.

2   With reference to Exercise 9 on page 23, find the range of the body weights of the 80 rats.

3   The records of 15 persons convicted of various crimes show that, respectively, 3, 4, 1, 3, 0, 2, 0, 1, 3, 4, 2, 0, 0, 3, and 4 of their grandparents were born in the United States. Find the standard deviation of these figures using
(a) the formula by which s is defined;
(b) the computing formula for s.

4   The following are the wind velocities reported at an airport at 6 P.M. on six consecutive days: 13, 8, 15, 11, 3, and 10. Find the variance of these figures using
(a) the formula by which s^2 is defined;
(b) the computing formula for s^2.

5   In four attempts it took a mechanic 14, 11, 16, and 20 minutes to perform a certain task.
(a) Use the computing formula for s to calculate the standard deviation of these data.
(b) Subtract 10 from each of the values and then use the computing formula to calculate the standard deviation of the resulting data. What general rule does this suggest for simplifying the calculation of s?

6   With reference to Exercise 4 on page 36, find the variance of the weight losses of the ten persons.

7   With reference to Exercise 6 on page 36, find the standard deviation of the IQ's of the twenty persons.

8   Find s^2 for the twelve means and the twelve medians obtained in part (a) of Exercise 14 on page 38. What is illustrated by the difference in the size of these two variances?

9   In a city in the Southwest, restaurants charge on the average $6.45 for a steak dinner (with a standard deviation of $0.40), $3.65 for a chicken dinner (with a standard deviation of $0.25), and $8.95 for a lobster dinner (with a standard deviation of $0.30). If a restaurant in this city charges $7.05 for a steak dinner,

$4.15 for a chicken dinner, and $9.35 for a lobster dinner, which of the three dinners is relatively most overpriced?   *The chicken dinner.*

10 The applicants to one state university have an average ACT mathematics score of 20.4 with a standard deviation of 3.1, while the applicants to another state university have an average ACT mathematics score of 21.1 with a standard deviation of 2.8. If a student scores 25 on this test, with respect to which of these two universities is he or she relatively speaking in a better position? How about a student who scores 30 on this test?

11 Among two persons on a reducing diet, the first belongs to an age group for which the mean weight is 146 pounds with a standard deviation of 14 pounds, and the second belongs to an age group for which the mean weight is 160 pounds with a standard deviation of 17 pounds. If their respective weights are 178 pounds and 193 pounds, which of the two is more seriously overweight for his or her age group?

12 An airline's records show that its flights between two cities arrive on the average 4. 6 minutes late with a standard deviation of 1.4 minutes. At least what percentage of its flights between these two cities arrive anywhere between
   (a) 1.8 minutes late and 7.4 minutes late?   *75%*
   (b) 1.0 minutes early and 10.2 minutes late?   *94%*
   (c) 3.8 minutes early and 13.0 minutes late?   *97%*

13 A study of the nutritional value of a certain kind of bread shows that on the average one slice contains 0.260 milligram of thiamine (vitamin B_1) with a standard deviation of 0.005 milligram. Between what values must be the thiamine content of
   (a) at least $\frac{35}{36}$ of all slices of this bread?   *0.23 – 0.29*
   (b) at least $\frac{63}{64}$ of all slices of this bread?   *0. 22 – 0. 30 .*

14 If the weights of certain objects have a standard deviation of 0.1 ounce, this information alone does not really tell us whether there is a great amount of variability, or very little variability, among these weights. A standard deviation of 0.1 ounce would reflect a great amount of variability if we are weighing the eggs of tiny birds, but very little variability if we are weighing 100-pound bags of potatoes. What we need in a situation like this is a measure of **relative variation** such as the **coefficient of variation** which expresses the standard deviation as a percentage of the mean. It is given by

Coefficient of variation

$$V = \frac{s}{\bar{x}} \cdot 100 \quad or \quad V = \frac{\sigma}{\mu} \cdot 100$$

   (a) With reference to Exercise 4, find the coefficient of variation of the wind velocities.
   (b) With reference to Exercise 5, find the coefficient of variation of the amounts of time it took the mechanic to perform the given task.
   (c) To compare the precision of two micrometers, a laboratory technician studies recent measurements made with both instruments. The first

micrometer was recently used to measure the diameter of a ball bearing and several measurements had a mean of 4.98 mm and a standard deviation of 0.018 mm; the second was recently used to measure the unstretched length of a spring and several measurements had a mean of 2.56 in with a standard deviation of 0.012 in. Which of the two micrometers is relatively more precise? *Micrometer #1*

3.8

The Description of Grouped Data ★

Since published data are often available only in the form of a frequency distribution, we shall discuss briefly the calculation of statistical descriptions of grouped data.

As we have already seen, the grouping of data entails some loss of information. Each item loses its identity, so to speak; we only know how many items there are in each class, so we must be satisfied with approximations. In the case of the mean and the standard deviation, we can usually get good approximations by assigning to each item falling into a class the value of the class mark. For instance, to calculate the mean or the standard deviation of the grouped sulfur oxides emission data on page 14, we treat the three values falling into the class 5.0–8.9 as if they were all 6.95, the ten values falling into the class 9.0–12.9 as if they were all 10.95, ..., and the two values falling into the class 29.0–32.9 as if they were both 30.95. This procedure is usually quite satisfactory, since the errors which are thus introduced into the calculations will more or less "average out."

To give general formulas for the mean and the standard deviation of a distribution with k classes, let us designate the successive class marks x_1, $x_2, x_3, \ldots$, and x_k, and the corresponding class frequencies $f_1, f_2, f_3, \ldots$, and f_k. The total that goes into the numerator of the formula for the mean is the sum obtained by adding x_1 times f_1, x_2 times f_2, x_3 times $f_3, \ldots$, and x_k times f_k; namely, $x_1 f_1 + x_2 f_2 + x_3 f_3 + \cdots + x_k f_k = \sum x \cdot f$. In words, $\sum x \cdot f$ is the sum of the products obtained by multiplying each class mark by the corresponding class frequency. Thus, we can write

Mean of grouped data

$$\bar{x} = \frac{\sum x \cdot f}{n}$$

where $n = f_1 + f_2 + f_3 + \cdots + f_k = \sum f$ is the size of the sample (namely, the total number of items grouped). If the data constitute a population instead of a sample, we substitute μ for $\bar{x}$ in this formula, and N for n.

Similarly, the total that goes into the numerator of the formulas defining the sample variance and the sample standard deviation is the sum obtained by adding $(x_1 - \bar{x})^2$ times f_1, $(x_2 - \bar{x})^2$ times f_2, $(x_3 - \bar{x})^2$ times

$f_3, \ldots,$ and $(x_k - \bar{x})^2$ times f_k. Writing this sum as $\sum (x - \bar{x})^2 \cdot f$, we get

Standard deviation of grouped data

$$s = \sqrt{\frac{\sum (x - \bar{x})^2 \cdot f}{n - 1}}$$

This formula serves to define the standard deviation of grouped data, but in practice we use a computing formula analogous to the one on page 45. Replacing $\sum x$ by $\sum x \cdot f$ in the computing formula, and $\sum x^2$ by $\sum x^2 \cdot f$, we get

Computing formula for the standard deviation of grouped data

$$s = \sqrt{\frac{n(\sum x^2 \cdot f) - (\sum x \cdot f)^2}{n(n - 1)}}$$

To get corresponding formulas for the population standard deviation σ, we substitute μ for $\bar{x}$ and N for $n - 1$ in the first formula, and N for n in the numerator and N^2 for $n(n - 1)$ in the denominator of the second formula.

EXAMPLE   Let us calculate the mean and the standard deviation of the distribution of the sulfur oxides emission data on page 14. The first column of the table which follows contains the class marks (calculated on page 15, the second column contains the class frequencies, and the third and fourth columns contain the products $x \cdot f$ and $x^2 \cdot f$; the required totals are shown at the bottom of the table.

Class marks x	Frequencies f	$x \cdot f$	$x^2 \cdot f$
6.95	3	20.85	144.9075
10.95	10	109.50	1,199.0250
14.95	14	209.30	3,129.0350
18.95	25	473.75	8,977.5625
22.95	17	390.15	8,953.9425
26.95	9	242.55	6,536.7225
30.95	2	61.90	1,915.8050
	80	1,508.00	30,857.0000

Substituting $n = 80$ and the necessary sums into the formula for $\bar{x}$ and the computing formula for s, we obtain

$$\bar{x} = \frac{1,508}{80} = 18.85$$

and

$$s = \sqrt{\frac{80(30{,}857) - (1{,}508)^2}{80 \cdot 79}} = 5.55$$

These calculations were very tedious, and we went through them mainly to dramatize the simplification that can be attained by **coding** the class marks so that we can work with smaller numbers. Provided the class intervals are all equal, this coding consists of assigning the value 0 to one of the class marks (preferably at or near the center of the distribution), and representing all the class marks by means of successive integers.

EXAMPLE If a distribution has nine classes and the class mark of the middle class is assigned the value 0, the successive class marks of the distribution are assigned the values $-4, -3, -2, -1, 0, 1, 2, 3,$ and 4.

Of course, when we code the class marks like this, we must account for it in the formulas for the mean and the standard deviation. Referring to the new (coded) class marks as u's, we write

Mean of grouped data (with coding)

$$\bar{x} = x_0 + \frac{\sum u \cdot f}{n} \cdot c$$

where x_0 is the class mark in the original scale to which we assign 0 in the new scale, c is the class interval, n is the number of items grouped, and $\sum u \cdot f$ is the sum of the products obtained by multiplying each of the new class marks by the corresponding class frequency. Similarly, we write

Standard deviation of grouped data (with coding)

$$s = c \sqrt{\frac{n(\sum u^2 \cdot f) - (\sum u \cdot f)^2}{n(n - 1)}}$$

where $\sum u^2 \cdot f$ is the sum of the products obtained by multiplying the squares of the new class marks by the corresponding class frequencies. If a set of data constitutes a population rather than a sample, we make the same modifications in the formulas as on page 51.

EXAMPLE To demonstrate the simplification brought about by coding, let us recalculate the mean and the standard deviation of the distribution of the emission data. Arranging the work, as before, in a table, we get

Original class marks x	New class marks u	f	$u \cdot f$	$u^2 \cdot f$
6.95	-3	3	-9	27
10.95	-2	10	-20	40
14.95	-1	14	-14	14
18.95	0	25	0	0
22.95	1	17	17	17
26.95	2	9	18	36
30.95	3	2	6	18
		80	-2	152

where the class mark 18.95 is taken to be 0 in the new scale. (Of course, we could have used any class mark as the zero for the new scale, but the objective is to make the numbers, and the arithmetic, as simple as possible.)

Substituting $c = 4$, $x_0 = 18.95$, $n = 80$, $\sum u \cdot f = -2$, and $\sum u^2 \cdot f = 152$ into the above formulas for $\bar{x}$ and s, we get

$$\bar{x} = 18.95 + \frac{-2}{80} \cdot 4 = 18.85$$

and

$$s = 4 \sqrt{\frac{80(152) - (-2)^2}{80 \cdot 79}} = 5.55$$

These results are, as they should be, identical with the ones which we obtained earlier without coding.

Once a set of data has been grouped, we cannot find the exact value of the median because of the loss of information which results from the act of grouping, but we can still find the class into which the median must fall. So, we define the median of a distribution in a special way, as **the number which is such that half the total area of the rectangles of the histogram of the distribution lies to its left and the other half lies to its right** (see Figure 3.2). This definition is equivalent to the assumption that the values in the class containing the median are distributed evenly (that is, spread out evenly) throughout that class.

3.8 The Description of Grouped Data

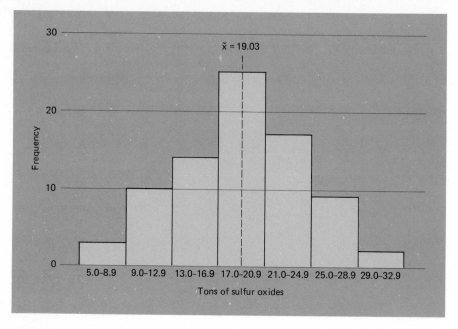

FIGURE 3.2

The median of the distribution of the emission data.

To find the dividing line between the two halves of a histogram (each of which represents $\frac{n}{2}$ of the items grouped), we must somehow count $\frac{n}{2}$ of the items starting at either end of the distribution.

EXAMPLE   For the distribution of the sulfur oxides emission data we have $n = 80$, and hence must count $\frac{80}{2} = 40$ items starting at either end. Counting from the bottom (that is, beginning with the smallest values), we find that $3 + 10 + 14 = 27$ of the values fall into the first three classes, and that $3 + 10 + 14 + 25 = 52$ of the values fall into the first four classes. Therefore, we must count $40 - 27 = 13$ more values beyond the 27 which fall into the first three classes, and on the assumption that the 25 values in the fourth class are spread evenly throughout that class, we can do this by adding $\frac{13}{25}$ of the class interval of 4 to 16.95, the lower boundary of the fourth class. This gives us

$$\tilde{x} = 16.95 + \frac{13}{25} \cdot 4 = 19.03$$

for the median of this distribution.

In general, if L is the lower boundary of the class into which the median must fall, f is its frequency, c is its class interval, and j is the number of items we still lack when we reach L, then the median of the distribution is

Median of grouped data

$$\tilde{x} = L + \frac{j}{f} \cdot c$$

If we prefer, we can find the median of a distribution by starting to count at the other end (beginning with the largest values) and subtracting an appropriate fraction of the class interval from the upper boundary U of the class into which the median must fall. The corresponding formula is

Alternate formula for the median of grouped data

$$\tilde{x} = U - \frac{j'}{f} \cdot c$$

where j' is the number of items we still lack when we reach U.

EXAMPLE Referring again to the distribution of the sulfur oxides emission data, we thus get

$$\tilde{x} = 20.95 - \frac{12}{25} \cdot 4 = 19.03$$

for the median, as before.

EXERCISES 1 The following is the distribution of the percentages of the students belonging to a certain ethnic group in 50 schools:

Percentage	Number of schools
0– 4	18
5– 9	15
10–14	9
15–19	7
20–24	1

Calculate $\bar{x}$ and s for this distribution
(a) without coding;
(b) with coding.

2 Calculate the median of the distribution of Exercise 1.

3 Find $\bar{x}$ and s^2 for the distribution of the weights of mineral specimens of Exercise 3 on page 22.

4 With reference to the distribution of the burning times obtained in part (a) of Exercise 8 on page 23, find
   (a) the mean;
   (b) the median;
   (c) the standard deviation.

5 With reference to the distribution of the body weights of rats obtained in part (a) of Exercise 9 on page 23, find
   (a) the mean;
   (b) the median;
   (c) the variance.

6 Referring to the distribution obtained in part (a) of Exercise 10 on page 23, calculate the mean and the standard deviation of this population of the amounts of money spent by a cab company on gasoline.

7 The three **quartiles** of a distribution Q_1, Q_2, and Q_3 are values which are such that 25 percent of the values are less than or equal to Q_1, 50 percent of the values are less than or equal to Q_2, and 75 percent are less than or equal to Q_3. To calculate the quartiles of a distribution we can use either of the two formulas for the median of a distribution on page 55. For Q_2 (which *is* the median) we count $\dfrac{n}{2}$ of the items starting at either end; for Q_1 we usually count $\dfrac{n}{4}$ of the items starting with the smallest values; and for Q_3 we usually count $\dfrac{n}{4}$ of the items starting with the largest values.
   (a) Find Q_1 and Q_3 for the distribution of the emission data on page 14.
   (b) Find Q_1, Q_2, and Q_3 for the distribution of Exercise 1.
   (c) Find Q_1, Q_2, and Q_3 for the distribution of the weights of mineral specimens of Exercise 3 on page 22.

8 Given the quartiles of a distribution (see Exercise 7), we can use as an alternative measure of variation the **interquartile range** $Q_3 - Q_1$; it is the length of the interval which contains the middle 50 percent of the data. Some research workers prefer to use the **semi-interquartile range**, also called the **quartile deviation,** which is given by the formula $\dfrac{Q_3 - Q_1}{2}$. A corresponding measure of relative variation, called the **coefficient of quartile variation,** is given by the ratio of the semi-interquartile range to the mean of Q_1 and Q_3 multiplied by 100; its formula can be written as $\dfrac{Q_3 - Q_1}{Q_3 + Q_1} \cdot 100$.
   (a) Use the results of part (a) of Exercise 7 to calculate the interquartile range and the coefficient of quartile variation for the distribution of the emission data.
   (b) Use the results of part (b) of Exercise 7 to calculate the semi-interquartile range and the coefficient of quartile variation for the distribution of Exercise 1.

(c) Use the results of part (c) of Exercise 7 to calculate the interquartile range and the coefficient of quartile variation for the distribution of the weights of the mineral specimens.

9 The procedure we use to find the median of a distribution (or its quartiles as in Exercise 7) can also be used to determine other **fractiles** or **quantiles**; that is, other values at or below which a given part of a set of data must fall. For instance, there are the nine **deciles** $D_1, D_2, \ldots$, and D_9, which are such that 10 percent of the data are less than or equal to D_1, 20 percent of the data are less than or equal to D_2, and so forth; and there are the 99 **percentiles** $P_1, P_2, \ldots$, and P_{99}, which are such that 1 percent of the data are less than or equal to P_1, 2 percent of the data are less than or equal to P_2, and so forth. It should be clear from this that D_5 and P_{50} are both equal to Q_2 or the median, and that P_{25} equals Q_1 and P_{75} equals Q_3.

(a) Find D_2, D_8, P_5, and P_{95} for the distribution of the emission data on page 14.

(b) Find $D_1, D_9, P_1, P_2, P_{98}$, and P_{99} for the following distribution of the grades obtained by 700 students in a geography test:

Grade	Number of students
10–24	64
25–39	100
40–54	132
55–69	187
70–84	165
85–99	52

10 To study the effect of grouping on the calculation of the mean and the standard deviation of a set of data, calculate

(a) the mean of the raw (ungrouped) emission data on page 14;

(b) their standard deviation.

For part (b) it will be necessary to use a calculator on which sums of squares can be accumulated, or a statistical calculator which gives the value of s directly. Compare the values obtained here with the corresponding values obtained for the distribution of the data; namely, $\bar{x} = 18.85$ and $s = 5.55$.

3.9

Some Further Descriptions ★

So far we have discussed statistical descriptions which come under the general heading of "measures of location" and "measures of variation." Actually, there is no limit to the number of ways in which statistical data can be described, and statisticians continually develop new methods of describing characteristics of numerical data that are of interest in particular problems. In this section we shall consider briefly the problem of describing the over-all shape of a distribution.

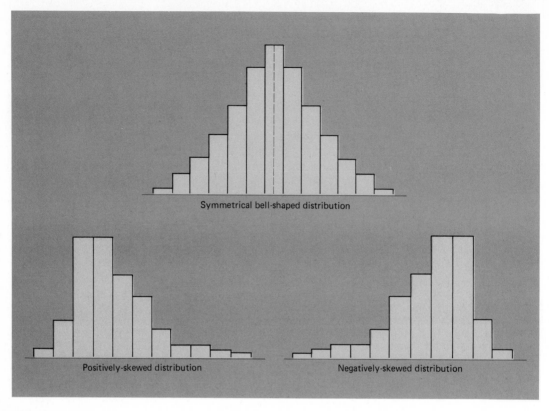

FIGURE 3.3
Bell-shaped distributions.

Although frequency distributions can take on almost any shape or form, most of the distributions we meet in practice can be described fairly well by one or another of a few standard types. Among these, foremost in importance is the aptly described symmetrical **bell-shaped distribution** shown in Figure 3.3. Indeed, there are theoretical reasons why, in many cases, distributions of actual data can be expected to follow its pattern. The other two distributions of Figure 3.3 can still, by a stretch of the imagination, be called bell-shaped, but they are not symmetrical. Distributions like these, having a "tail" on one side or the other, are said to be **skewed**; if the tail is on the left we say that they are **negatively skewed** and if the tail is on the right we say that they are **positively skewed.** Distributions of incomes or wages are often positively skewed because of the presence of some relatively high values that are not offset by correspondingly low values.

There are several ways of measuring the extent to which a distribution is skewed. A relatively easy one is based on the fact that for a perfectly symmetrical bell-shaped distribution such as the one of Figure 3.3, the values of the median and the mean coincide. Since the presence of some relatively high values that are not offset by correspondingly low values will tend to make the mean greater than the median (and the presence of some relatively low values that are not offset by correspondingly high values will tend to make the mean less than the median), we can use this relationship between the mean and the median to define a relatively simple measure of the extent to which a distribution is skewed. It is called the **Pearsonian coefficient of skewness**, and it is given by

Pearsonian coefficient of skewness

$$SK = \frac{3(mean - median)}{standard\ deviation}$$

For a perfectly symmetrical distribution the value of SK is 0, and in general its values must fall between -3 and 3.

EXAMPLE Substituting into this formula the values obtained for the mean, the median, and the standard deviation of the distribution of the sulfur oxides emission data (namely, $\bar{x} = 18.85$, $\tilde{x} = 19.03$, and $s = 5.55$), we get

$$SK = \frac{3(18.85 - 19.03)}{5.55} = -0.01$$

This value is small enough for us to say that the distribution is nearly symmetrical.

Two other kinds of distributions which sometimes arise in practice are the **reverse J-shaped** and **U-shaped** distributions shown in Figure 3.4. As can be seen from the diagram, the names of these distributions quite literally describe their shape. Examples of such distributions may be found in Exercises 4 and 5 on page 60.

FIGURE 3.4

Reverse J-shaped and U-shaped distributions.

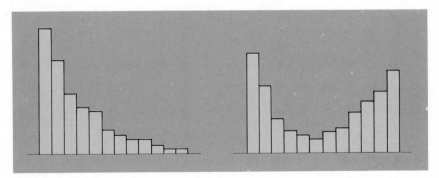

1 Use the results of Exercises 1 and 2 on pages 55 and 56 to calculate the Pearsonian coefficient of skewness for the distribution of the percentages.

2 Use the results of Exercise 4 on page 56 to calculate the Pearsonian coefficient of skewness for the distribution of the burning times.

3 Use the results of Exercise 5 on page 56 to calculate the Pearsonian coefficient of skewness for the distribution of the body weights of the rats.

4 Roll a pair of dice 120 times and construct a distribution showing how many times there were 0 sixes, how many times there was 1 six, and how many times there were 2 sixes. Draw a histogram of this distribution and describe its shape.

5 If a coin is flipped five times, the result may be represented by means of a sequence of H's and T's (for example, HHTTH), where H stands for *heads* and T for *tails*. Having obtained such a sequence of H's and T's, we can then check after each successive flip whether the number of heads exceeds the number of tails. For example, for the sequence HHTTH, heads is ahead after the first flip, after the second flip, after the third flip, not after the fourth flip, but again after the fifth flip; altogether, it is ahead four times. Repeat this experiment 50 times, and construct a histogram showing in how many cases heads was ahead altogether 0 times, 1 time, 2 times, . . . , and 5 times. Explain why the resulting distribution should be U-shaped.

3.10

Technical Note (Summations)

In the abbreviated notation introduced on page 30, $\sum x$ does not make it clear which, or how many, values of x we have to add. This is taken care of by the more explicit notation

$$\sum_{i=1}^{n} x_i = x_1 + x_2 + \cdots + x_n$$

where it is made clear that we are adding the x's whose subscripts i are 1, 2, . . . , and n. Generally, we shall not use the more explicit notation in this text to simplify the over-all appearance of the formulas, assuming that it is clear in each case what x's we are referring to and how many there are.

Using the $\sum$ notation, we shall also have occasion to write such expressions as $\sum x^2, \sum xy, \sum x^2 f, \ldots$, which (more explicitly) represent the sums

$$\sum_{i=1}^{n} x_i^2 = x_1^2 + x_2^2 + x_3^2 + \cdots + x_n^2$$

$$\sum_{j=1}^{m} x_j y_j = x_1 y_1 + x_2 y_2 + \cdots + x_m y_m$$

$$\sum_{i=1}^{n} x_i^2 f_i = x_1^2 f_1 + x_2^2 f_2 + \cdots + x_n^2 f_n$$

Working with two subscripts, we shall also have the occasion to evaluate **double summations** such as

$$\sum_{j=1}^{3} \sum_{i=1}^{4} x_{ij} = \sum_{j=1}^{3} (x_{1j} + x_{2j} + x_{3j} + x_{4j})$$

$$= x_{11} + x_{21} + x_{31} + x_{41} + x_{12} + x_{22} + x_{32} + x_{42}$$

$$+ x_{13} + x_{23} + x_{33} + x_{43}$$

To verify some of the formulas involving summations that are stated but not proved in the text, the reader will find it helpful to use the following rules:

Rules for summations

$$Rule\ A:\ \sum_{i=1}^{n} (x_i \pm y_i) = \sum_{i=1}^{n} x_i \pm \sum_{i=1}^{n} y_i$$

$$Rule\ B:\ \sum_{i=1}^{n} k \cdot x_i = k \cdot \sum_{i=1}^{n} x_i$$

$$Rule\ C:\ \sum_{i=1}^{n} k = k \cdot n$$

The first of these rules states that the summation of the sum (or difference) of two terms equals the sum (or difference) of the individual summations, and it can be extended to the sum or difference of more than two terms. The second rule states that we can, so to speak, factor a constant out of a summation, and the third rule states that the summation of a constant is simply *n* times that constant. All of these rules can be proved by actually writing out in full what each of the summations represents.

EXERCISES

1 Write each of the following in full; that is, without summation signs:

(a) $\sum_{i=1}^{6} x_i$;

(b) $\sum_{i=3}^{5} y_i$;

(c) $\sum_{i=1}^{3} x_i y_i$;

(d) $\sum_{j=1}^{8} x_j f_j$;

(e) $\sum_{i=3}^{7} x_i^2$;

(f) $\sum_{j=1}^{4} (x_j + y_j)$.

2 Write each of the following as summations:

(a) $z_1 + z_2 + z_3 + z_4 + z_5$; $\sum_{i=1}^{5} z$

(b) $x_5 + x_6 + x_7 + x_8 + x_9 + x_{10} + x_{11} + x_{12}$; $\sum_{i=5}^{12} x$

(c) $x_1 f_1 + x_2 f_2 + x_3 f_3 + x_4 f_4 + x_5 f_5 + x_6 f_6$; $\sum_{i=1}^{6} x_i f_i$

(d) $y_1^2 + y_2^2 + y_3^2$; $\sum_{i=1}^{3} y^2$

(e) $2x_1 + 2x_2 + 2x_3 + 2x_4 + 2x_5 + 2x_6 + 2x_7$;

(f) $(x_2 - y_2) + (x_3 - y_3) + (x_4 - y_4)$;

(g) $(z_2 + 3) + (z_3 + 3) + (z_4 + 3) + (z_5 + 3)$;

(h) $x_1 y_1 f_1 + x_2 y_2 f_2 + x_3 y_3 f_3 + x_4 y_4 f_4$.

3 Given $x_1 = 1, x_2 = 3, x_3 = -2, x_4 = 4, x_5 = -1, x_6 = 2, x_7 = 1$, and $x_8 = 2$, find

(a) $\sum_{i=1}^{8} x_i$; *10*

(b) $\sum_{i=1}^{8} x_i^2$. *40*

4 Given $x_1 = 3, x_2 = 4, x_3 = 5, x_4 = 6, x_5 = 7, f_1 = 3, f_2 = 7, f_3 = 10, f_4 = 5$, and $f_5 = 2$, find

(a) $\sum_{i=1}^{5} x_i$;

(c) $\sum_{i=1}^{5} x_i f_i$;

(b) $\sum_{i=1}^{5} f_i$;

(d) $\sum_{i=1}^{5} x_i^2 f_i$.

5 Given $x_1 = 2, x_2 = -3, x_3 = 4, x_4 = -2, y_1 = 5, y_2 = -3, y_3 = 2$, and $y_4 = -1$, find

(a) $\sum_{i=1}^{4} x_i$; *1*

(d) $\sum_{i=1}^{4} y_i^2$; *39*

(b) $\sum_{i=1}^{4} y_i$; *3*

(e) $\sum_{i=1}^{4} x_i y_i$. *29*

(c) $\sum_{i=1}^{4} x_i^2$; *33*

6 Use the more explicit summation notation to rewrite
   (a) the formula for the mean on page 30;
   (b) the formula for the mean of a distribution on page 50;
   (c) the formula for the weighted mean given in Exercise 16 on page 38;

7 Given $x_{11} = 3, x_{12} = 1, x_{13} = -2, x_{14} = 2, x_{21} = 1, x_{22} = 4, x_{23} = -2, x_{24} = 5, x_{31} = 3, x_{32} = -1, x_{33} = 2$, and $x_{34} = 3$, find

(a) $\sum_{i=1}^{3} x_{ij}$ separately for $j = 1, 2, 3$, and 4;

(b) $\sum_{j=1}^{4} x_{ij}$ separately for $i = 1, 2$, and 3.

8 With reference to the data of Exercise 7, evaluate the double summation $\sum_{i=1}^{3} \sum_{j=1}^{4} x_{ij}$ using
   (a) the results obtained in part (a) of Exercise 7;
   (b) the results obtained in part (b) of Exercise 7.

9 Show that $\sum_{i=1}^{n} (x_i - \bar{x}) = 0$ for any set of x's whose mean is $\bar{x}$.

10 Is it true in general that $\left(\sum_{i=1}^{n} x_i \right)^2 = \sum_{i=1}^{n} x_i^2$? (*Hint:* Check whether the equation holds for $n = 2$.)

BIBLIOGRAPHY

An informal discussion of the ethics involved in choosing among measures of location is given in

> HUFF, D., *How to Lie with Statistics*. New York: W. W. Norton & Company Inc., 1954.

A proof that division by $n - 1$ makes the sample variance an unbiased estimator of the population variance may be found in most textbooks on mathematical statistics; for instance, in

> FREUND, J. E., *Mathematical Statistics, 2nd ed.* Englewood Cliffs, N.J.: Prentice-Hall, Inc., 1972.

Some information about the effects of grouping on the calculation of x and s may be found in many of the older textbooks on statistics; for example, in

> MILLS, F. C., *Introduction to Statistics*. New York: Holt, Rinehart and Winston, Inc., 1956.

Further Exercises for Chapters

1,2, and 3

1 At 20 major intersections in a certain city there were 2, 0, 4, 1, 1, 3, 4, 0, 0, 2, 5, 1, 0, 3, 1, 2, 4, 0, 2, and 1 accidents during the Fourth of July weekend.
    (a) Find the mean and the median of these sample data.
    (b) Find the standard deviation and the range of these sample data.

2 The ages of the applicants for a certain job are grouped into a distribution having the classes 15–24, 25–34, 35–44, 45–54, and 55–64. What are the class boundaries if
    (a) the ages are rounded to the nearest year?
    (b) each applicant's age is given as of his or her last birthday?

3 If a student calculates his grade point index (that is, averages his grades) by counting A, B, C, D, and F, respectively, as 1, 2, 3, 4, and 5, what does this assume about the nature of such grades?

4 The mean weight of the 46 members of a football team is 212 pounds. If none of the players weighs less than 165 pounds, at most how many of them can weigh 250 pounds or more?

5 Use the formula of Exercise 17 on page 39 and the figures in the following table to determine the average weekly earnings of the 163, 900 workers employed in manufacturing, mining, and construction in Arizona in 1976:

	Number of workers	Average weekly earnings
Manufacturing	99,900	204.73
Mining	23,600	285.48
Construction	40,400	330.22

6 Chebyshev's theorem may be restated as follows: For any set of data and any constant k greater than 1, at most $\frac{1}{k^2}$ of the data can differ from the mean by k standard deviations or more.

(a) At most what percentage of the values in any set of data can differ from the mean by 7 standard deviations or more?

(b) Certain mass-produced metal shafts have a mean diameter of 28.00 mm (as required by specifications) with a standard deviation of 0.02 mm. At most what percentage of the shafts have a diameter which is off by 0.05 mm or more?

(c) Among a large number of persons tested, the mean reaction time to a visual stimulus is 0.45 second with a standard deviation of 0.04 second. At most what percentage of these persons took 0.65 second or more to react to the stimulus?

7 The following are the numbers of deer observed in 120 sections of land in a wildlife count:

11	14	21	15	16	14	18	11	17	11	17	12
18	8	9	22	12	16	20	33	15	21	18	13
21	0	16	12	20	17	13	20	10	16	5	10
13	19	0	2	14	17	11	18	16	13	12	6
5	19	10	6	15	1	26	8	18	19	1	14
15	10	16	14	29	17	4	18	20	10	16	9
17	6	15	14	22	7	7	13	19	0	15	17
8	12	13	21	8	11	19	1	14	4	19	16
2	16	11	18	10	28	15	24	8	20	6	7
12	7	18	3	14	19	14	0	23	7	13	2

Group these figures into a distribution having the classes 0–4, 5–9, 10–14, 15–19, 20–24, 25–29, and 30–34, and

(a) draw a histogram of this distribution;

(b) calculate $\bar{x}$;

(c) calculate the median;

(d) calculate s^2.

8 A rectangle has sides a and b. If we want to construct a square of side x, which "average" of a and b must be chosen for x so that

(a) the perimeter of the square will equal that of the rectangle?

(b) the area of the square will equal that of the rectangle?

9 Are the following nominal, ordinal, interval, or ratio data? Explain your answers.

(a) Elevations above sea level.

(b) Responses to the question whether (in the downtown area of a large city) living conditions are "getting much worse," "getting a little worse," "staying the same," "getting a little better," or "getting much better."

(c) Ages of second-hand cars.

(d) Data on eye color.

10 A technician working for a consumers' rating service found that four car doors made by company A withstood, respectively, 165, 172, 157, and 146 hours of continuous torture tests, while four car doors made by company B withstood 170, 182, 153, and 155 hours of continuous torture tests. Which of the following conclusions can be obtained from these data by purely descriptive methods and which require a statistical inference; namely, a generalization? Explain your answers.

(a) The car door which lasted the shortest period of time was made by company A.

(b) All car doors made by company B will withstand this kind of torture test for at least 150 hours.

(c) Each of the values obtained for the four car doors made by company B exceeds the smallest value obtained for the four car doors made by company A.

(d) Since the largest value obtained for the car doors made by company B is much larger than all the other values, it was probably recorded incorrectly.

11 Generalizing the argument used in the examples on page 31, it can be shown that for any set of non-negative data with the mean $\bar{x}$, the fraction of the data that are greater than or equal to any positive constant k cannot exceed $\dfrac{\bar{x}}{k}$.

Use this result, called **Markov's theorem,** to answer the following questions:

(a) If the mean breaking strength of certain linen threads is 32.5 ounces, at most what fraction of these threads can have a breaking strength of 40.0 ounces or more?

(b) If the diameters of the citrus trees in an orchard have a mean of 15.8 cm, at most what fraction of these trees can have a diameter of at least 25.0 cm?

12 With reference to Exercise 14 on page 25, what is the modal means of transportation by which the tourists arrived at the resort city?

13 In a popularity poll, 40 prominent coaches are asked to list their first, second, . . . , and tenth choices of basketball teams in the nation. Each first choice is counted ten points, each second choice nine points, . . . , and each tenth choice one point, and the teams are then ranked nationally on the basis of the total number of points which they receive.

(a) Are the final rankings nominal, ordinal, interval, or ratio data?

(b) What about the point totals which the teams receive?

14 An environmental engineer obtained the following data on the concentration of mercury (in parts per million) at ten locations along a stream: 0.062, 0.071, 0.067, 0.068, 0.066, 0.062, 0.068, 0.067, 0.060, and 0.065.

(a) Find $\bar{x}$ and s.

(b) Find the median and the range.

15 In two major golf tournaments one professional golfer finished second and ninth, while another finished sixth and fifth. Comment on the argument that since $2 + 9 = 6 + 5$, the over-all performance of the two golfers in these two tournaments was equally good.

16 Construct a pie chart (see Exercise 13 on page 25) to represent the following distribution:

*Number of B.A. degrees
in engineering conferred in
the United States in 1970–71*

Aeronautical	2,443
Chemical	3,579
Civil	6,526
Electrical	12,198
Mechanical	8,858
Industrial	3,171
General	2,864

17 Having kept records for many years, Mrs. Brown knows that it takes her on the average 47 minutes to prepare dinner; the standard deviation is 4.6 minutes. If she always starts preparing dinner at 10 minutes to 5, at least what percentage of the time has she had dinner ready before 6?

18 The following is the distribution of the numbers of mistakes 150 students made in translating a certain passage from Spanish to English:

Number of mistakes	Number of students
20–22	6
23–25	62
26–28	38
29–31	25
32–34	16
35–37	3

(a) Convert this distribution into a cumulative "less than" distribution and present it graphically in the form of an ogive.
(b) Calculate $\bar{x}$ and s.
(c) Calculate the median and the two quartiles Q_1 and Q_3.
(d) Calculate the coefficient of variation.
(e) Calculate the Pearsonian coefficient of skewness.

19 The dean of a university has complete records of how many A's, B's, C's, etc., each faculty member gave to his or her students during the academic year 1977–78. Give one example each of a situation in which the dean of this university would look upon these data
(a) as a population;
(b) as a sample.

20 A student took six readings of the direction from which the wind was blowing at a certain location, getting $9°$, $349°$, $350°$, $4°$, $18°$, and $350°$. (These angles are measured clockwise with $0°$ being due north.) Averaging these figures he gets a

mean of 180°, which would seem to indicate that on the average the wind blew from the south. Explain the fallacy of this argument and give a more appropriate way of presenting (and then averaging) these readings.

21 Find the mode (if it exists) of each of the following blood pressure readings:
    (a) 144, 145, 146, 146, 148, 146, 146, 145, 147, 145, 144, 147, 143;
    (b) 142, 146, 149, 145, 143, 146, 141, 146, 149, 147, 146, 149, 149;
    (c) 167, 151, 175, 160, 178, 144, 152, 148, 156, 169, 143, 177, 161.

22 To group sales invoices ranging from $15.00 to $40.00, a clerk uses the following classification: $15.00–$19.99, $20.00–$23.99, $25.00–$29.99, $30.00–$35.99, and $35.00–$39.99. Explain where difficulties might arise.

We cannot predict the outcome of a football game unless we know at least what teams are going to play, and we cannot predict which television program will get the highest rating in a given week unless we know at least what shows will be on the air. More generally, we cannot make intelligent predictions or decisions unless we know at least what is possible— in other words, we must know what is possible before we can judge what is probable. Thus, the first two sections of this chapter will be devoted to "what is possible" in a given situation, and then we shall learn how to judge also "what is probable."

Possibilities and Probabilities

4.1
Counting

In contrast to the complexity of many of the methods used nowadays in science, in business, and even in everyday life, the simple process of counting still plays an important role. We still have to count 1, 2, 3, 4, 5, . . . , for example, to determine how many persons take part in a demonstration, the size of the response to a questionnaire, the number of damaged cases in a shipment of wines from Portugal, or when preparing a report showing how many times the temperature in Phoenix, Arizona, went over 100 degrees in a given month. Sometimes, the process of counting can be simplified by using mechanical devices (for instance, when counting spectators passing through turnstiles), or by performing counts indirectly (for instance, by subtracting the serial numbers of invoices to determine the total number of sales). At other times, the process of counting can be simplified greatly by means of special mathematical techniques, such as the ones given below.

In the study of "what is possible," there are essentially two kinds of problems. There is the problem of listing everything that can happen in a given situation, and then there is the problem of determining how many different things can happen (without actually constructing a complete list). The second kind of problem is especially important, because there are many situations in which we do not need a complete list, and hence, can save ourselves a great deal of work. Although the first kind of problem may seem straightforward and easy, this is not always the case.

EXAMPLE   Suppose we want to know whether three law school graduates will pass the bar examination in Massachusetts on the first, second, or third try; specifically, we are interested only in how many of them pass the examination on each try. Clearly, there are many possibilities. All three might pass the examination on the first try; one might pass on the first try, another on the third try, while the other law school graduate fails all three times; one might pass on the first try and the other two on the second try; and all three of them might fail all three times. Continuing this way carefully, we might determine, correctly, that there are altogether 20 possibilities.

To handle this problem systematically, it would help to draw a **tree diagram** such as that of Figure 4.1. This diagram shows that for the first try there are four possibilities (four branches) corresponding to 0, 1, 2, or 3 of the law school graduates passing the bar examination. For the second try there are four branches emana-

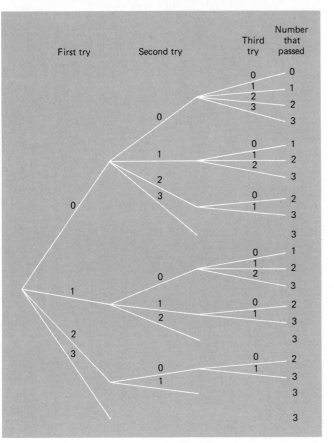

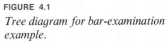

FIGURE 4.1

Tree diagram for bar-examination example.

ting from the top branch, three from the second branch, two from the third branch, and none from the bottom branch. Evidently, there are still four possibilities (0, 1, 2, or 3) when none of them passes on the first try, but only three possibilities (0, 1, or 2) when one passes on the first try, two possibilities (0 or 1) when two pass on the first try, and there is no need to go on when all three of them pass on the first try. The same sort of reasoning applies also to the third try, and (going from left to right) we find that there are altogether 20 different paths along the "branches" of the tree. In other words, 20 different things can happen in this situation. It can also be seen from this diagram that in ten of the cases all three of the law school graduates pass the Massachusetts bar examination (sooner or later) on the first three tries, in six of the cases only two of them pass the bar examination on the first three tries, in three of the cases only one of them passes on the first three tries, and in one case none of them passes on the first three tries.

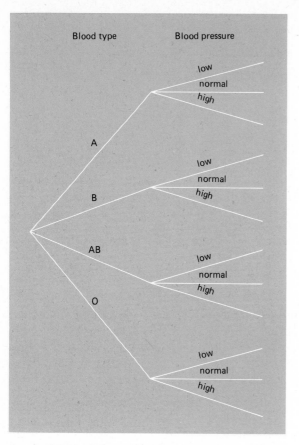

Blood type   Blood pressure

low
normal
high

A

low
normal
high

B

AB

low
normal
high

O

low
normal
high

FIGURE 4.2
*Tree diagram for classification of patients
in medical study.*

EXAMPLE   Suppose that in a medical study patients are classified according to whether they have blood type A, B, AB, or O, and also according to whether their blood pressure is low, normal, or high. To find the number of ways in which a patient can, thus, be classified, we have only to look at the tree diagram of Figure 4.2. Clearly, the answer is 12. Starting at the top, the first path along the "branches" of the tree corresponds to a patient having blood type A and low blood pressure, the second path corresponds to a patient having blood type A and normal blood pressure, . . . , and the twelfth path corresponds to a patient having blood type O and high blood pressure.

    The answer we got in the second example is $4 \cdot 3 = 12$; namely, the product of the number of ways in which a patient can be classified according to blood type and the number of ways in which a patient can be classified according to blood pressure. Generalizing from the example, let us state the following rule:

Multiplication of choices

> *If a choice consists of two steps, of which the first can be made in m ways and for each of these the second can be made in n ways, then the whole choice can be made in m · n ways.*

To prove this, we have only to draw a tree diagram similar to that of Figure 4.2. First there are *m* branches corresponding to the possibilities in the first step, and then there are *n* branches emanating from each of these branches to represent the possibilities in the second step. This leads to *m · n* paths along the branches of the tree diagram, and hence to *m · n* possibilities.

EXAMPLE   If a restaurant offers nine kinds of sandwiches, which it serves with coffee, Sanka, tea, milk, hot chocolate, or a soft drink, there are $9 \cdot 6 = 54$ different ways in which one can order a sandwich and a drink.

EXAMPLE   If a history department schedules four lecture sections and sixteen discussion groups for a course in modern European history, there are $4 \cdot 16 = 64$ ways in which a student can sign up for one of the lecture sections and one of the discussion groups in this course.

By using appropriate tree diagrams, we can easily generalize the above rule so that it will apply to choices involving more than two steps. For *k* steps, where *k* is a positive integer, we get the following rule:

Multiplication of choices (generalized)

> *If a choice consists of k steps, of which the first can be made in n_1 ways, for each of these the second can be made in n_2 ways, . . . , and for each of these the kth can be made in n_k ways, then the whole choice can be made in $n_1 \cdot n_2 \cdot \ldots \cdot n_k$ ways.*

We simply keep multiplying the number of ways in which the different steps can be made.

EXAMPLE   If a new-car buyer has the choice of four body styles, three different engines, and ten colors, he or she can order one of these cars in $4 \cdot 3 \cdot 10 = 120$ different ways. Furthermore, if there are the options of ordering the car with or without air conditioning, with or without an automatic transmission, with or without power brakes, and with or without bucket seats, the total number of possibilities becomes $4 \cdot 3 \cdot 10 \cdot 2 \cdot 2 \cdot 2 \cdot 2 = 1,920$.

4.1 Counting

73

EXAMPLE If a test consists of ten multiple choice questions, with each permitting four possible answers, there are

$$4 \cdot 4 \cdot 4 \cdot 4 \cdot 4 \cdot 4 \cdot 4 \cdot 4 \cdot 4 \cdot 4 = 1,048,576$$

ways in which a student can check off one answer to each question. Only in one of these cases all answers will be correct and in

$$3 \cdot 3 \cdot 3 \cdot 3 \cdot 3 \cdot 3 \cdot 3 \cdot 3 \cdot 3 \cdot 3 = 59,049$$

cases all answers will be wrong.

4.2

Permutations and Combinations

The generalized rule for the multiplication of choices is often applied when several selections are made from one set and we are concerned with the order in which they are made.

EXAMPLE If a panel of judges must select the winner, the first runner-up, and the second runner-up from among the twenty entries in an art exhibit, they can make their choice in $20 \cdot 19 \cdot 18 = 6,840$ different ways, for after each choice there is one less to choose from for the next selection.

EXAMPLE If the 52 members of a local union must choose a president, a vice-president, a secretary, and a treasurer, they can choose these four officers in $52 \cdot 51 \cdot 50 \cdot 49 = 6,497,400$ different ways.

In general, if r objects are selected from a set of n objects, any particular arrangement (order) of these objects is called a **permutation.**

EXAMPLE Wyoming, Idaho, and New Mexico is a permutation (a particular ordered arrangement) of three of the eight Mountain states, and *aei* and *uoa* are two different permutations of three of the five vowels *a, e, i, o,* and *u.*

EXAMPLE If we were asked to list all possible permutations of two of the four Greek letters α, β, π, and σ, we would write

$$\alpha\beta \qquad \alpha\pi \qquad \alpha\sigma \qquad \beta\pi \qquad \beta\sigma \qquad \pi\sigma$$
$$\beta\alpha \qquad \pi\alpha \qquad \sigma\alpha \qquad \pi\beta \qquad \sigma\beta \qquad \sigma\pi$$

Since products of consecutive integers arise in many problems relating to permutations and other kinds of special arrangements or selections, it is convenient to introduce here what is called the **factorial notation.** The product of all positive integers less than or equal to the positive integer n is called "n factorial" and denoted by $n!$. Thus, $1! = 1, 2! = 2 \cdot 1 = 2, 3! = 3 \cdot 2 \cdot 1 = 6, 4! = 4 \cdot 3 \cdot 2 \cdot 1 = 24, 5! = 5 \cdot 4 \cdot 3 \cdot 2 \cdot 1 = 120$, and in general $n! = n(n-1)(n-2) \cdot \ldots \cdot 3 \cdot 2 \cdot 1$. Also, to make various formulas more generally applicable, we shall let $0! = 1$ by definition.

Now, to develop a rule for counting the total number of permutations of r objects selected from n distinct objects (such as the eight Mountain states, the five vowels, or the four Greek letters), we observe that the first selection is made from the whole set of n objects, the second selection is made from the $n - 1$ objects which remain after the first selection has been made, the third selection is made from the $n - 2$ objects which remain after the first two selections have been made, $\ldots$, and the rth and final selection is made from the $n - (r - 1) = n - r + 1$ objects which remain after the first $r - 1$ selections have been made. Direct application of the generalized rule for the multiplication of choices shows that the total number of permutations, which we shall denote $_nP_r$, is $n(n-1)(n-2) \cdot \ldots \cdot (n-r+1)$. Multiplying this product by $\dfrac{(n-r)!}{(n-r)!}$ and making use of the fact that $n(n-1)(n-2) \cdot \ldots \cdot (n-r+1) \cdot (n-r)! = n!$, we can also write the formula for $_nP_r$ as $\dfrac{n!}{(n-r)!}$. To summarize:

Number of permutations of n objects taken r at a time

The number of permutations of r objects selected from a set of n distinct objects is

$$_nP_r = n(n-1)(n-2) \cdot \ldots \cdot (n-r+1)$$

or, in factorial notation,

$$_nP_r = \frac{n!}{(n-r)!}$$

The first formula is generally easier to use, but the one in factorial notation is easier to remember and more easily programmed for solution on a digital computer.

EXAMPLE For the number of permutations of $r = 4$ objects selected from a set of $n = 10$ distinct objects (say, the number of ways in which the first prize, the second prize, the third prize, and the fourth prize, can be awarded to the ten entries in an essay contest), the first formula yields

$$_{10}P_4 = 10 \cdot 9 \cdot 8 \cdot 7 = 5,040$$

and the second formula yields

$$_{10}P_4 = \frac{10!}{6!} = \frac{10 \cdot 9 \cdot 8 \cdot 7 \cdot 6!}{6!} = 5{,}040$$

Essentially, the work is the same, but the second formula required an extra step based on the fact that

$$n! = n \cdot (n - 1)!$$
$$= n(n - 1) \cdot (n - 2)!$$
$$= n(n - 1)(n - 2) \cdot (n - 3)!$$

and so forth.

EXAMPLE   For the number of permutations of $r = 0$ objects selected from a set of $n = 50$ distinct objects, we have to use the second formula and we get

$$_{50}P_0 = \frac{50!}{50!} = 1$$

This result may be trivial, but it shows that the factorial notation makes the formula for the number of permutations more generally applicable.

To find the formula for the number of permutations of n distinct objects taken all together, we substitute $n = r$ into the second formula for $_nP_r$, getting $\frac{n!}{(n - n)!} = \frac{n!}{0!} = n!$ (since $0! = 1$ by definition). We have thus shown that

Number of permutations of n objects taken all together

$$_nP_n = n!$$

EXAMPLE   The number of ways in which eight teaching assistants can be assigned to eight sections of a course in College Algebra is $8! = 40{,}320$, and (barring ties) the number of ways in which twelve bowlers can place in a tournament is $12! = 479{,}001{,}600$.

There are many problems in which we want to know the number of ways in which r objects can be selected from a set of n objects, but we do not care about the order in which the selection is made. For instance, we may want to know in how many ways a committee of four can be selected from among the 64 members of a union, or the number of ways in which the IRS can choose five of 36 tax returns for a special audit. To develop a formula which applies to problems like these, let us first examine the following 24 permutations of three of the first four letters of the alphabet:

abc	acb	bac	bca	cab	cba
abd	adb	bad	bda	dab	dba
acd	adc	cad	cda	dac	dca
bcd	bdc	cbd	cdb	dbc	dcb

If we are not concerned with the order in which three letters are chosen from the four letters a, b, c, and d, there are only four ways in which the selection can be made. They are abc, abd, acd, and bcd; namely, the values shown in the first column. Each row of the table merely contains the $3! = 6$ different permutations of the letters shown in the first column. In general, there are $r!$ permutations of any r objects we select from a set of n distinct objects, and hence the $_nP_r$ permutations of r objects selected from a set of n objects contain each set of r objects $r!$ times. (In our example, the 24 permutations of three letters selected from among the first four letters of the alphabet contain each set of three letters $3! = 6$ times.) Therefore, to write a formula for the number of ways in which r objects can be selected from a set of n distinct objects, also called the number of **combinations** of n objects taken r at a time and denoted $\binom{n}{r}$, we must divide $_nP_r$ by $r!$, and we get

Number of combinations of n objects taken r at a time

> *The number of ways in which r objects can be selected from a set of n distinct objects is*
>
> $$\binom{n}{r} = \frac{n(n-1)(n-2)\cdot\ldots\cdot(n-r+1)}{r!}$$
>
> *or, in factorial notation,*
>
> $$\binom{n}{r} = \frac{n!}{r!(n-r)!}$$

Like the two formulas for $_nP_r$, the first of these formulas is generally easier to use, and the one in factorial notation is easier to remember and more easily programmed for solution on a digital computer.

For $n = 0$ to $n = 20$, the values of $\binom{n}{r}$ may be read from Table VIII at the end of the book, where they are referred to as **binomial coefficients.** The reason for this is explained in Exercise 25 on page 82.

EXAMPLE   The number of ways in which a person can choose three books from a list of eight books (the number of combinations of eight objects taken three at a time) is

$$\binom{8}{3} = \frac{8!}{3!5!} = \frac{8 \cdot 7 \cdot 6 \cdot 5!}{3!5!} = \frac{8 \cdot 7 \cdot 6}{3 \cdot 2 \cdot 1} = 56$$

and the number of ways in which a social scientist can select six of the 20 largest cities in the United States to be included in a survey is

$$\binom{20}{6} = \frac{20 \cdot 19 \cdot 18 \cdot 17 \cdot 16 \cdot 15}{6!} = 38{,}760$$

Also, the number of ways in which a committee of four can be selected from among the 64 members of a union is

$$\binom{64}{4} = \frac{64 \cdot 63 \cdot 62 \cdot 61}{4!} = 635{,}376$$

Note that the values obtained for $\binom{8}{3}$ and $\binom{20}{6}$ can be checked in Table VIII, but the value obtained for $\binom{64}{4}$ cannot.

EXAMPLE   If the director of a research laboratory has to select two chemists from among seven applicants and three physicists from among nine applicants, she can select the two chemists in $\binom{7}{2}$ ways, the three physicists in $\binom{9}{3}$ ways, and by the multiplication of choices she can fill all five positions in

$$\binom{7}{2} \cdot \binom{9}{3} = 21 \cdot 84 = 1{,}764 \text{ ways}$$

In this case we looked up the binomial coefficients in Table VIII.

When r objects are selected from a set of n distinct objects, $n - r$ of the n objects are left; thus, there are as many ways of leaving (or selecting)

$n - r$ objects from a set of n distinct objects as there are of selecting r objects. Symbolically,

Rule for binomial coefficients

$$\binom{n}{r} = \binom{n}{n - r} \qquad for \ r = 0, 1, 2, \ldots, n$$

and in Exercise 26 on page 82 the reader will be asked to verify this result algebraically. The above rule has many applications—sometimes it serves to simplify calculations and sometimes it is needed in connection with the use of Table VIII.

EXAMPLE   To determine $\binom{100}{97}$ without having to evaluate the product $100 \cdot 99 \cdot 98 \cdot \ldots \cdot 4$, we write

$$\binom{100}{97} = \binom{100}{3} = \frac{100 \cdot 99 \cdot 98}{3!} = 161,700$$

EXAMPLE $\binom{19}{13}$ cannot be looked up directly in Table VIII, but making use of the fact that $\binom{19}{13} = \binom{19}{6}$, we look up $\binom{19}{6}$ instead, getting 27,132.

EXERCISES

1   Suppose that in a baseball World Series (in which the winner is the first team to win four games) the National League champion leads the American League champion three games to two. Construct a tree diagram to show the number of ways in which these teams may win or lose the remaining game or games.

2   A person with $2 in his pocket bets $1, even money, on the flip of a coin, and he continues to bet $1 so long as he has any money. Draw a tree diagram to show the various things that can happen during the first three flips of the coin. In how many of the cases will he be
   (a) exactly $1 ahead?
   (b) exactly $1 behind?

3   A student can study 0, 1, or 2 hours for a statistics test on any given night. Draw a tree diagram to show that there are six different ways in which she can study altogether 4 hours for the test on three consecutive nights.

4   There are four routes, A, B, C, and D, between a person's home and the place where he works, but route A is one-way so that he cannot take it on the way to work, and route D is one-way so that he cannot take it on the way home.
   (a) Draw a tree diagram showing the various ways he can go to and from work.
   (b) Draw a tree diagram showing the various ways he can go to and from work, but does not go by the same route both ways.

Permutations and Combinations

5 A student has to take two different courses in the humanities, one each during the two semesters of her freshman year. Draw a tree diagram to show the various ways in which she can make her choice, if during the first semester they offer introductory courses in philosophy, fine arts, history, and literature, and during the second semester they offer the same courses in philosophy and history, but not those in the other subjects.

6 If the NCAA has applications from four universities for hosting its intercollegiate wrestling championships in 1982 and 1983, in how many ways can they select the sites for these championship meets
   (a) if they are not to be held at the same university?
   (b) if they may be held at the same university?

7 In a political science survey, voters are classified into six categories according to their income and into four categories according to their education. In how many different ways can a voter thus be classified?

8 In an optics kit there are six concave lenses, four convex lenses, two prisms, and two mirrors. In how many different ways can one choose a concave lens, a convex lens, a prism, and a mirror from this kit?

9 A psychologist preparing three-letter nonsense words for use in a memory test chooses the first letter from among the consonants q, w, x, and z; the second letter from among the vowels e, i, and u; and the third letter from among the consonants c, f, p, and v.
   (a) How many different three-letter nonsense words can he construct?
   (b) How many of these nonsense words will begin with the letter w?
   (c) How many of these nonsense words will end either with the letter f or the letter p?

10 A true–false test consists of 15 questions. In how many ways can a student check off his or her answers to these questions?

11 In an ice cream parlor a customer can order a sundae with any one of 14 different kinds of ice cream, with any one of six different kinds of syrup, with or without whipped cream, and with or without nuts. In how many different ways can a customer order one of these sundaes?

12 In an example on page 74 we listed Wyoming, Idaho, and New Mexico as a permutation of three of the eight Mountain states. How many different permutations like these are there altogether?

13 The price of a European tour includes four stopovers to be selected from among nine cities. In how many different ways can one plan such a tour
   (a) if the order of the stopovers matters?
   (b) if the order of the stopovers does not matter?

14 In an English class, the students are given the choice of eight different essay topics. In how many different ways can four students each choose a topic
   (a) if no two students may choose the same topic?
   (b) if there is no restriction on the choice of topics?

15 In how many ways can a television director schedule six different commercials during the six time slots allocated to commercials during the telecast of the first period of a hockey game?

16 The number of ways in which n distinct objects can be arranged in a circle is $(n - 1)!$.
  - (a) Present an argument to justify this formula.
  - (b) In how many ways can six persons be seated at a round table (if we care only who sits on whose left or right side).
  - (c) In how many ways can a window dresser display four tennis rackets in a circular arrangement?

17 If among n objects r are alike, and the others are all distinct, the number of permutations of these n objects taken all together is $\dfrac{n!}{r!}$.
  - (a) How many permutations are there of the letters in the word "class"?
  - (b) In how many ways (according to manufacturer only) can five cars place in a stock-car race, if three of the cars are Fords, one is a Chevrolet, and one is a Dodge?
  - (c) In how many ways can the television director of Exercise 15 fill the six time slots allocated to commercials, if she has four different commercials, of which a given one is to be shown three times, while each of the others is to be shown once?
  - (d) Present an argument to justify the formula given in this exercise.

18 If among n objects r_1 are identical, another r_2 are identical, and the rest (if any) are all distinct, the number of permutations of these n objects taken all together is $\dfrac{n!}{r_1! \cdot r_2!}$.
  - (a) How many permutations are there of the letters in the word "greater"?
  - (b) In how many ways can the television director of Exercise 15 fill the six time slots allocated to commercials, if she has only two different commercials, each of which is to be shown three times?
  - (c) Generalize the formula so that it applies if among n objects r_1 are identical, another r_2 are identical, another r_3 are identical, and the rest (if any) are all distinct. In how many ways can the television director of Exercise 15 fill the six time slots allocated to commercials, if she has three different commercials, each of which is to be shown twice?
  - (d) In its cookbook section, a bookstore has four copies of *The New York Times Cookbook*, two copies of *The Joy of Cooking*, five copies of the *Better Homes and Gardens Cookbook*, and one copy of *The Secret of Cooking for Dogs*. If these books are sold one at a time, in how many different sequences can they be sold?

19 Calculate the number of ways in which a motel chain can choose 2 of 12 locations for the construction of new motels.

20 Calculate the number of ways in which the IRS can choose 5 of 36 tax returns for a special audit.

21 A true–false test consists of 16 questions. Calculate the numbers of ways in which a student can mark this test and get
  - (a) 8 right and 8 wrong;
  - (b) 10 right and 6 wrong;
  - (c) 3 right and 13 wrong.

22 Among the eight nominees for two vacancies on a school board are four men and four women. In how many ways can these vacancies be filled
  (a) with any two of the eight nominees?
  (b) with any two of the female nominees?
  (c) with one of the male nominees and one of the female nominees?

23 A shipment of 15 alarm clocks contains one that is defective. In how many ways can an inspector choose three of the alarm clocks for inspection so that
  (a) the defective alarm clock is not included?
  (b) the defective alarm clock is included?

24 Suppose that among the 15 alarm clocks of Exercise 23 there are two defectives. In how many ways can the inspector choose three of the alarm clocks for inspection so that
  (a) none of the defective alarm clocks is included?
  (b) only one of the defective alarm clocks is included?
  (c) both of the defective alarm clocks are included?

25 The quantity $\binom{n}{r}$ is called a binomial coefficient because it is, in fact, the coefficient of $a^{n-r}b^r$ in the binomial expansion of $(a + b)^n$. Verify this for $n = 2, 3$, and 4, by expanding $(a + b)^2$, $(a + b)^3$, and $(a + b)^4$ and comparing the coefficients with the values corresponding to $n = 2, 3$, and 4 in Table VIII.

26 Verify the formula $\binom{n}{r} = \binom{n}{n - r}$ on page 79 by expressing the binomial coefficients in terms of factorials.

27 A table of binomial coefficients is easy to construct by following the pattern shown below, which is called **Pascal's triangle.**

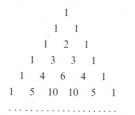

In this arrangement, each row begins with a 1, ends with a 1, and each other entry is given by the sum of the nearest two entries in the row immediately above.
  (a) Construct the next three rows of Pascal's triangle and verify that they are the same as the rows corresponding to $n = 6, 7$, and 8 in Table VIII.
  (b) Express each binomial coefficient in terms of factorials to verify the identity $\binom{n + 1}{r} = \binom{n}{r} + \binom{n}{r - 1}$, on which the construction of Pascal's triangle is based.

4.3

The Classical Probability Concept and the Frequency Interpretation

So far we have studied only what is possible in a given situation. In some instances we listed all possibilities and in others we merely determined how many different possibilities there are. Now we shall go one step further and judge also what is probable and what is improbable. The most common way of measuring the uncertainties connected with events (say, the outcome of a presidential election, the side effects of a new serum, the durability of an exterior paint, or the total number of points we may roll with a pair of dice) is to assign them **probabilities,** or to specify the **odds** at which it would be fair to bet that the events will occur.

Historically, the oldest way of measuring probabilities applies when all possible outcomes are equally likely, as is presumably the case in many games of chance. We can then say that

The classical probability concept

> *If there are n equally likely possibilities, of which one must occur and s are regarded as favorable, or as a "success," then the probability of a "success" is given by the ratio $\frac{s}{n}$.*

In the application of this rule, the terms "favorable" and "success" are used rather loosely—what is favorable to one player is unfavorable to his opponent, and what is a success from one point of view is a failure from another. Thus, the terms "favorable" and "success" can be applied to any particular kind of outcome, even if "favorable" means that a television set does not work, or "success" means that someone catches the flu. This usage dates back to the days when probabilities were quoted only in connection with games of chance.

EXAMPLE   Following this rule and assuming in each case that the possibilities are all equally likely, we find that the probability of drawing an ace from an ordinary deck of playing cards is $\frac{4}{52}$ (there are 4 aces among the 52 cards), the probability of getting heads with a balanced coin is $\frac{1}{2}$, and the probability of rolling either a 1 or a 2 with a die is $\frac{2}{6}$.

EXAMPLE   Somewhat more complicated is the problem of determining the probability that two cards drawn from an ordinary deck of 52 playing cards will both be black. According to what we learned in the preceding section, the total number of possibilities is $\binom{52}{2} = \frac{52 \cdot 51}{2} = 1,326$, the number of favorable possibilities is $\binom{26}{2} = \frac{26 \cdot 25}{2} = 325$ since half of the 52 cards are black and the others are red, and it follows that the probability of getting two black cards is

$$\frac{\binom{26}{2}}{\binom{52}{2}} = \frac{325}{1,326} = 0.245$$

rounded to three decimals.

Although equally likely possibilities are found mostly in games of chance, the classical probability concept applies also in a great variety of situations where gambling devices are used to make *random selections*—say, when offices are assigned to research assistants by lot, when laboratory animals are chosen for an experiment (perhaps, by the method which we shall discuss in Chapter 10) so that each one has the same chance of being selected, when each family in a township has the same chance of being included in a sample survey, or when machine parts are chosen for inspection so that each part produced has the same chance of being selected.

A major shortcoming of the classical probability concept is its limited applicability, for there are many situations in which the possibilities that arise cannot all be regarded as equally likely. This would be the case, for example, when we are concerned with the question whether it will rain on the next day; when we wonder whether a person will get a raise; when we want to predict the outcome of an election or the score of a baseball game; or when we want to judge whether a stock market index will go up or down.

Among the various probability concepts, most widely held is the frequency interpretation, according to which

The frequency interpretation of probability

> *The probability of an event (happening or outcome) is the proportion of the time that events of the same kind will occur in the long run.*

EXAMPLE   If we say that the probability is 0.78 that a jet from San Francisco to Phoenix will arrive on time, we mean that such flights arrive on time 78 percent of the time. Also, if the Weather Service predicts that there is a 40 percent chance for rain (namely, that the probability is 0.40 that it will rain), they mean that under the same weather conditions it will rain 40 percent of the time. More generally, we say that an event has a probability of, say, 0.90, in the same sense in which we might say that our car will start in cold weather 90 percent of the time. We cannot guarantee what will happen on any particular occasion—the car may start and then it may not—but if we kept records over a long period of time, we should find that the proportion of "successes" is very close to 0.90.

In accordance with the frequency interpretation of probability, we *estimate* the probability of an event by observing what fraction of the time similar events have occurred in the past.

EXAMPLE If data kept by a government agency show that (over a period of time) 468 of 600 jets from San Francisco to Phoenix arrived on time, we estimate the probability that any one flight from San Francisco to Phoenix will arrive on time as $\frac{468}{600} = 0.78$. Similarly, if 504 of 813 automatic dishwashers sold by a large retailer required repairs within the warranty year, we estimate the probability that any one automatic dishwasher sold by the retailer will require repairs within the warranty year as $\frac{504}{813} = 0.62$ (rounded to two decimals).

When probabilities are thus estimated, it is only reasonable to ask whether the estimates are any good. To answer this question, we refer to a remarkable theorem called the **Law of Large Numbers**; informally, it may be stated as follows:

The Law of Large Numbers

> *If a situation is repeated again and again, the proportion of successful outcomes will tend to approach the constant probability that any one of the outcomes will be a success.*

EXAMPLE If we repeatedly flip a balanced coin, observe the accumulated proportion of heads, say, after every fifth flip, and plot the results graphically as in Figure 4.3, we should find that although they

FIGURE 4.3
Graph illustrating the law of large numbers.

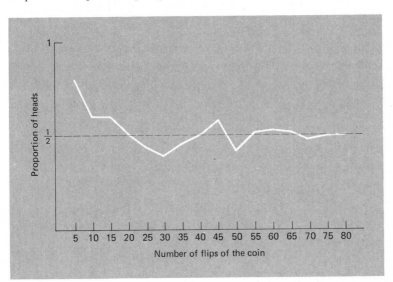

fluctuate, the proportions come closer and closer to $\frac{1}{2}$, the probability of heads for each flip of the coin.

Theoretical support for the Law of Large Numbers may be found in Exercise 15 on page 180.

In the frequency interpretation, the probability of an event is defined in terms of what happens to similar events in the long run, so let us examine briefly whether it is at all meaningful to talk about the probability of an event which can occur only once.

EXAMPLE Can we assign a probability to the event that a certain Ms. Barbara Smith will be able to leave the hospital within four days after having an appendectomy, or to the event that a certain major-party candidate will win an upcoming gubernatorial election? If we put ourselves in the position of Ms. Smith's doctor, we could check medical records, discover that patients left the hospital within four days after having an appendectomy in, say, 32 percent of hundreds of reported cases, and apply this figure to Ms. Smith. This may not be of much comfort to Ms. Smith, but it does provide a meaning for a probability statement about her leaving the hospital within four days—the probability that she will leave the hospital within four days after having her appendectomy is 0.32.

This illustrates the fact that when we make a probability statement about a specific (nonrepeatable) event, the frequency interpretation of probability leaves us no choice but to refer to a set of similar events. As can well be imagined, however, this can easily lead to complications, since the choice of "similar" events is often neither obvious nor straightforward.

EXAMPLE With reference to Ms. Smith's appendectomy, we might consider as "similar" only cases in which the patients were of the same sex, only cases in which the patients were also of the same age as Ms. Smith, or only cases in which the patients were also of the same height and weight as Ms. Smith.

Ultimately, the choice of "similar" events is a matter of personal judgment, and it is by no means contradictory, therefore, that we can arrive at different probabilities, all valid, concerning the same event.

EXAMPLE With regard to the other question asked earlier on this page, the one concerning the gubernatorial election, suppose we ask the persons who have conducted a poll "how sure" they are that the given candidate will win. If they say they are "99 percent sure" (that is, if they assign a probability of 0.99 to the candidate's winning the election), this is not meant to imply that the candidate would win 99 percent of the time if he ran for office a great number of times. Rather, it means that their conclusion is based on a method which "works" 99 percent of the time. It is in this sense that many of the probabilities we use in statistics to express our faith in predictions or decisions are simply **success ratios** that apply to the methods we employ.

4.4

Probabilities and Odds

To explain what we mean by the "odds" that an event will occur, let us first establish a relationship between the probability that an event will occur and the probability that it will not occur. In the classical probability concept, if there are s "successes" among n equally likely possibilities, then there are $n - s$ "failures," the probabilities of "success" and "failure" are, respectively, $\frac{s}{n}$ and $\frac{n - s}{n}$, and simple algebra shows that the sum of these two probabilities is 1. So far as the frequency interpretation is concerned, a 0.62 probability for rain implies that under certain conditions it will rain 62 percent of the time; thus, it will not rain 38 percent of the time, and the probability that it will not rain is 0.38. Again, the sum of the two probabilities is 1. Based on these observations, let us state the following rule of probability:

The probability that an event will not occur

> *If the probability that an event (happening or outcome) will occur is p and the probability that it will not occur is q, then $p + q = 1$ and $q = 1 - p$.*

EXAMPLE On page 83 we showed that the probability of drawing an ace from an ordinary deck of 52 playing cards is $\frac{4}{52}$; now we can add that the probability of not getting an ace is $1 - \frac{4}{52} = \frac{48}{52}$. (Of course, this result could also have been obtained directly by observing that there are 48 nonaces among the 52 cards.) Also, on page 85 we estimated as 0.78 the probability that a jet from San Francisco to Phoenix will arrive on time; it follows that the probability is $1 - 0.78 = 0.22$ that such a jet will not arrive on time.

Now let us give a definition of "odds." If an event is twice as likely to occur than not to occur, we say that the odds are 2 to 1 that it will occur. If an event is three times as likely not to occur than to occur, we say that the odds are 3 to 1 that it will not occur. In general, the **odds** that an event will occur are given by the ratio of the probability that the event will occur to the probability that it will not occur. Symbolically,

Formula relating odds to probabilities

> *If the probability of an event is p, the odds for its occurrence are a to b, where*
>
> $$\frac{a}{b} = \frac{p}{1 - p}$$

It is customary to express odds as a ratio of two positive integers having no common factor.

EXAMPLE   If the probability of an event is $\frac{5}{7}$, then the odds for its occurrence are $\frac{5}{7}$ to $1 - \frac{5}{7} = \frac{2}{7}$, or 5 to 2. Also, if the probability of an event is 0.65, then the odds for its occurrence are 0.65 to $1 - 0.65 = 0.35$, 65 to 35, or better, 13 to 7.

If an event is more likely not to occur than to occur, it is customary to quote the odds that it will not occur rather than the odds that it will occur.

EXAMPLE   If the probability of an event is 0.20, then the odds for its occurrence are 0.20 to $1 - 0.20 = 0.80$, or 1 to 4, and we say instead that the odds against the occurrence of the event are 4 to 1.

In gambling, the word "odds" is also used to denote the ratio of the wager of one party to that of another. For instance, if a gambler says that he will give 2 to 1 odds on the occurrence of an event, he means that he is willing to bet $2 against $1 (or perhaps $200 against $100) that the event will occur. If such **betting odds** actually equal the odds that the event will occur, we say that the betting odds are **fair.**

EXAMPLE   Betting on the flip of a balanced coin (heads or tails) is an even-money bet; regardless of whether we bet on heads or tails, the probability of winning is $\frac{1}{2}$ and fair odds are 1 to 1.

EXAMPLE   Records show that $\frac{1}{12}$ of the trucks weighed at a certain check point in Nevada carry too heavy a load. If someone offered to bet $20 against $2 that the next truck weighed at this check point will not carry too heavy a load, would these odds be fair? The answer is "No." Since the probability is $1 - \frac{1}{12} = \frac{11}{12}$ that the truck will not carry too heavy a load, the odds are 11 to 1, and the bet would have been fair if the person had offered to bet $22 against $2 that the next truck weighed at the check point will not carry too heavy a load. As it stands, the $20 against $2 bet favors the person who is offering the bet.

4.5
Subjective Probabilities

Our discussion of odds and betting odds was intended to lay the groundwork for an alternative point of view about probabilities, which is currently gaining in favor. According to this point of view, probabilities are interpreted as **personal** or **subjective evaluations.** Such probabilities measure one's belief with regard to the uncertainties that are involved, and they apply especially when there is little or no direct evidence, so that there really is no choice but to consider collateral (indirect) information, "educated guesses," and perhaps intuition and other subjective factors.

EXAMPLE   If a businessman "feels" that the odds for the success of a new restaurant are 3 to 2, this means that he would be willing to bet (or consider it fair to bet) $300 against $200, or perhaps $3,000 against $2,000, that the new restaurant will be a success. In this way he would be expressing his belief regarding the uncertainties connected with the success of the restaurant, and it may be based on business conditions in general, the opinion of an expert consultant, or his own subjective, perhaps optimistic, evaluation of the whole situation.

To convert odds like these into corresponding subjective probabilities, we refer to the equation $\frac{a}{b} = \frac{p}{1 - p}$ (which we gave on page 88 to express the relationship between probabilities and odds), and solve for p. Cross multiplying, we get $a(1 - p) = bp$, and then solving for p, we get $a = ap + bp$, $a = (a + b)p$, and finally $p = \frac{a}{a + b}$. Thus, we have shown that

Formula relating probabilities to odds

> If the odds are a to b that an event will occur, the probability of its occurrence is
>
> $$p = \frac{a}{a + b}$$

This formula holds regardless of whether the odds are arrived at subjectively as in the above example, or objectively (say, by referring to past occurrences in accordance with the frequency interpretation).

EXAMPLE   The businessman who is willing to give odds of 3 to 2 that the new restaurant will succeed is actually assigning its success a (personal) probability of $\frac{3}{3 + 2} = 0.60$.

EXAMPLE If a high school senior feels that the odds are 7 to 2 that she will be accepted by the college of her choice, then her (personal) probability that she will be accepted is $\frac{7}{7+2} = \frac{7}{9}$. The odds that she will not be accepted are 2 to 7, the probability that she will not be accepted is $\frac{2}{2+7} = \frac{2}{9}$, and as should have been expected, the sum of these two probabilities is equal to 1.

EXERCISES

1   When one card is drawn from a well-shuffled deck of 52 playing cards, what are the probabilities of getting
    (a) a red queen?
    (b) a queen, king, or ace of any suit?
    (c) a red card?
    (d) a 6, 7, 8, or 9?

2   If H stands for heads and T for tails, the eight possible outcomes in three successive flips of a coin are HHH, HHT, HTH, THH, HTT, THT, TTH, and TTT. Assuming that these eight possibilities are equally likely, what are the respective probabilities of getting 0, 1, 2, or 3 heads?

3   A bowl contains 11 red beads, 10 white beads, 25 blue beads, and 4 black beads. If one of these beads is drawn at random, what are the probabilities that it will be
    (a) red?
    (b) white or blue?
    (c) black?
    (d) neither white nor black?

4   Referring to Exercise 23 on page 82 and assuming that the three alarm clocks are chosen at random (namely, that each possible set of three of the clocks has the same chance of being selected), find the probability that the defective alarm clock will be included among the ones chosen by the inspector.

5   Assuming that the selection is random (namely, that each set of four of the twelve cities has the same chance of being selected), what are the probabilities that a survey conducted in four of the twelve largest cities in the United States will
    (a) include New York City?
    (b) include Los Angeles and Philadelphia?
    (c) not include Detroit?
    These four cities are among the twelve largest cities in the United States.

6   The records of a life insurance company show that 1,564 of the 1,840 policyholders whose policies were issued at age 25 were still alive at age 45.
    (a) Estimate the probability that a 25-year-old person who takes out a policy with this life insurance company will still be alive at age 45.
    (b) Can we use the same figures to estimate the probability that any 25-year-old person will still be alive at age 45? Explain.

7   In a radar check conducted on a Los Angeles freeway early in the morning, 214 of 856 cars were found to exceed the legal speed limit of 55 mph. Estimate

the probability that a car traveling on that freeway at that time of the day will be exceeding the legal speed limit of 55 mph.

8   Statistics compiled for a ski area in New Hampshire show that it has snowed there on Christmas Day in 35 of the last 63 years.
   (a) Estimate the probability that it will snow there on Christmas Day.
   (b) What are the odds that it will snow there on Christmas Day?
   (c) Would it be wise to bet $30 against $20 that it will snow there on Christmas Day?

9   Among the 102 times that Rita was late for work (over a number of years), her boss got there ahead of her 34 times.
   (a) Estimate the probability that if she is late for work her boss will already be there.
   (b) What are the odds that if she is late for work her boss will not yet be there?
   (c) If she offers a friend even money that if she is late for work her boss will already be there, who would be favored by this bet?

10   Convert each of the following probabilities to odds:
   (a) The probability that the last digit of a car's license plate is 2, 3, 4, 5, 6, or 7 is $\frac{3}{5}$.
   (b) If a secretary arbitrarily puts special delivery stamps on 3 of 6 letters, the probability that they will be on the three letters which are supposed to go by special delivery is $\frac{1}{20}$.
   (c) The probability of getting at least 3 heads in 6 flips of a balanced coin is $\frac{21}{32}$.

11   Convert each of the following probabilities to odds:
   (a) If a person has eight $1 bills, five $5 bills, and one $20 bill in her purse and randomly pulls out three of the bills, the probability that they will not all be $1 bills is $\frac{11}{13}$.
   (b) The probability of rolling "7 or 11" with a pair of balanced dice is $\frac{2}{9}$.
   (c) If a teacher randomly selects two of eight students to recite parts of a poem, the probability that Tom (who is one of the eight students) will be chosen is $\frac{1}{4}$.

12   If someone feels that 13 to 3 are fair odds that a construction job will be finished on time, what probability does he assign to the job's being finished on time?

13   A sportswriter feels that the odds are 5 to 1 that the home team will lose an upcoming basketball game. What subjective probability expresses his feelings about the home team's winning the game?

14   A used-car salesman offers his employer a bet of $55 against $25 that he will be able to sell a car to a difficult customer. If the employer feels that this is fair, what personal probability is he thus assigning to the used-car salesman's success with the difficult customer?

15   Convert the following odds to probabilities:
   (a) If an urn contains 22 black balls and 7 white balls, the odds for drawing two balls that are both black are 33 to 25.
   (b) On a tray there are six pieces of chocolate pie and four pieces of banana cream pie. If a waitress arbitrarily takes two of these pieces of pie and gives them to customers who ordered banana cream pie, the odds are 13 to 2 that she will make a mistake.

16 A stock broker is unwilling to bet $40 against $120 that the price of a certain stock will go up within a week. What does this tell us about the subjective probability he assigns to the stock's price going up within a week? (*Hint:* The answer should read "less than. . . .")

17 If a student is anxious to bet $25 against $5 that she will pass a certain course, what does this tell us about the personal probability she assigns to her passing the course? (*Hint:* The answer should read "greater than. . . .")

18 A television producer is willing to bet $1,500 against $1,000, but not $2,000 against $1000, that a new game show will be a success. What does this tell us about the probability which the producer assigns to the show's success?

19 Discuss the following assertion: If the weatherman says that the probability for rain is 0.30, whatever happens on that day cannot prove him right or wrong.

20 Discuss the following assertion: Since probabilities are measures of uncertainty, the probability we assign to a future event will always increase when we get more information.

21 The following illustrates how one's intuition can be misleading in connection with probabilities or odds: A box contains 100 beads, some red and some white. One bead will be drawn, and you are asked to call beforehand whether it is going to be red or white. At what odds would you be willing to bet on this game if
    (a) you have no idea how many of the beads are red and how many are white?
    (b) you are told that 50 of the beads are red and 50 are white?

22 Explain how one might assign a probability to the truth of testimony given at a trial, using
    (a) the frequency interpretation of probability;
    (b) subjective probabilities.

23 Suppose that someone flips a coin 100 times and gets 30 heads, which is short of the number of heads she might expect. Then she flips the coin another 100 times and gets 44 heads, which is again short of the number of heads she might expect. Can she accuse the Law of Large Numbers of "letting her down?" Explain.

BIBLIOGRAPHY   Informal introductions to probability, written essentially for the layman, may be found in

> LEVINSON, H. C., *Chance, Luck, and Statistics.* New York: Dover Publications, Inc., 1963.
> WEAVER, W., *Lady Luck—The Theory of Probability.* Garden City, N.Y.: Doubleday & Company, Inc., 1963.

and subjective probabilities, in particular, are discussed in

> BOREL, E., *Elements of the Theory of Probability.* Englewood Cliffs, N.J.: Prentice-Hall, Inc., 1965.
> KYBURG, H. E., JR., and SMOKLER, H. E., *Studies in Subjective Probability.* New York: John Wiley & Sons, Inc., 1964.

Some Rules of Probability

In the study of probability there are basically three kinds of questions: (1) What do we mean when we say that the probability of an event is, say, 0.50, 0.78, or 0.04? (2) How are the numbers we call probabilities determined, or measured, in actual practice? (3) What mathematical rules must probabilities obey?

The first and second kinds of questions have already been considered to some extent in Chapter 4: In the classical probability concept we are concerned with equally likely possibilities and count "favorable" cases; in the frequency interpretation we are concerned with proportions of "successes" in the long run and base our estimates on what happened in the past; and when it comes to subjective probabilities we are concerned with a measure of a person's belief and observe at what odds he or she would be willing to bet (or consider it fair to bet) that an event will occur.

In this chapter, after some preliminaries, we shall study the question of what rules probabilities must obey, or how they must "behave." Although these rules were originally developed for the "benefit" of gamblers, some familiarity with them can be of value to anyone. Businessmen must buy merchandise without knowing for sure whether it will sell, military strategists must commit men and equipment to the hazards of battle, we travel by car or plane without knowing for certain whether we will reach our destination, doctors risk their lives in combating disease, important messages are sent by mail without any assurance that they will be delivered on time, and so on—and in each case the study of probability makes it possible, or at least easier, to "live with the corresponding uncertainties."

5.1

Sample Spaces and Events

Since probabilities always refer to the occurrence or nonoccurrence of events, let us see first what is meant here by "event" and by the related terms "experiment," "outcome," and "sample space." For lack of a better term, it is customary in statistics to refer to any process of observation or measurement as an **experiment.** Using "experiment" in this unconventional way, an experiment may consist of counting how many "yes" votes a congressman has cast on certain critical issues; it may consist of the simple process of noting whether a house is occupied or empty, or whether a light is on or off; or it may consist of the very complicated process of obtaining and evaluating data to predict trends in the economy, to find the sources of social unrest, or to study the cause of a disease. The results one obtains from an experiment, whether they are instrument readings, counts, "yes" or "no" answers, or values obtained through extensive calculations, are called the **outcomes** of the experiment.

When we study the outcomes of an experiment, we usually identify the different possibilities with numbers or points so that we can treat all questions about them mathematically, without having to go through long verbal descriptions of what has taken place, is taking place, or will take place. This is precisely what we did on page 5, where we suggested that a person's marital status may be recorded as 1, 2, 3, or 4 depending on whether the person is single, married, widowed, or divorced. The use of points rather than numbers has the added advantage that it makes it easier to visualize the various possibilities, and perhaps discover some special features which several of the outcomes may have in common.

EXAMPLE Suppose that a Ford dealer has three new Mustangs in stock, and we are interested in how many of them a certain salesperson will sell in a given week. The most obvious way to represent the four outcomes is by means of the points shown in Figure 5.1.

Had we been interested in two salespersons, we could have used coordinates so that (0, 1), for example, represents the outcome that the first salesperson will sell none of the Mustangs and the second salesperson will sell one, and (1, 1) represents the outcome that each of the two salespersons will sell one of the Mustangs. The ten possible outcomes are (0, 0), (0, 1), (0, 2), (0, 3), (1, 0), (1, 1), (1, 2), (2, 0), (2, 1), and (3, 0), and the corresponding points are shown in Figure 5.2. Actually, we could have used ten points in any kind of pattern, but the use of coordinates makes it easy to identify each

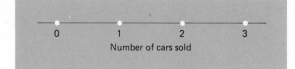

FIGURE 5.1

Outcomes for one-sales-person example.

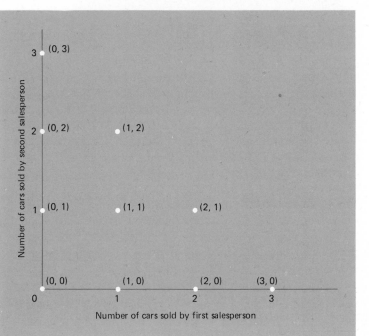

FIGURE 5.2

Outcomes for two-sales-persons example.

point with the corresponding outcome. Had we been interested in more than two salespersons trying to sell the same cars, things would have been much more complicated. For instance, for four salespersons there would be 35 possibilities, among which (0, 1, 1, 0) represents the case where the first and fourth salespersons will not sell any of the cars, while the second and third salespersons will each sell one.

It is customary to refer to a set of points which represents all the possible outcomes of an experiment as the **sample space** of the experiment and to denote it by the letter *S*. Thus, the four points of Figure 5.1 constitute the sample space for the case where one salesperson is trying to sell the three cars, and the ten points of Figure 5.2 constitute the sample space for the case where two salespersons are trying to sell the three cars.

In any discussion of an experiment, no matter how simple, it is always important to specify the sample space, and, as we shall see, this will depend on what we look upon as an individual outcome.

5.1 Sample Spaces and Events

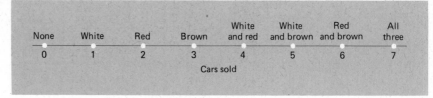

FIGURE 5.3

Outcomes for modified one-salesperson example.

EXAMPLE   Had the three Mustangs been distinguishable in the one-salesperson example (say, one was white, one was red, and one was brown), there would have been the eight outcomes shown in Figure 5.3, to which we arbitrarily assigned the numbers 0, 1, 2, 3, 4, 5, 6, and 7. Whether we use the sample space of Figure 5.3 in any given problem, or that of Figure 5.1, would depend entirely on whether the differences in color are of any relevance.

The sample spaces of Figures 5.1 and 5.3 are one-dimensional and the sample space of Figure 5.2 is two-dimensional; a corresponding sample space for the case where three salespersons are trying to sell the three cars would be three-dimensional and for four salespersons it would take a space of four dimensions. Although it may be useful to know the number of dimensions in which we picture the points of a sample space, it is more common to classify sample spaces according to the number of points which they contain. All the sample spaces mentioned so far have been **finite**, since they all consisted of a finite, or fixed, number of points.

EXAMPLE   Much larger, though still finite, sample spaces are the ones representing all possible 5-card poker hands one can deal with an ordinary deck of 52 playing cards, where there are $\binom{52}{5} = 2,598,960$ possibilities, and the one representing all the possible ways in which the 10 universities in the Big Ten Conference can finish the football season, where there are $10! = 3,628,800$ possibilities, without allowing for ties.

In the remainder of this chapter we shall consider only finite sample spaces, although in later chapters we shall consider also sample spaces that are infinite.

Having explained what we mean by a sample space, let us now state formally that when we speak of an **event,** we are speaking of a subset of a sample space. By subset we mean any part of a set, including the set as a whole and, trivially, a set called the **empty set,** which has no elements at all.

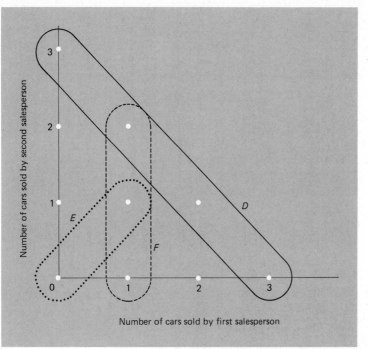

FIGURE 5.4

Sample space for two-salespersons example.

EXAMPLE   With reference to Figure 5.2, the point (0, 0) constitutes the event that neither salesperson will sell any of the cars, the points (1, 0) and (2, 1) together constitute the event that the first salesperson will sell one more car than the second salesperson, and the points (0, 0), (1, 0), (0, 1), and (1, 1) together constitute the event that neither salesperson will sell more than one car.

Still referring to the same example, suppose that *D* stands for the event that they will sell all three cars, *E* stands for the event that the two salespersons will sell equally many cars, and *F* stands for the event that the first salesperson will sell one car. We see from Figure 5.4, showing the same sample space as Figure 5.2, that *D* consists of the four points inside the solid line, *E* consists of the two points inside the dotted line, and *F* consists of the three points inside the dashed line.

In the preceding example, events *D* and *E* have no points in common and they are referred to as **mutually exclusive events.** Such events cannot both occur at the same time. It is also apparent from Figure 5.4 that events *D* and *F* are not mutually exclusive, and neither are events *E* and *F*.

In many probability problems we are interested in events that can be expressed in terms of two or more events by forming **unions, intersection,** and

complements. In general, the union of two events A and B, denoted $A \cup B$, is the event which consists of all the outcomes (points) contained in event A, in event B, or in both; the intersection of two events A and B, denoted $A \cap B$, is the event which consists of all the outcomes (points) contained in both A and B; and the complement of an event A, denoted A', is the event which consists of all the outcomes (points) of the sample space that are not contained in A. It is common practice to read $\cup$ as "or," $\cap$ as "and," and A' as "not A."

EXAMPLE   If A is the event that Mr. Jones is out of work and B is the event that his wife is out of work, then $A \cup B$ is the event that at least one of the two (Mr. Jones, Mrs. Jones, or both) is out of work, $A \cap B$ is the event that they are both out of work, and B' is the event that Mrs. Jones is not out of work.

EXAMPLE   With reference to Figure 5.4, $E \cup F$ is the event that the first salesperson will sell one car or neither of them will sell any of the cars, and it contains the points (0, 0), (1, 0), (1, 1), and (1, 2); $D \cap F$ is the event that the first salesperson will sell one car and the second salesperson will sell two cars, and it contains only the point (1, 2); and D' is the event that they will not sell all three of the cars. Also, since D and E are mutually exclusive, we write $D \cap E = \varnothing$, where the symbol $\varnothing$ denotes the empty set.

Sample spaces and events, particularly relationships among events, are often pictured by means of **Venn diagrams** such as those of Figures 5.5 and 5.6. In each case the sample space is represented by a rectangle, while events are represented by regions within the rectangle, usually by circles or parts of circles. The tinted regions of the four Venn diagrams of Figure 5.5 represent event X, the complement of event X, the union of two events X and Y, and the intersection of two events X and Y.

EXAMPLE   If X is the event that a certain high school senior applies for admission to the University of Tennessee and Y is the event that she applies to the University of North Carolina, then the region tinted in the first diagram of Figure 5.5 represents the event that she applies to the University of Tennessee, the region tinted in the second diagram represents the event that she does not apply to the University of Tennessee, the region tinted in the third diagram represents the event that she applies to the University of Tennessee and/or the University of North Carolina, and the region tinted in the fourth diagram represents the event that she applies to both of these universities.

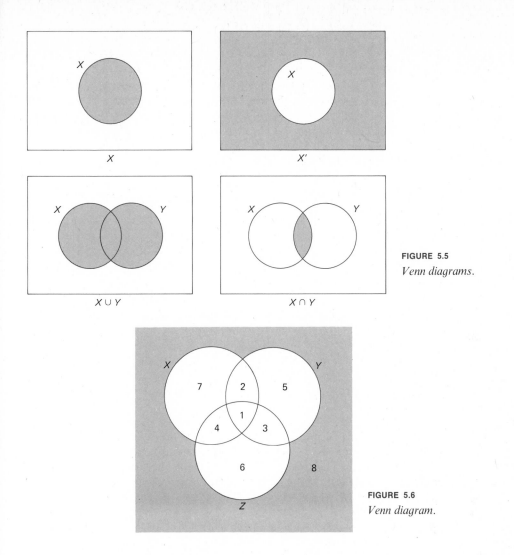

FIGURE 5.5
Venn diagrams.

FIGURE 5.6
Venn diagram.

When we deal with three events, X, Y, and Z, we draw the circles as in Figure 5.6. In this diagram, the circles divide the sample space into eight regions (which we numbered 1 through 8) and it is easy to determine whether each of the corresponding events is contained in X or in X', in Y or in Y', and in Z or in Z'.

EXAMPLE   Suppose that an engineer has designed a new kind of engine and that X represents the event that its gasoline consumption will be low, Y represents the event that its maintenance cost will be low, and Z represents the event that it can be sold at a profit. Looking at Figure 5.6, we see that region 4, for example, is contained in X and Z but not in Y, so that it represents the event that the gasoline consumption of the engine will be low and it can be sold at a profit, but the maintenance cost will not be low. Similarly, we see that region 5 represents the event that the maintenance cost will be low, but the gasoline consumption will not be low and the engine cannot be sold at a profit. It will be left to the reader to identify some of the other regions (and also some of their combinations) in Exercise 11 on page 102.

EXERCISES

1 With reference to the sample space of Figure 5.4, describe in words the events which are represented by the following sets of points:
   (a) (1, 1), (1, 2), and (2, 1);
   (b) (0, 2), (1, 2), and (0, 3);
   (c) (1, 0), (2, 0), (3, 0), and (2, 1).

2 With reference to the sample space of Figure 5.4, list the sets of points which constitute the following events:
   (a) One of the three cars will remain unsold.
   (b) One salesperson or the other will sell all three of the cars.
   (c) The second salesperson will not sell any of the cars.

3 Two professors and five graduate assistants are responsible for the supervision of a chemistry lab, and at least one professor and two graduate assistants must be present whenever the lab is in use.
   (a) Using two coordinates so that (1, 5), for example, represents the event that one of the professors and all five of the graduate assistants are present, and (2, 3) represents the event that both professors and three of the graduate assistants are present, draw a diagram (similar to that of Figure 5.2) showing the eight points of the corresponding sample space.
   (b) Describe in words the event which is represented by each of the following sets of points: the event K which consists of the points (1, 2) and (2, 3), the event L which consists of the points (1, 3) and (2, 2), and the event M which consists of the points (1, 5) and (2, 5).
   (c) With reference to part (b), list the points of the sample space which represent the event $K \cup L$, and describe it in words.
   (d) With reference to part (b), which of the pairs of events, K and L, K and M, and L and M are mutually exclusive?

4 A movie critic has two days in which to view some of the eight movies that have recently been released. She wants to see at least four of the movies, but not more than four on either day.
   (a) Using two coordinates so that (2, 3), for example, represents the event that she will see two of the movies on the first day and three on the second day, draw a diagram (similar to that of Figure 5.2) showing the 15 points of the corresponding sample space.
   (b) List the points of the sample space of part (a) which constitute the following events: event T that she will see only four of the movies, event

U that she will see more of the movies on the second day than on the first day, event V that she will see at least three of the movies on the first day, and event W that she will see as many movies on the first day as on the second day.

    (c) With reference to part (b), describe in words each of the events T', $U \cup W$, and $T \cap W$, and list the points which they contain.

    (d) With reference to part (b), which of the pairs of events, U and V, T and V, U and W, and T and W are mutually exclusive?

5   Among six applicants for an executive job, A is a college graduate, foreign born, and single; B is not a college graduate, foreign born, and married; C is a college graduate, native born, and married; D is not a college graduate, native born, and single; E is a college graduate, native born, and married; and F is not a college graduate, native born, and married. One of these applicants is to get the job, and the event that the job is given to a college graduate, for example, is denoted $\{A, C, E\}$. State in a similar manner the event that the job is given to

    (a) a single person;

    (b) a native-born college graduate;

    (c) a married person who is foreign born.

6   To construct sample spaces for experiments in which we deal with categorical data, we often code the various alternatives by assigning them numbers. For instance, if persons are asked whether they are for a candidate running for the state senate, undecided, or against him, we might assign these three alternatives the codes 1, 2, and 3.

    (a) Draw a suitable sample space showing the different ways in which one person can respond. Also, encircle by means of a solid line the event that the person is not for the candidate, and by means of a dotted line the event that the person is not against the candidate. Are these two events mutually exclusive?

    (b) Use two coordinates to represent, in order, the responses of two persons asked about the candidate, and draw a diagram (similar to that of Figure 5.2) which shows the nine points of the corresponding sample space.

    (c) Describe in words the event which is represented by each of the following subsets of the sample space of part (b): the event D, which consists of the points $(1, 1)$, $(1, 2)$, and $(1, 3)$; the event E, which consists of the points $(1, 3)$, $(2, 3)$, $(3, 3)$; the event F, which consists of the points $(1, 1)$, $(1, 3)$, $(3, 1)$, and $(3, 3)$; and the event G, which consists of the points $(1, 2)$ and $(2, 1)$.

    (d) With reference to part (c), describe in words the events which are denoted F', $D \cup E$, and $E \cap F$. Also list the points which comprise each of these three events.

    (e) If we use three coordinates to represent, in order, the responses of three persons asked about the candidate, what event is represented by the point $(3, 2, 1)$; what event is represented by the set of points $(1, 1, 1)$, $(2, 2, 2)$, and $(3, 3, 3)$; and what event is represented by the set of points $(3, 3, 1)$, $(3, 3, 2)$, and $(3, 3, 3)$?

7   If we code tails and heads as 0 and 1, we could let $(1, 0, 0)$ represent the event that we get heads, tails, tails (in that order) in three flips of a coin. Use this notation to list the eight possible outcomes for three flips of a coin, and draw the corresponding three-dimensional sample space.

8 Suppose that in the two-salespersons example on page 94 the three Mustangs had been distinguishable (say, one was white, one was red, and one was brown).

   (a) Draw a tree diagram to find the number of outcomes (that is, the number of ways, in which sales can be made).

   (b) Verify the result obtained in part (a) by arguing that one of three things must happen to each car (it will be sold by the first salesperson, it will be sold by the second salesperson, or it will remain unsold), and then using the formula for the multiplication of choices.

9 Which of the following pairs of events are mutually exclusive? Explain your answers.

   (a) Having rain and sunshine on July 4, 1982.

   (b) Being under 25 years of age and being President of the United States.

   (c) One person wearing a blue shirt and a red tie.

   (d) A driver getting a ticket for speeding and a ticket for going through a red light.

   (e) A person leaving Los Angeles by jet at 11:50 P.M. and arriving in New York City on the same day.

   (f) A baseball player getting a walk and hitting a home run in the same game.

   (g) A baseball player getting a walk and hitting a home run in the same time at bat.

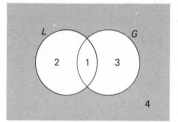

FIGURE 5.7
Venn diagram for Exercise 10.

10 In Figure 5.7, L is the event that a person arrested for car theft can afford to pay a lawyer and G is the event that he or she is found guilty of the crime. Explain in words what events are represented by regions 1, 2, 3, and 4 of the diagram.

11 With reference to the example on page 99 and Figure 5.6, explain in words what events are represented by the following regions of the Venn diagram:

   (a) region 1;

   (b) region 2;

   (c) region 6;

   (d) regions 1 and 3 together;

   (e) regions 2 and 7 together;

   (f) regions 4, 6, 7, and 8 together.

12 Suppose that a group of biologists plan a trip to study endangered species in the Lake Victoria region of Africa and that B is the event that they will run into bad weather, P is the event that they will have problems with local authorities, and E is the event that they will have difficulties with their photographic equipment. With reference to the Venn diagram of Figure 5.8, express in words what

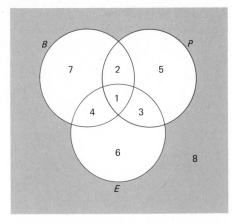

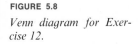

FIGURE 5.8

Venn diagram for Exercise 12.

events are represented by the following regions:

    (a) region 3;

    (b) region 7;

    (c) region 8;

    (d) regions 1 and 4 together;

    (e) regions 3 and 6 together;

    (f) regions 2, 5, 7, and 8 together.

13 With reference to Exercise 12 and the Venn diagram of Figure 5.8, list (by numbers) the regions or combinations of regions which represent the events that they will

    (a) run into bad weather and have difficulties with their photographic equipment, but no problems with local authorities;

    (b) not run into bad weather and have no difficulties with their photographic equipment, but problems with local authorities;

    (c) run into bad weather, but have no difficulties with their photographic equipment;

    (d) have no problems with local authorities and no difficulties with their photographic equipment.

14 Venn diagrams are often used to verify relationships among sets, subsets, or events, without requiring formal proofs based on the algebra of sets. We simply check whether the expressions which are supposed to be equal are represented by the same region of a Venn diagram. Use Venn diagrams to show that

    (a) $A \cup (A \cap B) = A$;

    (b) $(A \cap B) \cup (A \cap B') = A$;

    (c) $(A \cap B)' = A' \cup B'$ and also $(A \cup B)' = A' \cap B'$;

    (d) $(A \cup B) = (A \cap B) \cup (A \cap B') \cup (A' \cap B)$;

    (e) $A \cap (B \cup C) = (A \cap B) \cup (A \cap C)$.

5.2

The Postulates of Probability

               To turn now to the question of how probabilities must "behave," let us begin by stating the three basic postulates. To formulate these postulates and some of their immediate consequences, we shall continue the practice of denoting events by capital letters, and we shall write the probability of event A as $P(A)$, the probability of event B as $P(B)$, and so forth. As before, we shall denote the set of all possible outcomes, the sample space, by the letter S. As we shall formulate them here, the three postulates of probability apply only when the sample space S is finite.[†]

First two postulates of probability

> **1** *The probability of any event is a positive real number or zero; symbolically, $P(A) \geqslant 0$ for any event A.*
>
> **2** *The probability of any sample space is equal to 1; symbolically, $P(S) = 1$ for any sample space S.*

It is important to note that both of these postulates are compatible with all three probability concepts which we studied in Chapter 4. So far as the first postulate is concerned, the fraction $\dfrac{s}{n}$ is always positive or zero, so are percentages and proportions, and when the amounts a and b bet for and against the occurrence of an event are positive, the probability $\dfrac{a}{a+b}$ cannot be negative.

The second postulate states indirectly that certainty is identified with a probability of 1; after all, it is always assumed that one of the possibilities included in S must occur, and it is to this certain event that we assign a probability of 1. For equally likely outcomes, $s = n$ for the whole sample space and $\dfrac{s}{n} = \dfrac{n}{n} = 1$; and in the frequency interpretation, a probability of 1 implies that the event will occur 100 percent of the time, or in other words, that it is certain to occur. So far as subjective probabilities are concerned, the surer we are that an event will occur, the "better" odds we should be willing to give—say, 100 to 1, 1,000 to 1, or perhaps even 1,000,000 to 1. The corresponding probabilities are $\dfrac{100}{100+1}$, $\dfrac{1,000}{1,000+1}$, and $\dfrac{1,000,000}{1,000,000+1}$ (or approximately 0.99, 0.999, and 0.999999), and it can be seen that the surer we are that an event will occur, the closer its subjective probability will be to 1.

In actual practice, we also assign a probability of 1 to events which are "practically certain" to occur. For instance, we would assign a probability of 1 to the event that at least one person will vote in the next presidential election, and that among all new cars sold during any one model year at least one will be involved in an accident before it has been driven 12,000 miles.

[†] A modification that is required when this assumption of finiteness is dropped will be given on page 165.

The third postulate of probability is especially important, and it is not quite so "obvious" as the other two.

Third postulate of probability

3 *If two events are mutually exclusive, the probability that one or the other will occur equals the sum of their probabilities. Symbolically,*

$$P(A \cup B) = P(A) + P(B)$$

for any two mutually exclusive events A and B.

EXAMPLE If the probability that weather conditions will improve during a certain week is 0.62 and the probability that they will remain unchanged is 0.23, then the probability that they will either improve or remain unchanged is $0.62 + 0.23 = 0.85$. Similarly, if the probabilities that a student will get an *A* or a *B* in a course are, respectively, 0.13 and 0.29, then the probability that he will get either an *A* or a *B* is $0.13 + 0.29 = 0.42$.

This postulate is also compatible with the first two of the probability concepts of Chapter 4. In the classical concept, if s_1 of n equally likely possibilities constitute event *A* and s_2 others constitute event *B*, then these $s_1 + s_2$ equally likely possibilities constitute event $A \cup B$, and we have

$$P(A) = \frac{s_1}{n}, P(B) = \frac{s_2}{n}, P(A \cup B) = \frac{s_1 + s_2}{n}, \text{ and } P(A) + P(B) = P(A \cup B).$$

In accordance with the frequency interpretation, if one event occurs, say, 36 percent of the time, another event occurs 41 percent of the time, and they cannot both occur at the same time (that is, they are mutually exclusive), then one or the other will occur $36 + 41 = 77$ percent of the time; this satisfies the third postulate.

When it comes to subjective probabilities, the third postulate does not follow from our discussion in Chapter 4, but it is generally imposed as a **consistency criterion.** In other words, if a person's subjective probabilities "behave" in accordance with the third postulate, he is said to be **consistent;** otherwise, he is said to be **inconsistent** and his probability judgments cannot be taken seriously (see Exercises 7 and 8 on page 112).

EXAMPLE A stockbroker feels that the odds are 2 to 1 that the price of a given stock will not go up during the coming week, 5 to 1 that it will not remain the same, and 3 to 2 that it will go up or remain the same. The corresponding probabilities that the price of the stock will go up, remain the same, and go up or remain the same are $\frac{1}{3}$, $\frac{1}{6}$, and $\frac{3}{5}$, and since $\frac{1}{3} + \frac{1}{6} \neq \frac{3}{5}$, they are inconsistent.

By using the three postulates of probability, we can derive many further rules according to which probabilities must "behave"—some of them are easy to prove and some are not, but they all have important applications. Among the immediate consequences of the three postulates we find that *the probabilities that an event will occur and that it will not occur always add up to* 1, that *probabilities can never be greater than* 1, and that *an event which cannot occur has the probability* 0. Symbolically,

$$P(A) + P(A') = 1 \quad \text{for any event } A$$
$$P(A) \leqslant 1 \quad \text{for any event } A$$
$$P(\varnothing) = 0$$

So far as the first of these rules is concerned, we already showed in Chapter 4 that it is compatible with the classical probability concept and the frequency interpretation. Of course, this does not constitute a proof, but we shall leave that to the reader in Exercise 19 on page 113. The second of the rules simply expresses the fact that there cannot be more favorable outcomes than there are outcomes, that an event cannot occur more than 100 percent of the time, and that $\dfrac{a}{a + b}$ cannot exceed 1 when a and b are positive amounts bet for and against the occurrence of an event. The third of the rules expresses the fact that when an event is impossible there are $s = 0$ favorable outcomes, that an impossible event happens 0 percent of the time, and that a person would not be willing to bet at any odds on an event which cannot occur. In actual practice, we also assign 0 probabilities to events which are so unlikely that we are "practically certain" that they will not occur. For instance, we would assign a probability of 0 to the event that a monkey set loose on a typewriter will by chance type Plato's *Republic* word for word without a single mistake.

5.3
Addition Rules

The third postulate of probability applies only to two mutually exclusive events, but it can easily be generalized; repeatedly using this postulate, it can be shown that

Generalization of Postulate 3

If k events are mutually exclusive, the probability that one of them will occur equals the sum of their respective probabilities; symbolically

$$P(A_1 \cup A_2 \cup \cdots \cup A_k) = P(A_1) + P(A_2) + \cdots + P(A_k)$$

for any mutually exclusive events $A_1, A_2, \ldots,$ *and* A_k.

EXAMPLE If the probabilities that Mrs. F. will buy her new tennis dress at a sporting goods store, at a dress shop, or at a department store are, respectively, 0.18, 0.35, and 0.22, then the probability that she will buy it at one of these places is $0.18 + 0.35 + 0.22 = 0.75$.

EXAMPLE If the probabilities that a consumer testing service will rate a new antipollution device for cars very poor, poor, fair, good, very good, or excellent are, respectively, 0.07, 0.12, 0.17, 0.32, 0.21, and 0.11, then the probability that it will rate the device very poor, poor, fair, or good is $0.07 + 0.12 + 0.17 + 0.32 = 0.68$.

The job of assigning probabilities to all possible events connected with a given situation can be very tedious, to say the least.

EXAMPLE If there are only five individual outcomes (or points) in a sample space S, there are already $2^5 = 32$ different subsets or events. One subset (the empty set) consists of no outcomes at all, $\binom{5}{1} = 5$ consist of one outcome, $\binom{5}{2} = 10$ consist of two outcomes, $\binom{5}{3} = 10$ consist of three outcomes, $\binom{5}{4} = 5$ consist of four outcomes, and one (the sample space itself) consists of all five of the outcomes. As the number of individual outcomes in S increases slowly, the number of subsets increases rapidly. For instance, for a sample space with 20 individual outcomes, there are already more than a million different subsets or events; $2^{20} = 1,048,576$ to be exact.

Fortunately, it is seldom necessary to assign probabilities to all possible events, and the following rule (which is a direct application of the generalization of Postulate 3 given on page 106) makes it easy to determine the probability of any event on the basis of the probabilities assigned to the individual outcomes (points) of the corresponding sample space:

Rule for calculating the probability of an event

> *The probability of any event A is given by the sum of the probabilities of the individual outcomes comprising A.*

This rule is illustrated in Figure 5.9, where the dots represent the individual (mutually exclusive) outcomes.

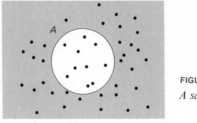

FIGURE 5.9

A sample space.

EXAMPLE If the probabilities are 0.12, 0.26, 0.43, 0.12, and 0.07 that a student will get an A, B, C, D, or F in a psychology course, then the probability that the student will get a B or a C is 0.26 + 0.43 = 0.69; the probability that the student will get a C, D, or F is 0.43 + 0.12 + 0.07 = 0.62; the probability that the student will get an A, B, C, or D is 0.12 + 0.26 + 0.43 + 0.12 = 0.93; and so forth.

EXAMPLE Referring again to the example dealing with the two salespersons and the three Ford Mustangs, suppose that the ten points of the sample space have the probabilities shown in Figure 5.10 (which is otherwise like Figures 5.2 and 5.4). If we are interested in the probability that the first salesperson will not sell any of the cars, we add the probabilities associated with the points (0, 0), (0, 1), (0, 2), and

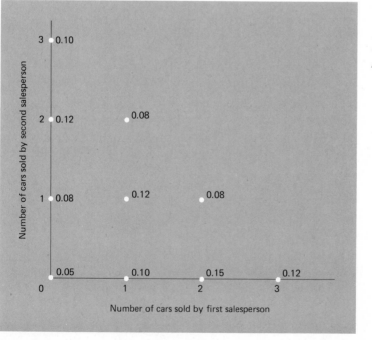

FIGURE 5.10

Sample space with probabilities for the two-salesperson example.

(0, 3), and we get $0.05 + 0.08 + 0.12 + 0.10 = 0.35$. Similarly, if we are interested in the probability that the two salespersons, between them, will sell two of the cars, we add the probabilities associated with the points (0, 2), (1, 1), and (2, 0), and we get $0.12 + 0.12 + 0.15 = 0.39$; and if we are interested in the probability that the second salesperson will sell at least two of the cars, we add the probabilities associated with the points (0, 2), (1, 2), and (0, 3), and we get $0.12 + 0.08 + 0.10 = 0.30$.

Since the third postulate of probability applies only to mutually exclusive events, it cannot be used, for example, to find the probability that at least one of two roommates will pass a final examination in Freshman English, the probability that a bird watcher will spot a roadrunner or a cactus wren, or the probability that a customer will buy a shirt or a sweater shopping at Bullock's. Both roommates can pass the examination, a bird watcher can spot both kinds of birds, and a customer can buy a shirt and sweater in the same store. To find a formula for $P(A \cup B)$ which holds regardless of whether the events A and B are mutually exclusive, let us consider the following example:

EXAMPLE Figure 5.11 concerns the appointment of a college president. The letter G stands for the event that the appointee will be a graduate of the given college, and the letter W stands for the event that the appointee will be a woman. It follows from the figures in the Venn diagram that

$$P(G) = 0.55 + 0.12 = 0.67$$

$$P(W) = 0.12 + 0.04 = 0.16$$

and

$$P(G \cup W) = 0.55 + 0.12 + 0.04 = 0.71$$

where we were able to add the respective probabilities because they pertain to mutually exclusive events (namely, to regions of the Venn diagram which do not overlap).

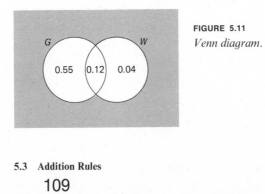

FIGURE 5.11
Venn diagram.

Had we erroneously used the third postulate of probability to calculate $P(G \cup W)$, we would have obtained $P(G \cup W) = 0.67 + 0.16 = 0.83$, which exceeds the correct value by 0.12. This error results from adding $P(G \cap W) = 0.12$ in twice, once in $P(G) = 0.67$ and once in $P(W) = 0.16$, and we could correct for it by subtracting 0.12 from 0.83. Symbolically, we could thus write

$$P(G \cup W) = P(G) + P(W) - P(G \cap W)$$

$$= 0.67 + 0.16 - 0.12$$

$$= 0.71$$

and this agrees, as it should, with the result obtained before.

Since the argument which we have presented holds for any two events A and B, we can now state the following **general addition rule,** which applies regardless of whether A and B are mutually exclusive events:

General addition rule

$$P(A \cup B) = P(A) + P(B) - P(A \cap B)$$

When A and B are mutually exclusive, $P(A \cap B) = 0$ (since by definition the two events cannot both occur at the same time), and the above formula reduces to that of the third postulate of probability. In this connection, the third postulate of probability is also referred to as the **special addition rule.**

EXAMPLE If the probabilities are, respectively, 0.20, 0.15, and 0.03 that a student will get a failing grade in chemistry, English, or both, then the probability is

$$0.20 + 0.15 - 0.03 = 0.32$$

that the student will fail at least one of these subjects.

EXAMPLE If the probabilities are, respectively, 0.87, 0.36, and 0.29 that a family (randomly chosen for a sample survey in a large metropolitan area) will own a color television set, a black and white set, or both kinds of sets, then the probability is

$$0.87 + 0.36 - 0.29 = 0.94$$

that such a family will own either kind of set.

EXERCISES   1 Suppose that in a study of juvenile delinquents in a certain city, D stands for the event that a delinquent has dropped out of school and W stands for the event

that a delinquent's parents are on welfare. State in words what probability is expressed by
- (a) $P(D')$;
- (b) $P(W')$;
- (c) $P(D \cap W)$;
- (d) $P(D \cup W')$;
- (e) $P(D' \cap W')$;
- (f) $P(D' \cup W)$.

2 If F is the event that a dishonest stockbroker is in financial difficulties, T is the event that he has tax problems, and Q is the event that he uses questionable sales practices, write in symbolic form the probabilities that a dishonest stockbroker
- (a) is in financial difficulties and uses questionable sales practices;
- (b) does not have tax problems but is in financial difficulties;
- (c) uses questionable sales practices and/or is in financial difficulties;
- (d) has neither financial difficulties nor tax problems.

3 Explain why there must be a mistake in each of the following statements:
- (a) The probability that a mineral sample will contain copper is 0.28 and the probability that it will not contain copper is 0.62.
- (b) The probability that Dan will pass the bar examination is 0.34 and the probability that he will not pass is -0.66.
- (c) The probability that the home team will win an upcoming football game is 0.77, the probability that it will tie the game is 0.08, and the probability that it will win or tie the game is 0.95.
- (d) The probability that a chemistry experiment will succeed is 0.73 and the probability that it will not succeed is 0.47.
- (e) The probability that it will rain is 0.67 and the probability that it will rain or snow is 0.55.
- (f) The probability that an ambulance service will receive more than ten calls on a certain day is 0.53 and the probability that it will receive more than twelve calls is 0.60.
- (g) The probability that a student will get a passing grade in English is 0.72 and the probability that she will get a passing grade in English and French is 0.85.
- (h) The probability that a person with a fever will get a shot from his doctor is 0.48, the probability that he will get some medication but no shot is 0.36, and the probability that he will get no medication and no shot is 0.12.

4 Given the mutually exclusive events C and D for which $P(C) = 0.31$ and $P(D) = 0.44$, find
- (a) $P(C')$;
- (b) $P(D')$;
- (c) $P(C \cap D)$;
- (d) $P(C \cup D)$;
- (e) $P(C' \cup D')$;
- (f) $P(C' \cap D')$.

(*Hint:* Draw a Venn diagram and fill in the probabilities associated with the various regions.)

5 If the probabilities that a certain missile will explode during lift-off or have its guidance system fail in flight are, respectively, 0.0002 and 0.0005, find the probabilities that such a missile will
  (a) not explode during lift-off;
  (b) either explode during lift-off or have its guidance system fail in flight;
  (c) not explode during lift-off nor have its guidance system fail in flight.

6 An economist claims that the odds are 2 to 1 that unemployment will go up and 3 to 1 that it will go down. Can these odds be right? Explain.

7 A high school principal feels that the odds are 7 to 5 against her getting a $1,000 raise and 11 to 1 against her getting a $2,000 raise. Furthermore, she feels that it is an even-money bet that she will get one of these raises or the other. Discuss the consistency of the corresponding subjective probabilities.

8 There are two Porsches in a race, and a reporter feels that the odds against their winning are, respectively, 3 to 1 and 4 to 1. To be consistent, what odds should he assign to the event that neither car will win?

9 The probabilities that the serviceability of some new X-ray equipment will be rated very difficult, difficult, average, easy, or very easy are, respectively, 0.11, 0.18, 0.34, 0.27, and 0.10. Find the probabilities that the serviceability of the new X-ray equipment will be rated
  (a) difficult or very difficult;
  (b) neither very difficult nor very easy;
  (c) average or worse;
  (d) average or better.

10 The probabilities that a TV station will receive 0, 1, 2, 3, . . . , 8, or at least 9 complaints after showing a controversial program are, respectively, 0.01, 0.03, 0.07, 0.15, 0.19, 0.18, 0.14, 0.12, 0.09, and 0.02. What are the probabilities that after showing such a program the station will receive
  (a) at most 4 complaints?
  (b) at least 6 complaints?
  (c) from 5 to 8 complaints?

11 A police department needs new tires for its patrol cars and the probabilities are 0.17, 0.22, 0.03, 0.29, 0.21, and 0.08 that it will buy Uniroyal tires, Goodyear tires, Michelin tires, General tires, Goodrich tires, or Armstrong tires. Find the probabilities that it will buy
  (a) Goodyear or Goodrich tires;
  (b) Uniroyal, General, or Goodrich tires;
  (c) Michelin or Armstrong tires;
  (d) Goodyear, General, or Armstrong tires.

12 With reference to Figure 5.10 find the probabilities that
  (a) the second salesperson will sell one of the cars;
  (b) neither salesperson will sell more than one car;
  (c) between them, the two salespersons will sell all three cars;
  (d) the first salesperson will sell at least one car.

13 With reference to Exercise 4 on page 100, suppose that each of the 15 points of the sample space has the probability $\frac{1}{15}$. What are the probabilities of events T, U, V, and W described in part (b) of that exercise?

14 Among the 68 doctors on the staff of a hospital, 59 carry malpractice insurance, 35 are surgeons, and 31 of the surgeons carry malpractice insurance. If one of

these doctors is chosen by lot to represent the hospital staff at an AMA convention (that is, each of the doctors has a probability of $\frac{1}{68}$ of being selected), what is the probability that the one chosen is not a surgeon and does not carry malpractice insurance?

15 Given two events G and H for which $P(G) = 0.46$, $P(H) = 0.35$, and $P(G \cap H) = 0.21$, find
   (a) $P(G')$;
   (b) $P(H')$;
   (c) $P(G \cup H)$;
   (d) $P(G' \cap H)$;
   (e) $P(G' \cup H)$;
   (f) $P(G' \cap H')$.
(*Hint:* Draw a Venn diagram and fill in the probabilities associated with the various regions.)

16 A geology professor has two graduate assistants helping her with her research. The probability that the older of the two will be absent on any given day is 0.08, the probability that the younger of the two will be absent on any given day is 0.06, and the probability that they will both be absent on any given day is 0.02. Find the probabilities that
   (a) either or both of the graduate assistants will be absent on any given day;
   (b) at least one of the two graduate assistants will not be absent on any given day;
   (c) only one of the two graduate assistants will be absent on any given day.

17 The probabilities that a person stopping at a gas station will ask to have his tires checked is 0.14, the probability that he will ask to have his oil checked is 0.27, and the probability that he will ask to have them both checked is 0.09. What are the probabilities that a person stopping at this gas station will have
   (a) either his tires or his oil checked?
   (b) neither his tires nor his oil checked?
(*Hint:* Draw a Venn diagram and fill in the probabilities associated with the various regions.)

18 A student artist who has entered an oil painting and a watercolor in a show feels that the probabilities are, respectively, 0.23, 0.19, and 0.08 that she will sell the oil painting, the watercolor, or both. What are the probabilities that she will sell
   (a) either of these works?
   (b) neither of these works?
   (c) the oil painting but not the watercolor?

19 The following is a proof of the fact that $P(A) \leq 1$ for any event A: By definition A and A' represent mutually exclusive events and $A \cup A' = S$ (since A and A' together constitute all the points of the sample space S). So, we can write $P(A \cup A') = P(S)$, and it follows that

$$P(A) + P(A') = P(S) \qquad \text{step 1}$$

$$P(A) + P(A') = 1 \qquad \text{step 2}$$

$$P(A) = 1 - P(A') \qquad \text{step 3}$$

$$P(A) \leq 1 \qquad \text{step 4}$$

State which of the three postulates of probability justify the first, second, and fourth steps of this proof; the third step is simple arithmetic. Note also that in step 2 we actually proved the first of the three rules on page 106.

20 Making use of the fact that $S \cup \varnothing = S$ for any sample space S, and S and $\varnothing$ are mutually exclusive (by default), prove the last of the three rules on page 106; namely, that $P(\varnothing) = 0$.

5.4
Conditional Probability

Difficulties can easily arise when we speak of, or ask for, the probability of an event without specifying the sample space on which the event is defined. For instance, if we ask for the probability that a lawyer makes more than \$40,000 per year, we may well get many different answers, and they can all be correct. One of them might apply to all lawyers in the United States, another to corporation lawyers, a third to lawyers employed by the federal government, another to lawyers handling only divorces, and so forth. Since the choice of the sample space (that is, the set of all possibilities under consideration) is by no means always self-evident, it is helpful to use the symbol $P(A|S)$ to denote the **conditional probability** of event A relative to the sample space S, or as we often call it "the probability of A given S." The symbol $P(A|S)$ makes it explicit that we are referring to a particular sample space S, and it is generally preferable to the abbreviated notation $P(A)$ unless the tacit choice of S is clearly understood. It is also preferable when we have to refer to different sample spaces in the same problem, as in the examples which follow.

EXAMPLE To elaborate on the idea of a conditional probability, suppose that a consumer research organization has studied the service under warranty provided by the 200 tire dealers in a large city, and that their findings are summarized in the following table:

	Good service under warranty	Poor service under warranty
Name-brand tire dealers	64	16
Off-brand tire dealers	42	78

Suppose, further, that a person randomly selects one of these tire dealers, where "randomly" means that each of the dealers has the same chance, a probability of $\frac{1}{200}$, of being selected. Now, if we let N denote the selection of a name-brand dealer and G the selection of a dealer who provides good service under warranty, we find that

$$P(N \cap G) = \frac{64}{200} = 0.32$$

$$P(N) = \frac{64 + 16}{200} = 0.40$$

and

$$P(G) = \frac{64 + 42}{200} = 0.53$$

where all these probabilities were calculated by means of the formula $\frac{s}{n}$ for equally likely possibilities.

Since the third of these probabilities is particularly disconcerting—there is almost a fifty–fifty chance of choosing a tire dealer who provides poor service under warranty—let us see what will happen if we limit the choice to name-brand dealers. Looking at the reduced sample space which consists of the first row of the table, we get

$$P(G|N) = \frac{64}{64 + 16} = 0.80$$

and this is quite an improvement over $P(G) = 0.53$, as might have been expected. Note that this conditional probability, 0.80, can also be written as

$$P(G|N) = \frac{64/200}{(64 + 16)/200} = \frac{P(N \cap G)}{P(N)}$$

which is the ratio of the probability of choosing a name-brand dealer who provides good service under warranty to the probability of choosing a name-brand dealer.

Generalizing from this example, let us now make the following definition of conditional probability, which applies to any two events A and B

belonging to a given sample space S:

Definition of conditional probability

> If $P(B)$ is not equal to zero, then the conditional probability of A relative to B, namely, the probability of A given B, is
>
> $$P(A|B) = \frac{P(A \cap B)}{P(B)}$$

EXAMPLE With reference to the tire dealers, we can also say on the basis of the table on page 114 that the probability that an off-brand tire dealer will give good service under warranty is

$$P(G|N') = \frac{P(G \cap N')}{P(N')} = \frac{42/200}{(42 + 78)/200} = 0.35$$

and the probability that a tire dealer who gives poor service under warranty is a name-brand dealer is

$$P(N|G') = \frac{P(N \cap G')}{P(G')} = \frac{16/200}{(16 + 78)/200} = 0.17$$

Of course, the fractions $\dfrac{42}{42 + 78}$ and $\dfrac{16}{16 + 78}$ could also have been obtained directly from the entries, and the row and column totals of the table.

Although we justified the formula for $P(A|B)$ by means of an example in which the possibilities were all equiprobable, this is not a requirement for its use. The only restriction is that $P(B)$ must not equal zero.

EXAMPLE If the probability that a communication system will have high selectivity is $P(A) = 0.54$, the probability that it will have high fidelity is $P(B) = 0.81$, and the probability that it will have both is $P(A \cap B) = 0.18$, then the probability that it will have high selectivity given that it has high fidelity is

$$P(A|B) = \frac{0.18}{0.81} = \frac{2}{9}$$

and the probability that it will have high fidelity given that it has high selectivity is

$$P(B|A) = \frac{0.18}{0.54} = \frac{1}{3}$$

EXAMPLE If the probability that a research project will be well planned is 0.60 and the probability that it will be well planned and well executed is 0.54, then the probability that it will be well executed given that it is well planned is $\dfrac{0.54}{0.60} = 0.90$.

EXAMPLE With reference to the sample space of Figure 5.10 on page 108, the probability that the first salesperson will sell one of the cars is $0.10 + 0.12 + 0.08 = 0.30$ and the probability that the first salesperson will sell one of the cars and the second salesperson will sell two of the cars is 0.08. Thus, the probability that the second salesperson will sell two of the cars given that the first salesperson sells one of the cars is $\dfrac{0.08}{0.30} = 0.267$.

To introduce another concept which is important in the study of probability, let us refer to the example on page 110, which dealt with the possibility that a student might get a failing grade in chemistry, English, or both.

EXAMPLE If we let C denote the event that the student will get a failing grade in chemistry, E that he will get a failing grade in English, and $P(C) = 0.20$, $P(E) = 0.15$, and $P(C \cap E) = 0.03$, as before, then the probability that he will get a failing grade in chemistry given that he gets a failing grade in English is

$$P(C|E) = \frac{P(C \cap E)}{P(E)} = \frac{0.03}{0.15} = 0.20$$

What is special, and interesting, about this result is that

$$P(C|E) = P(C) = 0.20$$

namely, that the probability of event C is the same regardless of whether or not event E has occurred (occurs, or will occur).

5.4 Conditional Probability

In general, if $P(A|B) = P(A)$, we say that event A is **independent** of event B, and since it can be shown that event B is independent of event A whenever event A is independent of event B, it is customary to say simply that **A and B are independent whenever one is independent of the other** (see also Exercise 5 on page 121). If two events A and B are not independent, we say that they are **dependent.**

5.5

Multiplication Rules

So far we have used the formula $P(A|B) = \dfrac{P(A \cap B)}{P(B)}$ only to calculate conditional probabilities, but if we multiply on both sides of the equation by $P(B)$, we get the following formula, called the **general multiplication rule,** which enables us to calculate the probability that two events will both occur:

General multiplication rule

$$P(A \cap B) = P(B) \cdot P(A|B)$$

In words, this formula states that the probability that two events will both occur is the product of the probability that one of the events will occur and the conditional probability that the other event will occur given that the first event has occurred (occurs, or will occur). As it does not matter which event is referred to as A and which is referred to as B, the above formula can also be written as

General multiplication rule

$$P(A \cap B) = P(A) \cdot P(B|A)$$

EXAMPLE   Suppose that we randomly select two shirts, one after the other, from a carton containing 12 shirts, three of which have blemishes. What is the probability that both of the shirts we pick will have blemishes? If we assume equal probabilities for each choice (which is, in fact, what we mean by the selections being "random"), the probability that the first shirt we pick will have blemishes is $\frac{3}{12}$, and the probability that the second shirt we pick will have blemishes is $\frac{2}{11}$. Clearly, there are only two shirts with blemishes among the 11 which remain after one shirt with blemishes has been picked. Hence, the probability of getting two shirts with blemishes is

$$\frac{3}{12} \cdot \frac{2}{11} = \frac{1}{22}$$

The same kind of argument leads to the result that the probability of getting two shirts without blemishes is

$$\frac{9}{12} \cdot \frac{8}{11} = \frac{12}{22}$$

and it follows, by subtraction, that the probability of getting one shirt with blemishes and one shirt without blemishes is

$$1 - \frac{1}{22} - \frac{12}{22} = \frac{9}{22}$$

When A and B are independent events, we can substitute $P(A)$ for $P(A|B)$ in the first of the two formulas for $P(A \cap B)$, or $P(B)$ for $P(B|A)$ in the second, and we obtain

Special multiplication rule

$$P(A \cap B) = P(A) \cdot P(B)$$

This rule tells us that the probability that two independent events will both occur is simply the product of their probabilities.

EXAMPLE The probability of getting two heads in two flips of a balanced coin is $\frac{1}{2} \cdot \frac{1}{2} = \frac{1}{4}$. Also, if 2 cards are drawn from an ordinary deck of 52 playing cards, and the first card is replaced before the second card is drawn, the probability of getting two aces in a row is

$$\frac{4}{52} \cdot \frac{4}{52} = \frac{1}{169}$$

However, if the first card is not replaced, the probability of getting two aces is

$$\frac{4}{52} \cdot \frac{3}{51} = \frac{1}{221}$$

for there are only three aces among the 51 cards which remain after one ace has been removed from the deck. This distinction is of basic importance in statistics, where we sometimes **sample with replacement** and sometimes **sample without replacement**.

EXAMPLE If the probability is 0.12 that a person will make a mistake in his or her income tax return, then the probability that two totally un-related persons (who do not use the same accountant or tax service) will both make a mistake is $(0.12)(0.12) = 0.0144$; also, if the probability is 0.27 that a person will name blue as his or her favorite color, then the probability that neither of two totally unrelated persons will name blue as their favorite color is $(0.73)(0.73) = 0.5329$.

The special multiplication rule can easily be generalized so that it applies to the occurrence of three or more independent events—again, we simply multiply together all their probabilities.

EXAMPLE The probability of getting three heads in three tosses of a balanced coin is

$$\frac{1}{2} \cdot \frac{1}{2} \cdot \frac{1}{2} = \frac{1}{8}$$

and the probability of first rolling four fives and then another number in five rolls of a balanced die is

$$\frac{1}{6} \cdot \frac{1}{6} \cdot \frac{1}{6} \cdot \frac{1}{6} \cdot \frac{5}{6} = \frac{5}{7,776}$$

For three or more dependent events the multiplication rule becomes somewhat more difficult, as is illustrated in Exercise 14 on page 122.

EXERCISES

1 Ms. Jones is looking for a job. If H is the event that she will find a job paying a high starting salary and G is the event that she will find a job with a good future, state in words what probabilities are expressed by
  (a) $P(G|H)$;
  (b) $P(H'|G)$;
  (c) $P(H|G')$;
  (d) $P(G'|H')$.

2 A guidance department gives students various kinds of tests. If I is the event that a student scores high in intelligence, A is the event that a student rates high on a social adjustment scale, and N is the event that a student displays neurotic tendencies, express each of the following probabilities in symbolic form:
  (a) The probability that a student who scores high in intelligence will display neurotic tendencies.
  (b) The probability that a student who does not rate high on the social adjustment scale will not score high in intelligence.
  (c) The probability that a student who displays neurotic tendencies will neither score high in intelligence nor rate high on the social adjustment scale.
  (d) The probability that a student who scores high in intelligence and rates high on the social adjustment scale will not display any neurotic tendencies.

3 There are 60 applicants for a job in the news department of a television station. Some of them are college graduates and some are not, some of them have at least three years' experience and some have not, with the exact breakdown being

	College graduates	Not college graduates
At least three years' experience	12	6
Less than three years' experience	24	18

If the order in which the applicants are interviewed by the station manager is random, G is the event that the first applicant interviewed is a college graduate, and T is the event that the first applicant interviewed has at least three years' experience, determine each of the following probabilities directly from the entries and the row and column totals of the table:

(a) $P(G)$;   .6 $(\frac{3}{5})$
(b) $P(T')$;   .7 $(\frac{7}{10})$
(c) $P(G \cap T)$;   .18 $(\frac{1}{5})$
(d) $P(G' \cap T')$;   .3 $(\frac{3}{10})$
(e) $P(T|G)$;   .333 $(\frac{1}{3})$
(f) $P(G'|T')$.   .43

4 Use the results of Exercise 3 to verify that

(a) $P(T|G) = \dfrac{P(G \cap T)}{P(G)}$;   (b) $P(G'|T') = \dfrac{P(G' \cap T')}{P(T')}$.

5 In the example on page 117 where $P(C) = 0.20$, $P(E) = 0.15$, and $P(C \cap E) = 0.03$, we showed that $P(C|E) = P(C)$; namely, that C is independent of E. Verify that

(a) $P(E|C) = P(E)$; namely, that E is also independent of C;
(b) $P(C|E') = P(C)$; namely, that C is also independent of E';
(c) $P(E|C') = P(E)$; namely, that E is also independent of C'.

6 Police records show that in a certain city the probability is 0.30 that a burglar will be caught, and 0.45 that, if caught, a burglar will be convicted. What is the probability that in this town a burglar will be caught and convicted?

7 The probability that a bus from Cleveland to Chicago will leave on time is 0.80, and the probability that it will leave on time and also arrive on time is 0.72.

(a) What is the conditional probability that if such a bus leaves on time it will also arrive on time?
(b) If the probability is 0.75 that such a bus will arrive on time, what is the conditional probability that if such a bus does not leave on time it will nevertheless arrive on time?

(*Hint:* Draw a Venn diagram and fill in the probabilities associated with the various regions.)

8   With reference to Exercise 17 on page 113, find the probabilities that
   (a) a person stopping at the gas station who has his tires checked will also have his oil checked;
   (b) a person stopping at the gas station who has his oil checked will also have his tires checked.

9   With reference to Exercise 18 on page 113, find the probabilities that the student artist will
   (a) sell the oil painting given that she is going to sell the watercolor;
   (b) sell the watercolor given that she is going to sell the oil painting?

10   With reference to the example on page 110, is family ownership of a color television set independent of family ownership of a black and white set?

11   If a zoologist has six male guinea pigs and nine female guinea pigs, and randomly selects two of them for an experiment, what are the probabilities that
   (a) both will be males?
   (b) both will be females?
   (c) there will be one of each sex?

12   In a fifth grade class of 18 boys and 12 girls, one pupil is chosen each week by lot to act as an assistant to the teacher. What is the probability that a girl will be chosen two weeks in a row if
   (a) the same pupil cannot serve two weeks in a row?
   (b) the restriction of part (a) is removed?

13   Find the probabilities of getting
   (a) six tails in a row in six flips of a balanced coin;
   (b) three spades in three draws from an ordinary deck of playing cards, if each card is replaced before the next card is drawn;
   (c) no threes in four rolls of a balanced die.

14   For three or more events which are not independent, the probability that they will all occur is obtained by multiplying the probability that one of the events will occur *times* the probability that a second of the events will occur given that the first event has occurred *times* the probability that a third of the events will occur given that the first two events have occurred, and so on. For instance, the probability of drawing without replacement three aces in a row from an ordinary deck of 52 playing cards is

$$\frac{4}{52} \cdot \frac{3}{51} \cdot \frac{2}{50} = \frac{1}{5,525}$$

   (a) A carton contains 12 shirts of which five have blemishes and the rest are good. What is the probability that if three of the shirts are randomly selected from the carton, they will all have blemishes?
   (b) If a person randomly picks four of the 15 gold coins a dealer has in stock, and six of the coins are counterfeits, what is the probability that the coins picked will all be counterfeits?
   (c) If five of a company's ten delivery trucks do not meet emission standards and three of them are chosen for inspection, what is the probability that none of the trucks chosen will meet emission standards?

(d) In a certain city, the probability that it will rain on a November day is 0.60, the probability that a rainy November day will be followed by another rainy day is 0.80, and the probability that a sunny November day will be followed by a rainy day is 0.30. What is the probability that it will rain, rain, not rain, and rain in this city on four consecutive November days?

5.6

Rule of Elimination ★

There are many situations in which the ultimate outcome depends on what happens in various intermediate stages. For instance, whether a patient will recover from a disease depends on the correctness of the diagnosis as well as the appropriateness of the treatment; whether an experiment will yield fruitful results depends on how well it is planned and how well it is executed; and the success of a manned space flight depends on the accuracy of preliminary calculations, the performance of all systems, and the competence of the astronauts.

EXAMPLE To illustrate the general technique used in problems of this kind, consider an election in which three candidates, Mr. Clark, Mr. Brown, and Ms. Jones, are running for the office of mayor of a certain city. The biggest campaign issue, that of raising the city sales tax, is strongly favored by Mr. Clark, strongly opposed by Ms. Jones, while Mr. Brown is uncommitted. In fact, it is felt that the probabilities for a raise in the city sales tax are 0.90, 0.50, and 0.05, depending on whether Mr. Clark, Mr. Brown, or Ms. Jones is elected. What we would like to know is the unconditional probability that the city sales tax will be raised, and to this end we must know something about the three candidates' chances of winning the election. Suppose, therefore, that the odds favor Mr. Clark by 3 to 2, but are 4 to 1 against Mr. Brown as well as Ms. Jones. These odds are consistent, since the corresponding probabilities are 0.60, 0.20, and 0.20, which add up to 1.

Referring to the tree diagram of Figure 5.12, we find that the probability associated with the first branch (namely, the probability that Mr. Clark will be elected and then raise the city sales tax) is $(0.60)(0.90) = 0.54$ in accordance with the general multiplication rule on page 118. Similarly, the probabilities associated with the other two branches of the tree diagram are $(0.20)(0.50) = 0.10$ and $(0.20)(0.05) = 0.01$, and since the three possibilities are mutually exclusive, we arrive at the result that the probability is $0.54 + 0.10 + 0.01 = 0.65$ that the city sales tax will be raised.

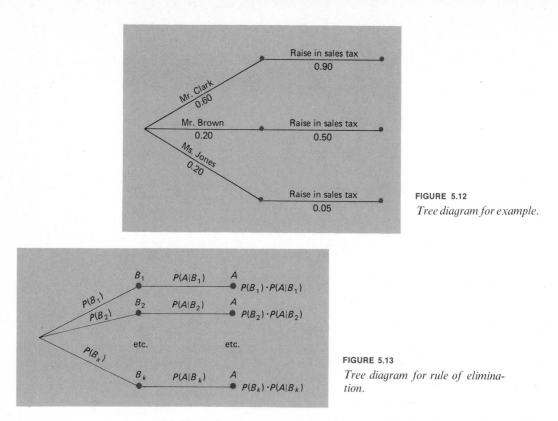

FIGURE 5.12
Tree diagram for example.

FIGURE 5.13
Tree diagram for rule of elimination.

This kind of argument can be used in general, when there are k mutually exclusive alternatives, B_1, B_2, ..., and B_k, which can all lead to event A. Referring to the tree diagram of Figure 5.13, we find that the probability associated with the first branch (namely, the probability of reaching A via B_1) is $P(B_1) \cdot P(A|B_1)$ in accordance with the general multiplication rule on page 118. Similarly, the probabilities associated with the other branches of the tree diagram are $P(B_2) \cdot P(A|B_2)$, ..., and $P(B_k) \cdot P(A|B_k)$, and since the k possibilities are mutually exclusive, we can write

Rule of elimination

> If B_1, B_2, ..., and B_k are mutually exclusive events of which one must occur, then for any event A
>
> $$P(A) = P(B_1) \cdot P(A|B_1) + P(B_2) \cdot P(A|B_2) + \cdots + P(B_k) \cdot P(A|B_k)$$

In the special case where there are only the two alternative B and B', we substitute B for B_1 and B' for B_2, and get

Special case of rule of elimination

> $$P(A) = P(B) \cdot P(A|B) + P(B') \cdot P(A|B')$$

EXAMPLE [†] To illustrate this special case, let us refer to a **randomized response technique** for getting answers to sensitive questions. Suppose, for instance, we want to determine what percentage of the students at a large university regularly smoke marijuana. We construct 20 flash cards, write "I smoke marijuana at least once a week" on 12 of the cards, where 12 is an arbitrary choice, and "I do not smoke marijuana at least once a week" on the others. Then, we let each student (in the sample interviewed) select one of the cards at random, and respond "Yes" or "No" without divulging the question. Strange as it may seem, this will enable us to estimate what percentage of the students at the given university smoke marijuana at least once a week.

If we let B denote the event that a student responds to "I smoke marijuana at least once a week" and B' denote the event that the student responds to "I do not smoke marijuana at least once a week," then $P(B) = \frac{12}{20} = 0.60$ and $P(B') = \frac{8}{20} = 0.40$. Furthermore, if we let A denote the event that a student answers "Yes" and p is the proportion of the students at the university who smoke marijuana at least once a week, then $P(A|B) = p$ and $P(A|B') = 1 - p$, since a "Yes" response to the second question is equivalent to a "No" response to the first question. Substituting all these values into the formula for the special case of the rule of elimination, we get

$$P(A) = P(B) \cdot P(A|B) + P(B') \cdot P(A|B')$$
$$= (0.60) \cdot p + (0.40) \cdot (1 - p)$$

Graphically, this is pictured in Figure 5.14. If we knew $P(A)$, we could calculate p, so let us suppose that in a random sample of 300 of the students, 126 answered "Yes" and 174 answered "No." Using $\frac{126}{300} = 0.42$ as an estimate of the probability $P(A)$ that a

---

[†] May be omitted without loss of continuity.

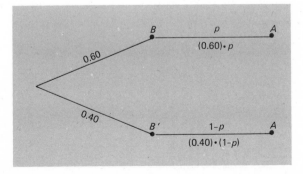

FIGURE 5.14
Tree diagram for example.

student will give a "Yes" answer, we get

$$0.42 = (0.60) \cdot p + (0.40) \cdot (1 - p)$$

and, hence, $0.42 = 0.60p + 0.40 - 0.40p$, $0.42 - 0.40 = 0.60p - 0.40p$, $0.02 = 0.20p$, and $p = \dfrac{0.02}{0.20} = 0.10$. Without knowing who answered what question, we estimate that ten percent of the students at the given university smoke marijuana at least once a week.

5.7

Bayes' Theorem ★

To consider problems which are closely related to those of the preceding section, suppose again that event A can be reached via any one of the intermediate steps $B_1, B_2, \ldots,$ or B_k, but instead of asking for $P(A)$, let us now ask for $P(B_i|A)$, where i can be 1, 2, ..., or k. $P(B_i|A)$ is the probability that event A, known to have occurred, was reached via the particular intermediate step B_i.

EXAMPLE   If a cannery has three assembly lines, I, II, and III, we may want to know the probability that an improperly sealed can (discovered in the final inspection of outgoing products) came from assembly line I. Here A is the event that in the final inspection a can is found to be improperly sealed, B_1 is the event that a can comes from assembly line I, and we are interested in the probability $P(B_1|A)$.

To find a formula for $P(B_i|A)$, let us use the definition of conditional probability and write

$$P(B_i|A) = \frac{P(A \cap B_i)}{P(A)}$$

Then, if we substitute $P(B_i) \cdot P(A|B_i)$ for $P(A \cap B_i)$ in accordance with the general multiplication rule, and substitute the expression given by the rule of elimination for $P(A)$, we arrive at the following result, which is named after the Methodist clergyman Thomas Bayes (1702–1761):

Bayes' theorem

If $B_1, B_2, \ldots,$ and B_k are mutually exclusive events of which one must occur, then

$$P(B_i|A) = \frac{P(B_i) \cdot P(A|B_i)}{P(B_1) \cdot P(A|B_1) + P(B_2) \cdot P(A|B_2) + \cdots + P(B_k) \cdot P(A|B_k)}$$

for $i = 1, 2, \ldots,$ or k.

In practice, the use of this theorem is facilitated by referring to a tree diagram like that of Figure 5.13 on page 124: $P(B_i|A)$ is the probability that event A, having occurred, was reached via the ith branch of the tree diagram, and its value is given by the ratio of the probability associated with the ith branch to the sum of the probabilities associated with all the branches of the tree diagram.

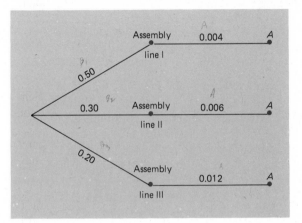

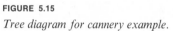

FIGURE 5.15

Tree diagram for cannery example.

EXAMPLE With reference to the cannery mentioned in the example on page 126, suppose that assembly lines I, II, and III account for 50 percent, 30 percent, and 20 percent of the total output, and quality control records show that 0.4 percent of the cans from assembly line I, 0.6 percent of the cans from assembly line II, and 1.2 percent of the cans from assembly line III are improperly sealed. Picturing this situation as in Figure 5.15, we find that the probabilities associated with the three branches of the tree diagram are, respectively, $(0.50)(0.004) = 0.0020$, $(0.30)(0.006) = 0.0018$, and $(0.20)(0.012) = 0.0024$, and it follows that the probability that an improperly sealed can came from assembly line I is

$$P(B_1|A) = \frac{0.0020}{0.0020 + 0.0018 + 0.0024} = 0.32$$

rounded to two decimals. Of course, if we had substituted directly into the formula for Bayes' theorem, all the calculations and the final result would have been the same.

EXAMPLE If a car fails a state's vehicular emission test, does this mean that the car actually emits excessive amounts of pollutants? Not necessarily, because there is always the possibility of an error due to faulty equipment calibration, or errors due to variations in temperature,

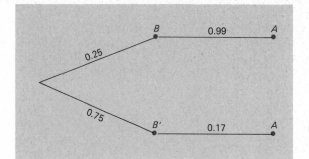

FIGURE 5.16
Tree diagram for emission test example.

humidity, or other factors. Suppose, for instance, that 25 percent of the cars in the given state emit excessive amounts of pollutants, the probability is 0.99 that a car emitting excessive amounts of pollutants will fail the test, and the probability is 0.17 that a car which does not emit excessive amounts of pollutants will fail the test. Picturing the situation as in Figure 5.16, we find that the probabilities associated with the two branches of the tree diagram are (0.25)(0.99) = 0.2475 and (0.75)(0.17) = 0.1275. Thus, the probability that a car which fails the test actually emits excessive amounts of pollutants is

$$\frac{0.2475}{0.2475 + 0.1275} = 0.66$$

and the probability that a car which fails the test does not emit excessive amount of pollutants is $1 - 0.66 = 0.34$. In view of the time it takes and the expense of having a car repaired and retested, a figure as high as this should be of concern. (See also Exercise 11 on page 130.)

EXERCISES

1   At an electronics plant, it is known from past experience that the probability is 0.86 that a new worker who has attended the company's training program will meet his production quota, and that the corresponding probability is 0.35 for a new worker who has not attended the company's training program. If 80 percent of all new workers attend the training program, what is the probability that a new worker will meet his production quota?

2   A hotel gets cars for its guests from three rental agencies, 20 percent from agency X, 40 percent from agency Y, and 40 percent from agency Z. If 14 percent of the cars from X, 4 percent from Y, and 8 percent from Z need tune-ups, what is the probability that a car needing a tune-up is delivered to one of the hotel's guests?

3   In a T-maze, a rat is given food if it turns left and an electric shock if it turns right. On the first trial there is a fifty–fifty chance that a rat will turn either way; then, if it receives food on the first trial, the probability that it will turn left on

the next trial is 0.68, and if it receives a shock on the first trial, the probability that it will turn left on the next trial is 0.84.

  (a) What is the probability that a rat will turn left on the second trial?
  (b) Assuming that the probability of a rat turning left on any given trial depends only on what it did on the preceding trial, find the probability that a rat will turn left on the third trial. (*Hint:* Draw a tree diagram and calculate, and then add, the probabilities associated with the four paths which lead to the rat turning left on the third trial.)

4 Two firms V and W consider bidding on a road-building job which may or may not be awarded depending on the amounts of the bids. Firm V submits a bid and the probability is $\frac{3}{4}$ that it will get the job provided firm W does not bid. The odds are 3 to 1 that W will bid, and if it does, the probability that V will get the job is only $\frac{1}{3}$.

  (a) What is the probability that V will get the job?
  (b) If V gets the job, what is the probability that W did not bid?

5 In a certain community, 8 percent of all adults over 50 have diabetes. If a doctor in this community correctly diagnoses 95 percent of all persons with diabetes as having the disease and incorrectly diagnoses 2 percent of all persons without diabetes as having the disease, what is the probability that an adult over 50 diagnosed by this doctor as having diabetes actually has the disease?

6 An importer, expecting a shipment of ginger roots from Indonesia, knows from past experience that the odds are 4 to 1 against such a shipment getting lost. He also knows from past experience that half the shipments that do not get lost arrive within a month. If the shipment of ginger roots has not arrived within a month, what are the odds that it will still come?

7 With reference to the example on page 127, what are the probabilities that the improperly sealed can came from

  (a) assembly line II?
  (b) assembly line III?

8 A mail-order house employs three stock clerks, P, Q, and R, who pull items from shelves and assemble them for subsequent verification and packaging. P makes a mistake in an order (gets a wrong item or the wrong quantity) one time in a hundred, Q makes a mistake in an order five times in a hundred, and R makes a mistake in an order three times in a hundred. Of all the orders delivered for verification, P, Q, and R fill, respectively, 30, 40, and 30 percent. If a mistake is found in a particular order, what is the probability that it was filled by Q?

9 (From Miller, I., and Freund, J. E., *Probability and Statistics for Engineers*, 2nd ed. Englewood Cliffs, N.J.: Prentice-Hall, Inc., 1977.) An explosion in an LNG storage tank in the process of being repaired could have occurred as the result of static electricity, malfunctioning electrical equipment, an open flame in contact with the liner, or purposeful action (industrial sabotage). Interviews with engineers who were analyzing the risks involved led to estimates that such an explosion would occur with probability 0.25 as a result of static electricity, 0.20 as a result of malfunctioning electric equipment, 0.40 as a result of an open flame, and 0.75 as a result of purposeful action. These interviews also yielded subjective estimates of the probabilities of the four causes of 0.30, 0.40, 0.15, and 0.15, respectively. Based on all this information, what is the most likely cause of the explosion?

10 Formally prove the special form of the rule of elimination given on page 124 by making use of part (b) of Exercise 14 on page 103 and the fact that events $A \cap B$ and $A \cap B'$ are mutually exclusive.

11 With reference to the example on page 128, suppose they can reduce to 0.03 the probability that a car which does not emit excessive amounts of pollutants will fail the test. Verify that this will lower to 0.08 the probability that a car which fails the test does not emit excessive amounts of pollutants.

BIBLIOGRAPHY

More detailed, though still elementary, treatments of probability may be found in

FREUND, J. E., *Introduction to Probability*. Encino, Calif.: Dickenson Publishing Co., Inc., 1973.

GOLDBERG, S., *Probability—An Introduction*. Englewood Cliffs, N.J.: Prentice-Hall, Inc., 1960.

MOSTELLER, F., ROURKE, R. E. K., and THOMAS, G. B., *Probability with Statistical Applications*, 2nd ed. Reading, Mass.: Addison-Wesley Publishing Co., Inc., 1970.

and some interesting applications in

MOSTELLER, F., KRUSKAL, W. H., LINK, R. F., PIETERS, R. S., and RISING, G. R., eds., *Statistics by Example: Weighing Chances*. Reading, Mass.: Addison-Wesley Publishing Co., Inc., 1973.

The randomized response technique illustrated in the example on page 125 was first proposed in

WARNER, S. L., "Randomized Response: A Survey Technique for Eliminating Evasive Answer Bias." *Journal of the American Statistical Association*, 1965.

Expectations
and
Decisions

When decisions are made in the face of uncertainty, they are seldom based on probabilities alone. In most cases we must also know something about the consequences (profits, losses, penalties, or rewards) to which we are exposed. For instance, if we must decide whether or not to buy a new car, knowing the chances that our old car will soon require repairs is not enough—to make an intelligent decision we would also have to know, among other things, the cost of the repairs and the trade-in value of our old car. To give another example, suppose that a building contractor has to decide whether to bid on a construction job which promises a profit of $120,000 with a probability of 0.20, or a loss of $27,000 (due, perhaps, to bad estimates, strikes, or the late delivery of materials) with a probability of 0.80. Clearly, the probability that the contractor will make a profit is not very high, but, on the other hand, the amount he stands to gain is much greater than the amount he stands to lose. Both of these examples demonstrate the need for a method of combining probabilities and consequences, and this is why we shall introduce here the concept of a **mathematical expectation.**

6.1

Mathematical Expectation

If an insurance agent tells us that in the United States a 45-year-old woman can expect to live 33 more years, this does not mean that anyone really expects a 45-year-old woman to live until her 78th birthday and then pass away the next day. Similarly, if we read that a person living in the United States can expect to eat 13.2 pounds of cheese and 307 eggs a year, or that a child in the age group from 6 to 16 can expect to visit a dentist 1.9 times a year, it must be obvious that the word "expect" is not being used in its colloquial sense. A child cannot go to the dentist 1.9 times, and it would be surprising, indeed, if we found somebody who has actually eaten 13.2 pounds of cheese and 307 eggs in a given year. So far as 45-year-old women are concerned, some will live another 15 years, some will another 20 years, some will live another 36 years, . . . , and the life expectancy of "33 more years" will have to be interpreted as an average, or as we shall call it here, a **mathematical expectation.**

Originally, the concept of a mathematical expectation arose in connection with games of chance, and in its simplest form it is the product of the amount a player stands to win and the probability that he or she will win.

EXAMPLE If we stand to win $4 if a balanced coin comes up heads and nothing if it comes up tails, our mathematical expectation is $4 \cdot \frac{1}{2} = \$2$.

EXAMPLE If we buy one of 1,000 raffle tickets issued for a prize (say, a television set) worth $480, the probability for each ticket is $\frac{1}{1,000}$ and our mathematical expectation is $480 \cdot \frac{1}{1,000} = \$0.48$ or 48 cents. Thus, it would be unwise to spend more than 48 cents for the ticket, unless, of course, the proceeds of the raffle go to a worthy cause (or the difference can be credited to whatever pleasure a person might derive from placing a bet).

So far we have considered only problems in which there is a single "payoff," namely, a single prize or a single payment. To see how the concept of a mathematical expectation can be generalized, let us consider the following change in the raffle of the preceding example:

EXAMPLE In addition to the television set worth $480, there is also a second prize (say, a tape recorder) worth $120 and a third prize (say, a small radio) worth $40. Now we can argue that 997 of the raffle tickets will not pay anything at all, one ticket will pay $480 (in merchandise), another will pay $120 (in merchandise), and a third will pay $40 (in merchandise); altogether, the 1,000 tickets will thus pay $480 + 120 + 40 = 640 (in merchandise), or on the average $0.64 per ticket. As before, this is the mathematical expectation for each ticket. Looking at the problem in a different way, we could argue that if the raffle were repeated many times, we would lose 99.7 percent of the time (or with probability 0.997) and win each of the prizes 0.1 percent of the time (or with probability 0.001). On the average we would thus win

$$0(0.997) + 480(0.001) + 120(0.001) + 40(0.001) = \$0.64$$

which is the sum of the products obtained by multiplying each amount by the corresponding proportion or probability.

Generalizing from this example, let us now make the following definition:

Mathematical expectation

> *If the probabilities of obtaining the amounts $a_1, a_2, \ldots,$ or a_k are, respectively, $p_1, p_2, \ldots,$ and p_k, then the mathematical expectation is*
>
> $$E = a_1 p_1 + a_2 p_2 + \cdots + a_k p_k$$

Each amount is multiplied by the corresponding probability, and the mathematical expectation, E, is given by the sum of all these products. In the $\sum$ notation, $E = \sum a \cdot p$.

So far as the a's are concerned, it is important to keep in mind that they are positive when they represent profits, winnings, or gains (namely, amounts which we receive), and that they are negative when they represent losses, penalties, or deficits (namely, amounts which we have to pay).

EXAMPLE   If we bet $5 on the flip of a balanced coin (that is, we either win $5 or lose $5 depending on the outcome), the amounts a_1 and a_2 are $+5$ and -5, the probabilities are $p_1 = 0.50$ and $p_2 = 0.50$, and the mathematical expectation is

$$E = 5(0.50) + (-5)(0.50) = 0$$

This is what the expectation should be in an **equitable game**; namely, in a game which does not favor either player.

EXAMPLE   Suppose that a businessman is interested in a piece of property for which the probabilities are 0.22, 0.36, 0.28, and 0.14 that he will be able to sell it at a profit of $2,500, that he will be able to sell it at a profit of $1,000, that he will break even, or that he will sell it at a loss of $1,500. If we substitute all these figures into the formula for E, we get

$$E = 2,500(0.22) + 1,000(0.36) + 0(0.28) + (-1,500)(0.14)$$

$$= \$700$$

and this is his expected profit.

Although we referred to the quantities $a_1, a_2, \ldots,$ and a_k as "amounts," they need not be cash winnings, losses, penalties, or rewards. When we said on page 132 that a child in the age group from 6 to 16 goes to the dentist 1.9 times a year, we referred to a result which was obtained by multiplying $0, 1, 2, 3, 4, \ldots,$ by the respective probabilities that a child in this age group will visit a dentist that many times a year.

EXAMPLE   The probabilities that on any one day the office of an airline at Sky Harbor Airport will get 0, 1, 2, 3, 4, 5, 6, 7, or 8 complaints about the handling of luggage are, respectively, 0.06, 0.21, 0.24, 0.18, 0.14, 0.10, 0.04, 0.02, and 0.01. To see how many complaints about the handling of luggage they can expect per day, we substitute into the formula and get

$$E = 0(0.06) + 1(0.21) + 2(0.24) + 3(0.18) + 4(0.14)$$

$$+ 5(0.10) + 6(0.04) + 7(0.02) + 8(0.01)$$

$$= 2.75$$

In all of the examples in this section we were given the values of a and p (or the values of the a's and p's) and calculated E. Now let us consider an example in which we are given values of a and E to arrive at some result about p, and also an example in which we are given values of p and E to arrive at some result about a.

EXAMPLE   A recent college graduate is faced with a decision which cannot wait; she must decide whether to accept a job paying $9,720 a year, or turn it down with the hope of getting another job paying $13,500 a year. If she turns down the $9,720 job, what does this tell us about the probability which she assigns to her getting the higher-paying job? Letting p denote this probability, we can argue that she attaches a mathematical expectation of $13,500p$ to the higher-paying job, and since she prefers this to $9,720, it follows that $13,500p > 9,720,$

and hence that $p > \dfrac{9,720}{13,500}$ or $p > 0.72$. This assumes, of course, that salary is the only factor which affects her decision.

EXAMPLE A friend claims that he would "give his right arm" for our ticket to an upcoming football game which has been sold out, but feels that $50 is too much. So, we offer him the following proposition: he pays us $12 (the actual price of the ticket), but he will get the ticket only if he draws a jack, a queen, a king, or an ace from an ordinary deck of 52 playing cards; otherwise, we keep the ticket and his $12. If he feels that this arrangement is fair, we can argue that the probability of his getting the ticket is $\frac{16}{52}$ (since there are four jacks, four queens, four kings, and four aces), the probability of his not getting the ticket is $1 - \frac{16}{52} = \frac{36}{52}$, and the mathematical expectation associated with the gamble is

$$\frac{16}{52} \cdot a + \frac{36}{52} \cdot 0 = \frac{4}{13} \cdot a$$

where a is the value he assigns to the ticket. Putting this mathematical expectation equal to $12, the amount he is willing to bet on the gamble, we get $\frac{4}{13} \cdot a = 12$ and $a = \dfrac{13 \cdot 12}{4} = 39$. Thus, he values the ticket at $39.

EXERCISES

1 If a charitable organization raises funds by selling 4,000 raffle tickets for a painting worth $3,000, what is the mathematical expectation of a person who buys one of the raffle tickets?

2 If someone were to give us $5.00 each time we roll a 5 or a 6 with a balanced die, how much should we pay him each time we roll a 1, 2, 3, or 4 to make the game equitable?

3 In the finals of a tennis tournament, the winner gets $30,000 and the loser gets $18,000. What are the two finalists' mathematical expectations if
   (a) they are evenly matched?
   (b) one of the players should be favored by odds of 2 to 1?

4 To introduce his new cars to the public, a dealer offers a first prize of $2,500 and a second prize of $1,000 to some lucky persons who come to his showroom, inspect the new cars, and write their comments and their names on a special card. The winners are to be drawn at random from these cards. What is each entrant's mathematical expectation, if 8,750 persons come to the dealer's showroom, inspect the new cars, and file such cards? Does this make it worthwhile to spend 50 cents on gasoline to drive to the dealer's showroom?

5 If the two league champions are evenly matched, the probabilities that a "best of seven" basketball play-off will take 4, 5, 6, or 7 games are $\frac{1}{8}$, $\frac{1}{4}$, $\frac{5}{16}$, and $\frac{5}{16}$. Under these conditions, how many games can we expect such a play-off to last?

6 A labor union wage negotiator feels that the odds are 3 to 1 that the union members will get a raise of 80 cents in their hourly wage, the odds are 17 to 3 against their getting a raise of 40 cents in their hourly wage, and the odds are 9 to 1 against their getting no raise at all. What is the expected raise in their hourly wage?

7 An importer is offered a shipment of pearls for $20,000, and the probabilities that he will be able to sell them for $24,000, $22,000, $20,000, or $18,000 are, respectively, 0.22, 0.47, 0.26, and 0.05. If she buys the pearls, what is her expected gross profit?

8 A police chief knows that the probabilities for 0, 1, 2, 3, 4, or 5 car thefts on any given day are 0.23, 0.34, 0.26, 0.12, 0.04, and 0.01. How many car thefts can he expect per day? It is assumed here that the probability of more than five car thefts is negligible.

9 A playwright is offered the option of either taking an immediate cash payment of $5,500 for her script, or gambling on the success of the play, in which case she will receive $20,000 if it is a success and only a token fee of $1,000 if it fails. What can we say about the probability which she assigns to the success of the play if she takes the immediate cash payment?

10 To handle a liability suit, a lawyer must decide whether to charge a straight fee of $1,200 or a contingent fee of $4,800 which he will get only if his client wins. What does the lawyer think about his client's chances if
    (a) he cannot make up his mind?
    (b) he prefers the straight fee of $1,200?
    (c) he prefers the contingent fee of $4,800?

11 Mr. Green has the choice of staying home and reading a good book or going to a party. If he goes to the party he might have a terrible time (to which he assigns a utility of 0), or he might have a wonderful time (to which he assigns a utility of 20 units). If he feels that the odds against his having a good time are 4 to 1 and he decides not to go, what can we say about the utility which he assigns to staying home and reading a good book?

12 Mr. Jones would like to beat Mr. Brown in an upcoming golf tournament, but his chances are nil unless he takes $500.00 worth of extra lessons, which (according to the pro at his club) will give him a fifty–fifty chance. If Mr. Jones assigns the utility U to his beating Mr. Brown and the utility $-\frac{1}{5}U$ to his losing to Mr. Brown, find U if Mr. Jones decides that it is just about worthwhile to spend the $500 on extra lessons.

6.2
Decision Making ★

When we are faced with uncertainties, mathematical expectations can often be used to great advantage in making decisions. In general, if we have to choose between two or more alternatives, it is considered "rational" to select the one with the "most promising" mathematical expectation: the one which maximizes expected profits, minimizes expected costs, maximizes expected tax advantages, minimizes expected losses, and so on. This approach to decision making has great intuitive appeal, but it is not without complications.

EXAMPLE   To illustrate some of these difficulties, suppose that a furniture manufacturer must decide whether to expand his plant capacity now or wait at least another year. His advisors tell him that if he expands now and economic conditions remain good, there will be a profit of $328,000 during the next fiscal year; if he expands now and there is a recession, there will be a loss (negative profit) of $80,000; if he waits at least another year to expand and economic conditions remain good, there will be a profit of $160,000; and if he waits at least another year and there is a recession, there will be a small profit of $16,000. Schematically, this information can be presented as follows:

	Expand now	*Delay expansion*
Economic conditions remain good	$328,000	$160,000
There is a recession	−$80,000	$16,000

Evidently, it will be advantageous to expand the plant capacity right away only if economic conditions remain good, and the furniture manufacturer's decision will, therefore, have to depend on the chances that this will be the case. Suppose, for instance, that he judges (on the basis of information available to him) that the odds are 2 to 1 that there will be a recession, and, hence, feels that the probability of a recession is $\frac{2}{2+1} = \frac{2}{3}$ and the probability that economic conditions will remain good is $1 - \frac{2}{3} = \frac{1}{3}$. He could then argue that if he expands his plant capacity right away, the expected profit is

$$328,000 \cdot \frac{1}{3} + (-80,000) \cdot \frac{2}{3} = \$56,000$$

and if the expansion is delayed, the expected profit is

$$160,000 \cdot \frac{1}{3} + 16,000 \cdot \frac{2}{3} = \$64,000$$

where both of these figures are mathematical expectations calculated in accordance with the formula on page 133. Since an expected profit of $64,000 is obviously preferable to an expected profit of $56,000, it stands to reason that the furniture manufacturer should delay expanding the plant. Or should he? What if he had been hasty in assessing the odds for a recession as 2 to 1? What if he had been better advised to use odds of 3 to 2? In that case the expected profit would be

$$328{,}000 \cdot \frac{2}{5} + (-80{,}000) \cdot \frac{3}{5} = \$83{,}200$$

if the plant capacity is expanded right away, and

$$160{,}000 \cdot \frac{2}{5} + 16{,}000 \cdot \frac{3}{5} = \$73{,}600$$

if expansion is delayed, and the decision would be reversed.

The last part of this example serves to illustrate that if mathematical expectations are to be used for making decisions, it is essential that one's appraisals of all relevant probabilities are close, if not correct. Not only that, but we must know the correct values of the "payoffs" which are associated with the various possibilities.

EXAMPLE   Would the furniture manufacturer's decision be the same, for instance, if the $328,000 profit in the table on page 137 is replaced by a $400,000 profit, or if the $80,000 loss is replaced by a $120,000 loss? It will be left to the reader to answer these questions in Exercise 6 on page 140.

The way in which we have studied the problem of the furniture manufacturer is called a **Bayesian analysis.** In this kind of analysis, probabilities are assigned to the alternatives about which uncertainties exist (the "states of nature," which in our example were business conditions remaining good or there being a recession); then we make whichever decision promises the greatest expected profit or the smallest expected loss.

EXERCISES   1   Ms. Cooper is planning to attend a convention in San Francisco, and she must send in her room reservation immediately. The convention is so large that the activities are held partly in hotel I and partly in hotel II, and Ms. Cooper does not know whether the particular session she wants to attend will be held at hotel I or hotel II. She is planning to stay only one day, which would cost her

$36.00 at hotel I and $32.40 at hotel II, but it will cost her an extra $6.00 for cab fare if she stays at the wrong hotel.

(a) Present all this information in a table like that on page 137.

(b) If Ms. Cooper feels that the odds are 3 to 1 that the session she wants to attend will be held at hotel I and she wants to minimize her expected cost, where should she make her reservation?

(c) If Ms. Cooper feels that the odds are 5 to 1 that the session she wants to attend will be held at hotel I and she wants to minimize her expected cost, where should she make her reservation?

2   A truck driver has to deliver a load of lumber to one of two construction sites, which are, respectively, 18 and 22 miles from the lumberyard, but he has misplaced the order telling him where the load of lumber should go. The two construction sites are 8 miles apart, and, to complicate matters, the telephone at the lumberyard is out of order.

(a) Construct a table (similar to the one on page 137) showing the total mileages he may have to drive depending on where he decides to go first and where the lumber should go.

(b) Where should he go first if he wants to minimize the distance he can expect to drive and he feels that the odds are 5 to 1 that the lumber should go to the construction site which is 22 miles from the lumberyard?

(c) Where should he go first if he wants to minimize the distance he can expect to drive and he feels that the odds are 2 to 1 that the lumber should go to the construction site which is 22 miles from the lumberyard?

(d) Show that it does not matter where he goes first when the odds are 3 to 1 that the lumber should go to the construction site which is 22 miles from the lumberyard.

3   With reference to the example on page 137, should the furniture manufacturer expand the plant capacity right away or delay expansion, if he wants to maximize the expected profit and feels that the odds are 7 to 4 that there will be a recession?

4   The management of a mining company must decide whether to continue an operation at a certain location. If they continue and are successful, they will make a profit of $2,250,000; if they continue and are not successful, they will lose $1,350,000; if they do not continue but would have been successful if they had continued, they will lose $900,000 (for competitive reasons); and if they do not continue and would not have been successful if they had continued, they will make a profit of $225,000 (because funds allocated to the operation remain unspent).

(a) Present all this information in a table like that on page 137.

(b) What decision would maximize the company's expected profit if the odds against success are 3 to 2?

(c) What decision would maximize the company's expected profit if the odds against success are 4 to 1?

5   A retailer has shelf space for 4 highly perishable items which are destroyed at the end of the day if they are not sold. The unit cost of the item is $3.00, the selling price is $6.00, and the profit is thus $3.00 per item sold. How many items should the retailer stock so as to maximize his expected profit, if he knows that the probabilities of a demand for 0, 1, 2, 3, or 4 items are, respectively, 0.10, 0.30, 0.30, 0.20, and 0.10?

6 With reference to the example on page 137, would the furniture manufacturer's decision remain the same if
  (a) the $328,000 profit is replaced by a $400,000 profit and the odds are 2 to 1 that there will be a recession?
  (b) the $80,000 loss is replaced by a $120,000 loss and the odds are 3 to 2 that there will be a recession?

7 With reference to the example on page 137, suppose that the furniture manufacturer has no idea about the chances that there will be a recession. Being a confirmed pessimist, however, he notes that if the plant capacity is expanded right away, there might be a loss of $80,000, whereas the decision to delay expansion will lead to a profit of at least $16,000. If he decides to delay expansion for this reason, he is said to be using the **maximin criterion**—he would be maximizing his company's minimum profit.
  (a) If the management of the mining company of Exercise 4 has no idea about the chances of success, what should they decide to do so as to maximize the company's minimum profit?
  (b) If Ms. Cooper of Exercise 1 has no idea about the chances that the session she wants to attend will be held at either hotel, where should she make her reservation so as to minimize her maximum cost? We refer to this as the **minimax criterion,** and we use it instead of the maximin criterion when we deal with losses or costs instead of profits or gains.
  (c) If the truck driver of Exercise 2 has no idea about the chances that the load of lumber should go to either construction site, where should he go first so as to minimize the maximum mileage?

8 If (in the absence of any knowledge about relevant probabilities) a decision is to be based on optimism (rather than on pessimism as in Exercise 7), we may use the **maximax criterion** (that is, maximize maximum gains or profits) or the **minimin criterion** that is, minimize minimum losses or costs).
  (a) If the furniture manufacturer of the example on page 137 has no idea about the chances that there will be a recession, what should he decide to do so as to maximize his company's maximum profit?
  (b) If Ms. Cooper of Exercise 1 has no idea about the chances that the session she wants to attend will be held at either hotel, where should she make her reservation so as to minimize her minimum cost?

9 With reference to the example in the text, suppose that the furniture manufacturer has the option of paying an infallible consultant $25,000 to find out for sure whether there will be a recession, before he decides whether or not to expand the capacity of his plant. This raises the question whether it is worthwhile to spend the $25,000. To answer it, let us take the furniture manufacturer's original 2 to 1 odds that there will be a recession. If he knew for sure whether or not there will be a recession, the right decision would yield his company a profit of $328,000 or a profit of $16,000 (as can be seen from the table on page 137), and since the corresponding probabilities are $\frac{1}{3}$ and $\frac{2}{3}$, the expected profit (with the help of the infallible expert) is

$$328,000 \cdot \frac{1}{3} + 16,000 \cdot \frac{2}{3} = \$120,000$$

This is called the **expected profit with perfect information.** On page 137 we showed that without the help of the infallible expert, the furniture manufacturer's expected profit was either $56,000 or $64,000, so that there is an improvement of at least $120,000 − $64,000 = $56,000, which makes the $25,000 fee well worthwhile. It is customary to refer to the amount by which perfect information will improve one's expectation, $56,000 in our example, as the **expected value of perfect information.**

(a) With reference to Exercise 1 and the odds of part (b), find the expected profit with perfect information and the expected value of perfect information. Would it be worthwhile to spend $1.20 on a long-distance call to find out for certain at which hotel the session (Ms. Cooper wants to attend) will be held?

(b) With reference to Exercise 4 and the odds of part (b), find the expected profit with perfect information and the expected value of perfect information. Would it be worthwhile to spend $200,000 to find out for sure whether the operation will be a success, before they decide whether or not it should be continued?

10 There are situations in which the criteria we have discussed are outweighed by other considerations. For instance, with reference to the example in the text, what may well be the furniture manufacturer's decision if

(a) he knows that he will be bankrupt unless he can make a profit of at least $200,000 during the next fiscal year?

(b) he knows that he will be bankrupt unless he can show a profit, no matter how small, during the next fiscal year?

Explain your answers.

6.3

Statistical Decision Problems ⋆

Modern statistics, with its emphasis on inference, may be looked upon as the art, or science, of decision making under uncertainty. This approach to statistics, called **decision theory,** dates back only to the middle of this century and the publication of John von Neumann and Oscar Morgenstern's *Theory of Games and Economic Behavior* in 1944 and Abraham Wald's *Statistical Decision Functions* in 1950. Since the study of decision theory is quite complicated mathematically, we shall limit our discussion here to an example in which the method of the preceding section is applied to a problem that is of a statistical nature.

EXAMPLE A government agency has appointed five teams to study racial discrimination, and on these teams there are, respectively, 1, 2, 5, 1, and 6 members known to favor liberal causes. These teams are randomly assigned to various cities, and we are asked to predict how many members known to favor liberal causes there will be on the team sent to our city. Thus faced with making a decision, we shall first have to see what the consequences may be. After all, if there are no penalties for being wrong, no rewards for being right or close, there is nothing at stake and it does not matter what we predict. For the sake of argument, let us investigate what we might do in each of the following situations:

1. *We are given a reward of $10 if our prediction is correct; otherwise, we are fined $3.*
2. *We are paid $5 for making the prediction, but fined an amount of money equal in dollars to the size of our error.*
3. *We are paid $10 for making the prediction, but fined an amount of money equal in dollars to the square of our error.*

In the first case, if our prediction is 1, the mode of the five numbers, we stand to make $10 with the probability $\frac{2}{5}$, lose $3 with the probability $\frac{3}{5}$, and hence can expect to gain

$$10 \cdot \frac{2}{5} + (-3) \cdot \frac{3}{5} = \$2.20$$

As can easily be verified, this is the best we can do; if we predict 2, 5, or 6, the expected gain is $-$0.40$ (or a loss of $0.40), and any other prediction entails a certain loss of $3. This illustrates the (perhaps, obvious) fact that if we have to pick the exact value on the nose and there is no reward for being close, the best prediction is the mode.

In the second case it is the median which yields the most promising prediction. If our prediction is 2, the median of the five numbers, the fine will be $1, $0, $3, or $4, depending on whether 1, 2, 5, or 6 of the members of the team sent to our city will be known to favor liberal causes. So, the expected fine is

$$1 \cdot \frac{2}{5} + 0 \cdot \frac{1}{5} + 3 \cdot \frac{1}{5} + 4 \cdot \frac{1}{5} = \$1.80$$

and the expected gain is $5.00 - $1.80 = $3.20. As can be verified, the expected fine would be greater for any number other than the median. For instance, if we predicted 3, the mean of the five numbers, the fine would be $2, $1, $2, or $3, depending on whether 1, 2, 5, or 6 of the members of the team will be known to favor liberal causes, and the expected fine would be

$$2 \cdot \frac{2}{5} + 1 \cdot \frac{1}{5} + 2 \cdot \frac{1}{5} + 3 \cdot \frac{1}{5} = \$2.00$$

The mean comes into its own right in the third case, where the fine increases more rapidly with the size of the error. If our prediction is 3, the mean of the five numbers, the fine will be $4, $1, $4, or $9, depending on whether 1, 2, 5, or 6 of the members of the team will be known to favor liberal causes, the expected fine is

$$4 \cdot \frac{2}{5} + 1 \cdot \frac{1}{5} + 4 \cdot \frac{1}{5} + 9 \cdot \frac{1}{5} = \$4.40$$

and the expected gain is $10.00 - $4.40 = $5.60. As can be verified, the expected fine would be greater for any number other than the mean, and in Exercise 2 on page 144 the reader will be asked to verify that it would be $5.40 if we predicted that 2 of the members of the team will be known to favor liberal causes.

The third case studied in this example plays an important role in statistics, as it ties in closely with the **method of least squares** which we shall study in Chapter 14. The idea of minimizing the squared errors is justifiable on the grounds that in actual practice the seriousness of an error often increases very rapidly with the size of the error, more rapidly than the magnitude of the error itself.

The greatest difficulty in applying the methods of this chapter to realistic problems in statistics is that we seldom know the exact values of all the risks that are involved; that is, we seldom know the exact values of the "payoffs" corresponding to the various eventualities. For instance, if the FDA must decide whether or not to release a new drug for general use, how can it put a cash value on the damage that might be done by not waiting for a more thorough analysis of possible side effects, or on the lives that might be lost by not making the drug available to the public right away? Similarly, if a faculty committee must decide which of several applicants should be admitted to a medical school or, perhaps, receive a scholarship, how can they possibly foresee all the consequences that might be involved?

The fact that we seldom have adequate information about relevant probabilities also provides obstacles to finding suitable decision criteria; without them, is it "reasonable" to base decisions, say, on pessimism or optimism as in Exercises 7 and 8 on page 140? Questions like these are difficult to answer, but their analysis at least serves the purpose of revealing some of the logic that underlies statistical thinking.

EXERCISES

1 The ages of seven finalists in a beauty contest are 18, 18, 18, 19, 21, 22, and 24, and we are to predict the age of the winner.

    (a) If we assign each finalist the same probability and there is a reward for being exactly right, none for being close, what prediction maximizes the expected reward?

(b) If we assign each finalist the same probability and there is a penalty proportional to the magnitude of the error, what prediction minimizes the expected penalty?

(c) Repeat part (b) for the case where the penalty is proportional to the square of the size of the error.

2 Verify that in the third case on page 143, the expected fine would be $5.40 if we predicted that 2 of the members of the team will be known to favor liberal causes.

3 Some of the used cars on a lot are priced at $895, some are priced at $1,395, some are priced at $1,795, and some are priced at $2,495. If we want to predict the price of the car which will be sold first, what prediction will minimize the maximum error? What name did we give to this statistic in Chapter 3?

4 Referring to the retailer of Exercise 5 on page 139, suppose he has no idea about the potential demand for the item. How many of the items should he stock so as to minimize the maximum loss to which he may be exposed? Discuss the reasonableness of this criterion in a problem of this kind.

BIBLIOGRAPHY   More detailed treatments of the subject matter of this chapter may be found in

BROSS, I. D. J., *Design for Decision*. New York: Macmillan Publishing Co., Inc., 1953.

JEFFREY, R. C., *The Logic of Decision*. New York: McGraw-Hill Book Co., 1965.

and in many textbooks on business statistics; for instance, in Chapter 6 of

FREUND, J. E., and WILLIAMS, F. J., *Elementary Business Statistics: The Modern Approach*, 3rd ed. Englewood Cliffs, N.J.: Prentice-Hall, Inc., 1977.

Further Exercises for Chapters

4, 5, and 6

1　In how many ways can ten welfare cases be assigned to three case workers, D, E, and F, so that D gets three cases, E gets three cases, and F gets four cases?

2　The probabilities that a person shopping at "The Bookstore" will buy 0, 1, 2, 3, or 4 books are 0.32, 0.44, 0.17, 0.06, and 0.01. How many books can a person shopping at this bookstore be expected to buy?

3　A department store which bills its charge-account customers once a month has found that if a customer pays promptly one month, the probability is 0.90 that he will also pay promptly the next month; however, if a customer does not pay promptly one month, the probability that he will pay promptly the next month is only 0.50.

    (a) What is the probability that a customer who pays promptly one month will also pay promptly the next three months?

    (b) What is the probability that a customer who does not pay promptly one month will also not pay promptly the next two months and then make a prompt payment the month after that?

    (c) What is the probability that a customer who pays promptly one month will also pay promptly the second month after that? (*Hint:* Draw a tree diagram, and/or use the rule of elimination.)

4　A small real-estate office has only three part-time salespersons.

    (a) Using two coordinates so that (2, 1), for example, represents the event that two of the salespersons are at work but only one is busy with a customer, and (3, 0) represents the event that all three salespersons are at work but none of them is busy with a customer, draw a diagram (similar to that of Figure 5.2) showing the ten points of the corresponding sample space.

    (b) List the points which comprise the event K that at least two of the sales-persons are busy with customers, the event L that only one of the salespersons is at work, and the event M that all the salespersons who are at work are busy with customers.

    (c) With reference to part (b), list the points of the sample space which represent M' and $M \cap L$, and describe in words the corresponding events.

    (d) With reference to part (b), which of the pairs of events, K and L, K and M, and L and M are mutually exclusive?

5 An old-fashioned string of Christmas-tree lights consists of eight bulbs connected in series, so that the whole string will not light up if any one of the eight bulbs is defective. If the probability is 0.90 for each bulb that it will work after a year's storage, what is the probability that the string will not light up after a year's storage?

6 A grab-bag contains 15 packages worth 70 cents apiece, 15 packages worth 40 cents apiece, and 10 packages worth 25 cents apiece. Is it worthwhile to pay 50 cents for the privilege of choosing one of these packages at random? Is it worthwhile to pay 40 cents?

7 The credit manager of a mortgage company figures that if an applicant for a certain-size home mortgage is a good risk and the company accepts him the company's profit will be $4,110, and if he is a bad risk and the company accepts him the company will lose $660. If the credit manager turns down the mortgage applicant, there will be no direct profit or loss either way.

    (a) Present all this information in a table like that on page 137.

    (b) If the credit manager feels that the probabilities are 0.10 and 0.90 that a certain loan applicant is a good risk or a bad risk, which decision would maximize the mortgage company's expected profit?

    (c) Would the credit manager's decision be the same if he feels that the probabilities are 0.20 and 0.80 that the loan applicant is a good risk or a bad risk?

8 On the faculty of a college there are three professors named Jones: Harry Jones, Mary Jones, and Elsa Jones. Draw a tree diagram to show the six ways in which the college's payroll department can distribute their paychecks so that each of them receives a check made out to one of the Joneses. In how many of these possibilities will

    (a) only one of them get the right check?

    (b) none of them get the right check?

9 In how many different ways can the manager of a baseball team arrange the batting order of the nine players in his starting line-up?

10 If each of the dots of Figure 5.9 represents an outcome having the same probability, what is the probability of event A?

11 An insurance company agrees to pay the promoter of a drag race $8,000 in case the event has to be cancelled because of rain. If the company's actuary feels that a fair net premium for this risk would be $1,280, what probability does she assign to the possibility that the race may have to be cancelled because of rain?

12 For a student enrolling as a freshman at a certain university, the probability is 0.25 that he or she will get a scholarship and 0.75 that he or she will graduate. If the probability is 0.80 that a student who gets a scholarship will graduate, what is the probability that a student who does not get a scholarship will graduate?

13 If we randomly arrange the letters in the word "nest," what are the odds that we will not get a meaningful word in the English language?

14 In a history course, the odds are 4 to 1 that a student who regularly does his homework will get a passing grade, and 3 to 1 that a student who does not regularly do his homework will not get a passing grade. If 60 percent of the students taking the course regularly do their homework, what is the probability that a student who gets a passing grade regularly did his homework?

15 As part of a promotional scheme in Arizona and New Mexico, a company distributing frozen foods will award a grand price of $100,000 to some person sending in his or her name on an entry blank, with the option of including a label from one of the company's products. A breakdown of the 225,000 entries received is shown in the following table:

	With label	*Without label*
Arizona	120,000	42,000
New Mexico	30,000	33,000

If the winner of the grand prize is chosen by lot, A represents the event that it will be won by an entry from Arizona, and L represents the event that it will be won by an entry which included a label, find each of the following probabilities:

(a) $P(A)$;

(b) $P(L)$;

(c) $P(A|L)$;

(d) $P(L|A)$;

(e) $P(A'|L')$;

(f) $P(L|A')$.

16 Suppose that in Exercise 15 the drawing is rigged so that by including a label each entry's probability of winning the grand prize is doubled. Recalculate the probabilities of parts (a) through (f).

17 Check each of the following to see whether it is true or false:

(a) $7! = 7 \cdot 6 \cdot 5!$;

(c) $\dfrac{10!}{10 \cdot 9 \cdot 8 \cdot 7} = 6!$;

(b) $2! + 2! = 4!$;

(d) $\dfrac{1}{2!} + \dfrac{1}{2!} = 1$.

18 To fill a number of vacancies, the personnel manager of a company must choose four secretaries from among 12 applicants and two file clerks from among six applicants. In how many ways can she make her choice?

19 If somebody claims that the odds are 33 to 17 that a certain shipment will arrive on time, what probability does he assign to the shipment's arriving on time?

20 A contractor has to choose between two jobs. The first job promises a profit of $120,000 with a probability of $\frac{3}{4}$ or a loss of $30,000 (due to strikes and other delays) with a probability of $\frac{1}{4}$; the second job promises a profit of $180,000 with a probability of $\frac{1}{2}$ or a loss of $45,000 with a probability of $\frac{1}{2}$.

(a) Which job should the contractor choose if he wants to maximize his expected profit?

(b) Which job would the contractor probably choose if his business is in fairly bad shape and he will go broke unless he can make a profit of at least $150,000 on his next job?

21 Suppose that the furniture manufacturer of the example on page 137 is the kind of person who always worries about missing out on a good deal. Looking at the table on page 137, he would argue that if he decides to expand his plant capacity now and there is a recession, he would have been better off by $96,000 if he had decided not to expand right away. This is the difference between $16,000 (his profit if he had made the right decision) and −$80,000 (his loss due to his making the wrong decision), and we refer to this difference as the **opportunity loss** associated with this situation.

    (a) What is his opportunity loss if he delays expansion and economic conditions remain good?

    (b) What are his opportunity losses in the other two cases?

    (c) With odds of 2 to 1 that there will be a recession, which decision will minimize his expected opportunity loss?

22 The probabilities that a doctor's answering service will receive 0, 1, 2, 3, 4, 5, 6, or 7 or more calls for him during the lunch hour are, respectively, 0.001, 0.007, 0.024, 0.055, 0.095, 0.131, 0.151, and 0.536. What are the probabilities that he will receive

    (a) fewer than five calls?

    (b) at least three calls?

    (c) anywhere from two calls to six calls, inclusive?

23 Some philosophers have argued that if we have absolutely no information about the likelihood of the different possibilities, it is reasonable to regard them all as equally likely. This is sometimes referred to as the "Principle of *Equal Ignorance*." Discuss the argument that human life either does or does not exist elsewhere in the universe, and since we really have no information one way or the other, the probability that human life exists elsewhere in the universe is $\frac{1}{2}$.

24 Among a company's replacement parts for a given assembly, 20 percent are defective and the rest are good, 60 percent were bought from external sources and the rest were made by the company itself, and of those bought from external sources 80 percent are good and the rest are defective. What is the probability that a replacement part, randomly selected from this stock, is

    (a) company-made and good?

    (b) either defective or bought?

    (c) neither company-made nor good?

    (d) bought, given that it is defective?

25 A carton of eggs contains two bad ones. In how many ways can we select three of the 12 eggs so that

    (a) none of the bad ones are included?

    (b) both of the bad ones are included?

    (c) one of the bad ones is included?

26 During a time of war, a country uses lie detectors to uncover security risks. As is well known, lie detectors are not infallible, so let us suppose that the probabilities are 0.10 and 0.05 that they will fail to detect a security risk and incorrectly label a person a security risk. If two percent of the persons who are given the lie detector test are security risks, what is the probability that a person labeled a security risk by a lie detector actually is a security risk?

27 Ms. Black feels that it is about a toss-up whether to accept $20 cash or to gamble on drawing a bead from an urn containing 15 white beads and 45 red beads,

where she is to receive $2 if she draws a white bead and a bottle of fancy perfume if she draws a red bead. What value, or utility, does she attach to the bottle of perfume?

28 A tennis player feels that the odds against her winning at Forest Hills are 3 to 1 and the odds against her winning at Forest Hills and at Wimbledon are 7 to 1. What are the odds that she will win at Forest Hills but not at Wimbledon?

29 Mrs. Jones, who lives in a suburb, plans to spend an afternoon shopping in downtown Manhattan, and she has some difficulty deciding whether or not to take along her raincoat. If it rains, she will be inconvenienced if she does not bring it along, and if it does not rain, she will be inconvenienced if she does. On the other hand, it will be convenient to have the raincoat if it rains, and she will be neither convenienced nor inconvenienced if she does not bring the raincoat and it does not rain. To express all this numerically, suppose that the "payoffs" in the following table are in units of inconvenience, so that the negative value reflects convenience:

	She takes the raincoat	She does not take the raincoat
It rains	−25	50
It does not rain	15	0

(a) What should Mrs. Jones do to minimize her expected inconvenience if she feels that the odds against rain are 6 to 1?

(b) What should Mrs. Jones do to minimize her expected inconvenience if she feels that the probabilities for rain and no rain are, respectively, 0.20 and 0.80?

(c) Does it matter whether or not Mrs. Jones takes her raincoat if she feels that the odds against rain are 5 to 1?

30 Mr. Brown is willing to bet Mrs. Brown $15 to her $5, but not $12 to her $3, that they will be late for the theater. What does this tell us about the probability which he assigns to their being late?

31 If 441 of 700 television viewers interviewed in a certain area feel that local news coverage is inadequate, estimate the probability that a television viewer randomly selected in this area will feel that local news coverage is inadequate.

32 On a weekday afternoon, a television station schedules five soap operas, four situation comedies, and two game shows. In how many different ways can one watch

(a) one of the soap operas, one of the situation comedies, and one of the game shows?

(b) three of the soap operas, two of the situation comedies, and one of the game shows?

33 Two friends are betting on repeated flips of a balanced coin. One has $5 at the start and the other has $3, and after each flip the loser pays the winner $1. If p is the probability that the one who starts with $5 will win his friend's $3 before he loses his own $5, explain why $3p - 5(1 - p)$ should equal 0, and then solve the equation $3p - 5(1 - p) = 0$ for p. Generalize this result to the case where two players start with a dollars and b dollars, respectively.

34 It is known from experience that in a certain industry 55 percent of all labor-management disputes are over wages, 10 percent are over working conditions, and 35 percent are over pension plans. Also, 40 percent of the disputes over wages are resolved without strikes, 70 percent of the disputes over working conditions are resolved without strikes, and 45 percent of the disputes over pension plans are resolved without strikes.

    (a) What percent of all labor-management disputes in this industry are resolved without strikes?

    (b) If a labor-management dispute in this industry is resolved without a strike, what is the probability that the dispute was not over wages?

35 Suppose that the probabilities are 0.75, 0.40, and 0.20 that a person vacationing in Virginia will visit Williamsburg, Monticello, or both.

    (a) What is the probability that such a person will visit Williamsburg but not Monticello?

    (b) What is the probability that such a person will visit Williamsburg, Monticello, or both?

    (c) What is the probability that such a person will visit Williamsburg given that he or she will visit Monticello?

36 The manufacturer of a new battery additive has to decide whether to sell his product for $1.00 a can, or for $1.25 with a "double-your-money-back-if-not-satisfied guarantee." How does he feel about the chances that a person will actually ask for double his money back if

    (a) he decides to sell the product for $1.00?

    (b) he decides to sell the product for $1.25 with the guarantee?

    (c) he cannot make up his mind?

37 The editor of a student newspaper wants to interview two professors chosen at random from among the twelve faculty members who serve on the student publications advisory committee. If nine of the faculty members favor a new student humor magazine while the other three are against it, find the probabilities that the editor will get

    (a) two opinions favoring the magazine;

    (b) two opinions opposed to the magazine;

    (c) one opinion favoring the magazine and one against it.

38 Suppose that a group of students is planning to visit a museum and that X is the event that they will see the ancient artifacts, Y is the event that they will see the modern paintings, and Z is the event that they will get very tired. With reference to the Venn diagram of Figure 5.6 on page 99, list (by number or numbers) the regions or combinations of regions which represent the events that the students will

    (a) see the ancient artifacts and the modern paintings, but get very tired;

    (b) see neither the ancient artifacts nor the modern paintings, and not get very tired;

(c) see the modern paintings and not get very tired;

(d) see the ancient artifacts or the modern paintings (or both), and get very tired.

39 A questionnaire sent through the mail as part of a market research study consists of eight questions, each allowing five different answers. In how many different ways can a person answer the eight questions?

40 How many different sums of money can be formed with one or more of the following coins: a dollar, a half-dollar, a quarter, a dime, a nickel, and a penny? (*Hint:* There are two possibilities for each coin depending on whether or not it is included.)

$$P(W/M) \quad \frac{p(W \cap M)}{P(M)} = \frac{.2}{.4} = .5$$

7

Uncertainties face us wherever we turn—we cannot be certain that an experiment will succeed, we cannot be certain that the weather will stay nice, we cannot be certain that the milk we buy in a sealed carton is fresh, we cannot be certain that a rocket will get off the ground, we cannot be certain that a marriage will last, we cannot be certain that an investment will pay off, and so on. Although life without any uncertainties would undoubtedly be very dull and most uncertainties cannot be eliminated, we shall see in this chapter that the potential size of chance fluctuations (namely, variations or differences brought about by chance) is quite often predictable; and sometimes it can even be controlled.

Probability Distributions

7.1
Random Variables

In most statistical problems we are interested only in one aspect, or at most in a few aspects, of the outcomes of experiments. For instance, a student taking a true-false test may be interested only in the number of questions he answers correctly, since this determines his grade; a geologist may be interested only in the age of a rock sample and not in its hardness; and a sociologist may be interested only in the socioeconomic status of a person interviewed in a survey and not in her age or weight. Also, an agronomist may be interested in determining not only the yield per acre of a new variety of corn but also the temperature at which it will germinate; and an automotive engineer may be interested in the brightness and the durability of the headlights proposed for a new model car and also in their projected cost.

In these five examples, the student, the geologist, the sociologist, the agronomist, and the automotive engineer are all interested in numbers that are associated with the outcomes of situations involving an element of chance, or more specifically, in values of **random variables.**

EXAMPLE To be more explicit, let us consider Figure 7.1, which (like Figure 5.10 on page 108) pictures the sample space for the example dealing with the two salespersons who hope to sell the three Ford Mustangs. Note, however, that we have added another number to each point—the number 0 to the point (0, 0); the number 1 to the points (1, 0) and (0, 1); the number 2 to the points (2, 0), (1, 1), and (0, 2); and the number 3 to the points (3, 0), (2, 1), (1, 2), and (0, 3). In this way we have associated with each point of the sample space the total number of cars which, between them, the two salespersons will sell.

Since "associating a number with each point of a sample space" is just another way of saying that we are defining a *function* over the points of the sample space, random variables are functions. Conceptually, though, most beginners find it easier to think of random variables simply as quantities which can take on different values depending on chance. For instance, the number of speeding tickets issued each day on the freeway between Indio and Blythe in California is a random variable, and so is the annual production

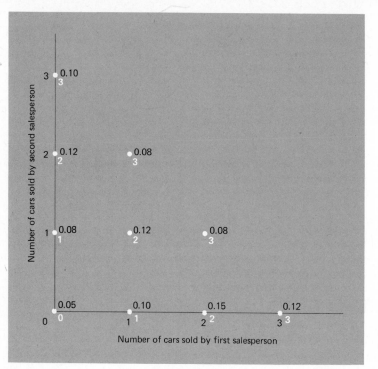

FIGURE 7.1

Sample space with values of random variable.

of coffee in Brazil, the number of persons visiting Disneyland each week, the wind velocity at Kennedy airport, the size of the audience at a baseball game, and the number of mistakes a person makes in typing a report.

7.2

Probability Distributions

In the study of random variables, we are usually interested mainly in the probabilities that they will take on the various values within their range. How such probabilities can be obtained, displayed, and summarized is the subject matter of this chapter, and to give an example, let us refer again to Figure 7.1.

EXAMPLE   Adding the probabilities associated with the respective points, we find that the random variable "the total number of cars which, between them, the two salespersons will sell" takes on the value 0 with probability 0.05, the value 1 with probability 0.10 + 0.08 = 0.18, the value 2 with probability 0.15 + 0.12 + 0.12 = 0.39, and the value 3 with probability 0.12 + 0.08 + 0.08 + 0.10 = 0.38. All this is summarized in the following table:

Number of cars sold	Probability
0	0.05
1	0.18
2	0.39
3	0.38

This table and the ones in the examples which follow serve to illustrate what we mean by a **probability function** or **probability distribution**—it is a correspondence which assigns probabilities to the values of a random variable.

EXAMPLE For the number of points rolled with a fair die, we assume that each face has the same probability and, hence, get the correspondence shown in the following table:

Number of points rolled with a die	Probability
1	$\frac{1}{6}$
2	$\frac{1}{6}$
3	$\frac{1}{6}$
4	$\frac{1}{6}$
5	$\frac{1}{6}$
6	$\frac{1}{6}$

EXAMPLE For four flips of a balanced coin there are the 16 equally likely possibilities HHHH, HHHT, HHTH, HTHH, THHH, HHTT, HTHT, THHT, HTTH, THTH, TTHH, HTTT, THTT, TTHT, TTTH, and TTTT, where H stands for heads and T for tails. Counting the number of heads in each case and using the formula $\frac{s}{n}$ for equiprobable possibilities, we obtain the following probability distribution for the total number of heads:

Number of heads	Probability
0	$\frac{1}{16}$
1	$\frac{4}{16}$
2	$\frac{6}{16}$
3	$\frac{4}{16}$
4	$\frac{1}{16}$

Whenever possible, we try to express probability distributions by means of mathematical formulas which enable us to calculate directly the probabilities associated with the various values of a random variable.

EXAMPLE   For the number of points we roll with a fair, or balanced die, we can write

$$f(x) = \frac{1}{6} \quad \text{for } x = 1, 2, 3, 4, 5, \text{ and } 6$$

where $f(1)$ represents the probability of rolling a 1, $f(2)$ represents the probability of rolling a 2, and so on, in the usual functional notation.[†]

EXAMPLE   For the number of heads in four flips of a balanced coin, observe that the numerators of the probabilities shown in the table on page 155—1, 4, 6, 4, and 1—are the same as the binomial coefficients $\binom{4}{0}, \binom{4}{1}, \binom{4}{2}, \binom{4}{3}$, and $\binom{4}{4}$. This is by no means coincidental, as we shall see later, and it enables us to write

$$f(x) = \frac{\binom{4}{x}}{16} \quad \text{for } x = 0, 1, 2, 3, \text{ and } 4$$

To conclude this preliminary discussion of probability distributions, let us point out the following two general rules which the values of any probability distribution must obey: **1. Since the values of probability distributions are probabilities, they must always be numbers between 0 and 1, inclusive. 2. Since a random variable has to take on one of its values, the sum of all the values of a probability distribution must always be equal to 1.**

7.3
The Binomial Distribution

There are many applied problems in which we are interested in the probability that an event will occur "x times out of n," or "x times in n trials" as we say in probability and statistics. For instance, we may be interested in the probability of getting 45 responses to 400 mail questionnaires, the probability that 5 of 12 mice will survive for a given length of time after the injection of a cancer-inducing substance, or the probability that 66 of 200 television viewers (interviewed by a rating service) will recall what products were advertised on a given program. To borrow again from the language of games of chance, we could

---

[†] Most of the time we shall write the probability that a random variable takes on the value x as $f(x)$, but we could just as well write $g(x)$, $h(x)$, $b(x)$, and so on.

say that in each of these examples we are interested in the probability of getting "x successes in n trials," or in other words, x successes and $n - x$ failures in n attempts.

In the problems which we shall study in this section, we shall always make the following assumptions: **1. The number of trials (attempts, or repetitions) is fixed. 2. The probability of a success is the same for each trial. 3. The trials are all independent; that is, what happens in any one trial does not affect the probability of a success in any other trial.** This means that the theory we shall develop will not apply, for example, if we are interested in the number of dresses a woman will try on before she finally buys one (where the number of trials is not fixed), if we check hourly whether traffic is congested at an important intersection (where the probability of a success is not the same for each trial), or if we are interested in a person's voting for the Democratic candidate in the 1968, 1972, 1976, and 1980 presidential elections (where the trials are not independent).

To solve problems which do meet the conditions listed in the preceding paragraph, we use a formula obtained in the following way: If p and $1 - p$ are the probabilities of a success and a failure on any given trial, then the probability of getting x successes and $n - x$ failures *in some specific order* is $p^x(1 - p)^{n-x}$; clearly, in this product of p's and $(1 - p)$'s there is one factor p for each success, one factor $1 - p$ for each failure, and the x factors p and $n - x$ factors $1 - p$ are all multiplied together by virtue of the generalized multiplication rule for more than two independent events. Since this probability applies to any point of the sample space which represents x successes and $n - x$ failures (in any specific order), we have only to count how many points of this kind there are, and then multiply $p^x(1 - p)^{n-x}$ by this number. Clearly, the number of ways in which we can select the x trials on which there is to be a success is $\binom{n}{x}$, the number of combinations of x objects selected from a set of n objects, and we have arrived at the following result:

Binomial distribution

> *The probability of getting x successes in n independent trials is*
>
> $$f(x) = \binom{n}{x} p^x(1 - p)^{n-x} \qquad for\ x = 0, 1, 2, \ldots, or\ n$$
>
> *where p is the constant probability of a success for each trial.*

It is customary to say here that the number of successes in n trials is a random variable having the **binomial probability distribution**, or simply the **binomial distribution.** The binomial distribution is called by this name because for $x = 0, 1, 2, \ldots$, and n, the values of the probabilities are the successive terms of the binomial expansion of $[(1 - p) + p]^n$.

7.3   **The Binomial Distribution**

EXAMPLE For the number of heads in four flips of a balanced coin, $n = 4$ and $p = \frac{1}{2}$, so the formula becomes

$$f(x) = \binom{4}{x}\left(\frac{1}{2}\right)^x\left(1 - \frac{1}{2}\right)^{4-x} = \binom{4}{x}\left(\frac{1}{2}\right)^4 = \frac{\binom{4}{x}}{16}$$

and this agrees with the result we gave on page 156.

EXAMPLE To give some numerical applications, let us first calculate the probability that two of five registered voters (randomly selected from official rolls) will vote in an election, given that for each one the probability is 0.70. Substituting $x = 2$, $n = 5$, $p = 0.70$, and $\binom{5}{2} = \frac{5 \cdot 4}{2!} = 10$ into the formula, we get

$$f(2) = \binom{5}{2}(0.70)^2(1 - 0.70)^{5-2}$$

$$= 10(0.70)^2(0.30)^3$$

$$= 0.1323$$

or approximately 0.13. Similarly, to find the probability that a cleaning fluid will remove six of eight spots, when the probability is 0.80 that it will remove any one spot, we substitute $x = 6$, $n = 8$, $p = 0.80$, and $\binom{8}{6} = \frac{8 \cdot 7 \cdot 6 \cdot 5 \cdot 4 \cdot 3}{6!} = 28$ into the formula, and get

$$f(6) = \binom{8}{6}(0.80)^6(0.20)^{8-6}$$

$$= 28(0.80)^6(0.20)^2$$

$$= 0.294$$

rounded to three decimals.

EXAMPLE To show how we calculate all the values of a binomial distribution, suppose that a customs official claims that 30 percent of all persons returning from a trip to Europe fail to declare one or more purchases on which there is duty. Assuming that this figure is correct, let us find the probabilities that among six randomly selected persons returning from Europe there will be 0, 1, 2, 3, 4, 5, or 6 who fail to declare one or more purchases on which there is duty. Substituting $n = 6$, $p = 0.30$, and, respectively, $x = 0, 1, 2, 3, 4, 5$, and 6 into the formula for the binomial distribution, we get

$$f(0) = \binom{6}{0}(0.30)^0(0.70)^6 = 0.118$$

$$f(1) = \binom{6}{1}(0.30)^1(0.70)^5 = 0.303$$

$$f(2) = \binom{6}{2}(0.30)^2(0.70)^4 = 0.324$$

$$f(3) = \binom{6}{3}(0.30)^3(0.70)^3 = 0.185$$

$$f(4) = \binom{6}{4}(0.30)^4(0.70)^2 = 0.060$$

$$f(5) = \binom{6}{5}(0.30)^5(0.70)^1 = 0.010$$

$$f(6) = \binom{6}{6}(0.30)^6(0.70)^0 = 0.001$$

where all the probabilities are rounded to three decimals. A histogram of this binomial distribution is shown in Figure 7.2.

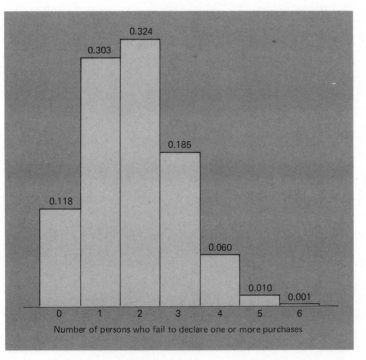

FIGURE 7.2

Histogram of binomial distribution with $n = 6$ and $p = 0.30$.

7.3   **The Binomial Distribution**

Probably, the reader is not especially interested in tourists smuggling in goods from abroad, but let us stress the importance of the binomial distribution as a **mathematical model** by pointing out that the results of the preceding example apply also if the probability is 0.30 that the energy cell of a watch will last a year under normal usage, and we want to know the probabilities that, among six of these cells, 0, 1, 2, 3, 4, 5, or 6 will last a year under normal usage; if the probability is 0.30 that an embezzler will be caught and brought to trial, and we want to know the probabilities that, among six embezzlers, 0, 1, 2, 3, 4, 5, or 6 will be caught and brought to trial; if the probability is 0.30 that the head of a household owns some life insurance, and we want to know the probabilities that, among six heads of households, 0, 1, 2, 3, 4, 5, or 6 will own some life insurance; or if the probability is 0.30 that a person having a given disease will live for another ten years, and we want to know the probabilities that, among six persons having the disease, 0, 1, 2, 3, 4, 5, or 6 will live another ten years. The argument we have presented here is precisely like the one we used in Section 1.1, where we tried to impress upon the reader the generality of statistical techniques.

In actual practice, binomial probabilities are seldom obtained by direct substitution into the formula. Sometimes we use approximations such as those discussed in Sections 7.5 and 8.5, but more often we refer to special tables such as Table VI at the end of this book or the more detailed tables listed in the Bibliography at the end of this chapter. Table VI is limited to the binomial probabilities for $n = 2$ to $n = 15$, and $p = 0.05$, 0.1, 0.2, 0.3, 0.4, 0.5, 0.6, 0.7, 0.8, 0.9, and 0.95.

EXAMPLE   Using Table VI, we can verify the results obtained in all of the preceding examples. Also using Table VI, we find, for instance, that if the probability is 0.40 that a divorced person will remarry within three years, then the probability that at most 3 of 10 divorced persons will remarry within three years is

$$0.006 + 0.040 + 0.121 + 0.215 = 0.382$$

and the probability that at least 7 will remarry within three years is

$$0.042 + 0.011 + 0.002 = 0.055$$

Similarly, if the probability is 0.10 that any one person will dislike the smell of a new deodorant, then the probability that none of 15 randomly selected persons will dislike it is 0.206, and the probability that 3 or more will dislike it is

$$0.129 + 0.043 + 0.010 + 0.002 = 0.184$$

The probabilities which were not added in these examples are all 0.0005 or less, and they are not shown in Table VI.

When we observe a value of a random variable having the binomial distribution—for instance, when we observe the number of heads in 25 flips of a coin, the number of seeds (in a package of 24 seeds) that germinate, the number of students (among 200 interviewed) who are opposed to a change in student activity fees, or the number of automobile accidents (among 20 investigated) that are due to drunk driving—we say that we are **sampling a binomial population.** This terminology is widely used in statistics.

7.4

The Hypergeometric Distribution ★

In Section 5.5 we spoke of sampling "with replacement" to illustrate the multiplication rule for independent events, and of sampling "without replacement" to illustrate the rule for dependent events. Since the binomial distribution applies only when the trials are all independent, it can be used when we sample with replacement, but not when we sample without replacement.

EXAMPLE To find the probability of drawing, with replacement, two aces in a row from an ordinary deck of 52 playing cards including four aces, we substitute $x = 2$, $n = 2$, and $p = \frac{4}{52} = \frac{1}{13}$ into the formula for the binomial distribution, and get

$$f(2) = \binom{2}{2}\left(\frac{1}{13}\right)^2\left(\frac{12}{13}\right)^{2-2}$$

$$= 1 \cdot \left(\frac{1}{13}\right)^2 \cdot 1$$

$$= \frac{1}{169}$$

as on page 119.

To introduce a probability distribution which applies when we sample without replacement, let us consider the following situation.

EXAMPLE A company ships automatic dishwashers in lots of 24. Before they are shipped, though, an inspector randomly selects four dishwashers from each lot, and the lot "passes" this inspection only if all four are in perfect condition; otherwise, each dishwasher is checked out individually (and repaired, if necessary) at considerable cost. Clearly, this kind of sampling inspection involves certain risks—it is possible for a lot to "pass" this inspection even though 12, or even 20, of the 24 dishwashers have serious defects, and it is possible for a lot to "fail" this inspection, even though only one of the dishwashers has a slight defect. This raises many questions. For instance, it may be of special interest to know the probability that a lot will "pass" the inspection when, say, four of the dishwashers are not in perfect condition. This means that we shall have to find the probability of 4 successes (perfect dishwashers) in 4 trials, but if sampling is without replacement (as it ordinarily is in sampling inspection), the formula for the binomial distribution cannot be used. Without replacement, the probability that the first dishwasher will be perfect is $\frac{20}{24}$, the probability that the second dishwasher will be perfect given that the first one was perfect is $\frac{19}{23}$, the probability that the third dishwasher will be perfect given that the first two were perfect is $\frac{18}{22}$, the probability that the fourth dishwasher will be perfect given that the first three were perfect is $\frac{17}{21}$, and the probability of choosing four perfect dishwashers in a row is

$$\frac{20}{24} \cdot \frac{19}{23} \cdot \frac{18}{22} \cdot \frac{17}{21} = \frac{5 \cdot 19 \cdot 17}{23 \cdot 22 \cdot 7} = \frac{1,615}{3,542}$$

or approximately 0.46.

In the preceding example, the result was obtained simply by applying the multiplication rule for dependent events. However, there are many problems of sampling without replacement where we proceed in a different way.

EXAMPLE   With reference to the dishwasher example, suppose we want to determine the probability that the inspector will find one defective and three perfect dishwashers in a random sample of four dishwashers taken without replacement from the lot of 24 dishwashers, in which there are four defectives and 20 are perfect. Now, we can argue that there are $\binom{4}{1} = 4$ ways in which the inspector can select one of the four defective dishwashers, $\binom{20}{3} = 1,140$ ways in which he can select three of the 20 perfect ones, and, hence, $\binom{4}{1} \cdot \binom{20}{3} = 4 \cdot 1,140 = 4,560$ ways in which he can select one defective dishwasher and three in perfect condition. Since there are altogether $\binom{24}{4} = 10,626$ equally likely ways (by virtue of the random sampling) in which the inspector can choose four dishwashers from the entire lot, we find that the probability of his getting one defective dishwasher and three perfect ones is

$$\frac{\binom{4}{1} \cdot \binom{20}{3}}{\binom{24}{4}} = \frac{4,560}{10,626} = 0.43$$

rounded to two decimals.

To generalize the method used in the preceding example, suppose that n objects are to be chosen from a set of a objects of one kind (successes) and b objects of another kind (failures), and that we are interested in the probability of getting "x successes and $n - x$ failures." Arguing as before, we can say that the x successes can be chosen in $\binom{a}{x}$ ways, the $n - x$ failures can be chosen in $\binom{b}{n - x}$ ways, and, hence, x successes and $n - x$ failures can be chosen in $\binom{a}{x} \cdot \binom{b}{n - x}$ ways. Also, n objects can be chosen from the whole set of $a + b$ objects in $\binom{a + b}{n}$ ways, and if we regard all these possibilities as equally likely, it follows that the probability of getting "x successes in n trials" is

Hypergeometric distribution

$$f(x) = \frac{\binom{a}{x} \cdot \binom{b}{n - x}}{\binom{a + b}{n}} \qquad \textit{for } x = 0, 1, 2, \ldots, \textit{or } n$$

This is the formula for the **hypergeometric distribution**, and we should add that it applies only when x does not exceed a and $n - x$ does not exceed b; clearly, we cannot very well get more successes (or failures) than there are in the whole set.

EXAMPLE   To consider another application of the hypergeometric distribution, suppose that as part of an air pollution survey an inspector decides to examine the exhaust of 8 of a company's 16 trucks. He suspects that 5 of the trucks emit excessive amounts of pollutants, and he wants to know the probability that if his suspicion is correct, his sample will catch at least 3 of the 5 trucks. The probability he wants to know is given by $f(3) + f(4) + f(5)$, where each term in this sum is to be calculated by means of the formula for the hypergeometric distribution with $a = 5$, $b = 11$, and $n = 8$. Substituting these values together with $x = 3$, then $x = 4$, and $x = 5$, we get

$$f(3) = \frac{\binom{5}{3} \cdot \binom{11}{5}}{\binom{16}{8}} = \frac{10 \cdot 462}{12,870} = 0.359$$

$$f(4) = \frac{\binom{5}{4} \cdot \binom{11}{4}}{\binom{16}{8}} = \frac{5 \cdot 330}{12,870} = 0.128$$

$$f(5) = \frac{\binom{5}{5} \cdot \binom{11}{3}}{\binom{16}{8}} = \frac{1 \cdot 165}{12,870} = 0.013$$

and the probability that the inspector will catch at least three of the five trucks which emit excessive amounts of pollutants is $0.359 + 0.128 + 0.013 = 0.500$.

As we have pointed out, the binomial distribution does not apply when sampling is without replacement, but sometimes it can be used as an approximation.

EXAMPLE   Let us return to the example on page 161, where we found the probability that a lot will "pass" the inspection (that is, all four of the dishwashers inspected will be in perfect condition) when four of the 24 dishwashers are not in perfect condition. To find the probability of "four successes (perfect dishwashers) in four trials," we might be tempted to argue that since 20 of the 24 dishwashers are in perfect condition, the probability of a success is $\frac{20}{24} = \frac{5}{6}$, and hence the probability of "four successes in four trials" is

$$f(4) = \binom{4}{4}\left(\frac{5}{6}\right)^4 \left(1 - \frac{5}{6}\right)^{4-4} = 1 \cdot \left(\frac{5}{6}\right)^4 \cdot 1 = \frac{625}{1,296}$$

or approximately 0.48, in accordance with the formula for the binomial distribution.

This result would be correct if each dishwasher is replaced before the next one is chosen for inspection, but if sampling is without replacement, the figure which we obtained must be looked upon as an approximation. In this case, 0.48 is not a bad approximation, since the correct value is 0.46, as we showed on page 162.

It is generally agreed that the approximation illustrated by this example is "safe" so long as n constitutes less than five percent of $a + b$, which was actually not the case for $n = 4$ and $a + b = 24$. The main advantages of this approximation are that the binomial distribution has been tabulated much more extensively than the hypergeometric distribution, and that it is generally easier to use.

7.5

The Poisson Distribution ⋆

If n is large and p is small, binomial probabilities are often approximated by means of the formula

Poisson
distribution

$$f(x) = \frac{(np)^x \cdot e^{-np}}{x!} \qquad \text{for } x = 0, 1, 2, 3, \ldots$$

which is that for the **Poisson distribution.** Here e is the number 2.71828... used in connection with natural logarithms, and the values of e^{-np} may be obtained from Table XII at the end of the book. For the Poisson distribution, x can take on the infinite set of values $0, 1, 2, 3, \ldots$, but this poses no problems, since the probabilities become negligible (very close to zero) after the first few values of x.[†]

EXAMPLE   To illustrate this approximation, suppose that a fire insurance company has 1,860 policyholders and that the probability is $\frac{1}{600}$ that any one of the policyholders will file at least one claim in any given year. To find the probabilities that in a given year 0, 1, 2, 3, 4, ... of the policyholders will file at least one claim, we cannot, for practical reasons, use the formula for the binomial distribution. Instead, we substitute $n = 1,860$ and $p = \frac{1}{600}$, and, hence, $np = 1,860 \cdot \frac{1}{600} = 3.1$ into the formula for the Poisson distribution, getting

$$f(0) = \frac{(3.1)^0 \cdot e^{-3.1}}{0!} = 0.045 \qquad f(1) = \frac{(3.1)^1 \cdot e^{-3.1}}{1!} = 0.140$$

$$f(2) = \frac{(3.1)^2 \cdot e^{-3.1}}{2!} = 0.216 \qquad f(3) = \frac{(3.1)^3 \cdot e^{-3.1}}{3!} = 0.223$$

for the first four values of x. Here $e^{-3.1} = 0.045$ is the only value that we need to look up in Table XII. Continuing this way, we obtain the values shown in the following table, where we stopped with $x = 10$ since the probabilities for $x = 11, x = 12, \ldots$, are all

[†] As formulated in Chapter 5, the postulates of probability apply only when the sample space is finite. When the sample space is **countably infinite** (that is, when there are as many outcomes as there are whole numbers), as is the case here, the third postulate has to be modified so that for any sequence of mutually exclusive events $A_1, A_2, A_3, \ldots$,

$$P(A_1 \cup A_2 \cup A_3 \cup \cdots) = P(A_1) + P(A_2) + P(A_3) + \cdots$$

0.000 rounded to three decimals:

Number of policyholders filing claims	Probability
0	0.045
1	0.140
2	0.216
3	0.223
4	0.173
5	0.108
6	0.056
7	0.025
8	0.010
9	0.003
10	0.001

A histogram of this distribution is shown in Figure 7.3.

FIGURE 7.3

Histogram of Poisson distribution with np = 3.1.

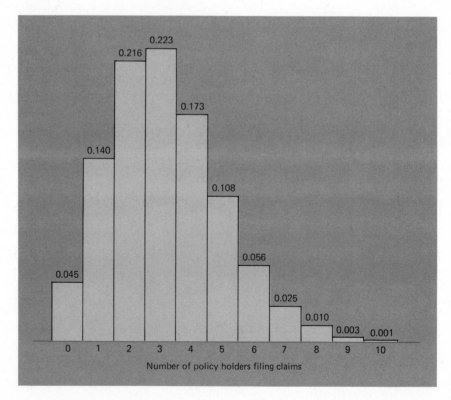

The Poisson distribution also has many important applications which have no direct connection with the binomial distribution. In that case, np is replaced by the parameter λ (Greek lowercase *lambda*) and we calculate the probability of getting x "successes" by means of the formula

Poisson distribution (with parameter λ)

$$f(x) = \frac{\lambda^x \cdot e^{-\lambda}}{x!} \qquad for\ x = 0, 1, 2, 3, \ldots$$

where the parameter λ is interpreted as the expected, or average, number of successes.

This formula applies to many situations where we can expect a fixed number of "successes" per unit time (or for some other kind of unit), say, when a bank can expect to receive six bad checks per day, when 1.6 accidents can be expected per day at a busy intersection, when eight small pieces of meat can be expected in a frozen meat pie, when 5.6 imperfections can be expected per roll of cloth, when 0.14 complaint per passenger can be expected by an airline, and so on.

EXAMPLE   If a bank receives on the average $\lambda = 6$ bad checks per day, the probability that it will receive $x = 4$ bad checks on any given day is

$$f(4) = \frac{6^4 \cdot e^{-6}}{4!} = \frac{(1{,}296)(0.0025)}{24} = 0.135$$

where the value of e^{-6} was read from Table XII. Also, if $\lambda = 1.6$ accidents can be expected at an intersection on any given day, the probability that there will be $x = 3$ accidents is

$$f(3) = \frac{1.6^3 \cdot e^{-1.6}}{3!} = \frac{(4.096)(0.202)}{6} = 0.138$$

EXERCISES

1   In each case determine whether the given values can be looked upon as the values of a probability distribution of a random variable which can take on only the values 1, 2, 3, and 4, and explain your answers:
(a) $f(1) = 0.26,\ f(2) = 0.26,\ f(3) = 0.26,\ f(4) = 0.26$;
(b) $f(1) = \frac{1}{9},\ f(2) = \frac{2}{9},\ f(3) = \frac{1}{3},\ f(4) = \frac{1}{3}$;
(c) $f(1) = 0.15,\ f(2) = 0.28,\ f(3) = 0.29,\ f(4) = 0.28$;
(d) $f(1) = 0.33,\ f(2) = 0.37,\ f(3) = -0.03,\ f(4) = 0.33$;
(e) $f(1) = \frac{1}{4},\ f(2) = \frac{1}{8},\ f(3) = \frac{1}{16},\ f(4) = \frac{1}{32}$.

2 Determine whether the following can be probability distributions (defined in each case only for the given values of x) and explain your answers:

(a) $f(x) = \dfrac{1}{5}$ for $x = 0, 1, 2, 3, 4, 5$;

(b) $f(x) = \dfrac{x + 1}{14}$ for $x = 1, 2, 3, 4$;

(c) $f(x) = \dfrac{x - 2}{5}$ for $x = 1, 2, 3, 4, 5$;

(d) $f(x) = \dfrac{x^2}{30}$ for $x = 0, 1, 2, 3, 4$.

3 A multiple-choice test consists of six questions and three answers to each question (of which only one is correct). If a student answers each question by rolling a balanced die and checking the first answer if he gets a 1 or a 2, the second answer if he gets a 3 or a 4, and the third answer if he gets a 5 or a 6, find the probabilities of getting
   (a) exactly three correct answers;
   (b) no correct answers;
   (c) at least five correct answers.

4 In a large government agency, illness is given one times in nine as the reason for all absences from work. What is the probability that three of four absences (randomly selected from the agency's records) were claimed to be due to illness?

5 If 40 percent of the mice used in an experiment will become very aggressive within one minute after having been administered an experimental drug, find the probability that exactly four of 12 mice which have been administered the drug become very aggressive within one minute, using
   (a) the formula for the binomial distribution;
   (b) Table VI.

6 If the probability is 0.10 that a hockey playoff game will go into overtime, find the probability that at most one of five playoff games will go into overtime, using
   (a) the formula for the binomial distribution;
   (b) Table VI.

7 In a certain city, incompatibility is given as the legal reason for 60 percent of all divorce cases. Find the probability that three of the next six divorce cases filed in this city will claim incompatibility as the reason, using
   (a) the formula for the binomial distribution;
   (b) Table VI.

8 Suppose that a civil service examination is designed so that 70 percent of all persons with an IQ of 90 can pass it. Use Table VI to find the probabilities that among 15 persons with an IQ of 90 who take the test
   (a) at most six will pass;
   (b) at least 12 will pass;
   (c) from eight through 12, inclusive, will pass.

9 A social scientist claims that only 40 percent of all high school seniors capable of doing college work actually go to college. Use Table VI to find the probabilities that among 14 high school seniors capable of doing college work
   (a) exactly four will go to college;
   (b) at most eight will go to college;

(c) at least five will go to college;

(d) from four through nine, inclusive, will go to college.

10 A quality control engineer wants to check whether (in accordance with specifications) 95 percent of the electronic components shipped by his company are in good working condition. To this end, he randomly selects 12 from each large lot ready to be shipped and passes the lot if they are all in good working condition; otherwise each of the components in the lot is checked. Assuming that the lots are so large that (for all practical purposes) we can use the binomial distribution, use Table VI to find the probabilities that he will commit the error of

(a) holding a lot for further inspection even though 95 percent of the components are in good working condition;

(b) passing a lot even though only 90 percent of the components are in good working condition;

(c) passing a lot even though only 80 percent of the components are in good working condition;

(d) passing a lot even though only 70 percent of the components are in good working condition.

11 A study shows that 60 percent of all patients coming to a certain medical building have to wait in their doctor's waiting room for at least 45 minutes. Use Table VI to find the probabilities that among ten patients coming to this medical building 0, 1, 2, 3, ..., or 10 have to wait in their doctor's waiting room for at least 45 minutes. Draw a histogram of this probability distribution.

12 In some situations where the binomial distribution applies, we are interested in the probabilities that the first success will occur on any given trial. For this to happen on the xth trial, it must be preceded by $x - 1$ failures for which the probability is $(1 - p)^{x-1}$, and it follows that the probability that the first success will occur on the xth trial is

Geometric distribution

$$f(x) = p(1 - p)^{x-1} \qquad for\ x = 1, 2, 3, 4, \ldots$$

This distribution is called the **geometric distribution** (because its successive values constitute a geometric progression) and it should be observed that, as in the case of the Poisson distribution, there is a countable infinity of possibilities. Using the formula, we find, for example, that in repeatedly rolling a balanced die, the probability that the first 6 will occur on the fifth roll is $\frac{1}{6}\left(\frac{5}{6}\right)^{5-1} = \frac{625}{7,776}$ or approximately 0.08.

(a) When taping a television commercial, the probability that a certain actor will get his lines straight on any one take is 0.40. What is the probability that this actor will get his lines straight for the first time on the fourth take?

(b) Suppose the probability is 0.25 that any given person will believe a rumor about the private life of a certain politician. What is the probability that the fifth person to hear the rumor will be the first one to believe it?

(c) The probability is 0.70 that a child exposed to a certain contagious disease will catch it. What is the probability that the third child exposed to the disease will be the first one to catch it?

13 Among the 15 applicants for a job, nine have college degrees. If two of the applicants are randomly chosen for interviews, find the probabilities that
    (a) neither has a college degree;
    (b) only one has a college degree;
    (c) both have college degrees.

14 Among the 20 cities which a political party is considering for its next three annual conventions, seven are in the western part of the United States. If, to avoid arguments, the selection is left to chance, what is the probability that none of these conventions will be held in the western part of the United States?

15 What is the probability that an IRS auditor will catch only two income tax returns with illegitimate deductions, if she randomly selects six returns from among 18 returns of which eight contain illegitimate deductions?

16 To pass a quality control inspection, two batteries are chosen from each lot of 12 car batteries, and the lot is passed only if neither battery has any defects; otherwise, each of the batteries in the lot is checked. If the selection of the batteries is random, find the probabilities that a lot will
    (a) pass the inspection when one of the 12 batteries is defective;
    (b) fail the inspection when three of the batteries are defective;
    (c) fail the inspection when six of the batteries are defective.

17 A secretary is supposed to send three of nine letters by special delivery. If he gets them all mixed up and randomly puts special delivery stamps on three of the letters, what are the probabilities that he will put
    (a) all the special delivery stamps on the wrong letters?
    (b) all the special delivery stamps on the right letters?

18 A panel of prospective jurors includes 240 whites and 60 blacks. Since the jury of 12 chosen from this panel does not include any blacks, the defense attorney suspects bias and would like to know the probability that there should be at least two blacks on the jury if the selection were random. Use the binomial approximation to the hypergeometric distribution (and Table VI) to find this probability.

19 A shipment of 100 burglar alarms contains four that are defective. If three of these burglar alarms are randomly selected and shipped to a customer, find the probability that she will get exactly one bad unit using
    (a) the formula for the hypergeometric distribution;
    (b) the binomial approximation to the hypergeometric distribution.

20 It is known from experience that 1.4 percent of the calls received at a switchboard are wrong numbers. Use the Poisson approximation to the binomial distribution to determine the probability that among 200 calls received at the switchboard, three are wrong numbers.

21 Records show that the probability is 0.00002 that a car will have a flat tire while driving over a certain bridge. Use the Poisson approximation to the binomial distribution to find the probability that among 20,000 cars driven over this bridge, not more than one will have a flat tire.

22 If 1.2 percent of the inhabitants of a border city are illegal immigrants, what is the approximate probability that in a random sample of 300 of them, two are illegal immigrants?

23 In a certain geographical region the number of persons who become seriously ill each year from eating a certain poisonous plant is a random variable having

the Poisson distribution with $\lambda = 1.8$. What are the probabilities of 0, 1, or 2 such illnesses in a given year?

24 Suppose that in the inspection of a fabric produced in continuous rolls the number of imperfections spotted by an inspector during a 5-minute period is a random variable having the Poisson distribution with $\lambda = 2.4$. Find the probabilities that during a 5-minute period an inspector will find
    (a) no imperfections;   (d) three imperfections;
    (b) one imperfection;   (e) at least four imperfections.
    (c) two imperfections;

25 If the number of javelinas (wild pigs) seen on a day's field trip in the Sonora desert is a random variable having the Poisson distribution with $\lambda = 3.2$, find the probabilities that on such a field trip one will see
    (a) no javelinas;   (c) two javelinas;
    (b) one javelina;   (d) at least three javelinas.

26 The binomial distribution applies only when there are two possible outcomes for each trial. If there are k possible (mutually exclusive) outcomes for each trial, and the probabilities of these outcomes are $p_1, p_2, \ldots,$ and p_k, then in n independent trials the probability of x_1 outcomes of the first kind, x_2 outcomes of the second kind, $\ldots,$ and x_k outcomes of the kth kind is

Multinomial distribution

$$\frac{n!}{x_1! x_2! \cdot \ldots \cdot x_k!} p_1^{x_1} \cdot p_2^{x_2} \cdot \ldots \cdot p_k^{x_k}$$

This distribution is called the **multinomial distribution.** Using this formula, we find, for example, that if on Saturday nights Channel 12 has 50 percent of the viewing audience, Channel 10 has 30 percent of the viewing audience, and Channel 3 has 20 percent of the viewing audience, the probability that among eight television viewers randomly selected on a Saturday night, five will be watching Channel 12, two will be watching Channel 10, and one will be watching Channel 3 is

$$\frac{8!}{5! \cdot 2! \cdot 1!} (0.50)^5 (0.30)^2 (0.20) = 0.0945$$

    (a) The probabilities are 0.40, 0.50, and 0.10 that in city driving a certain kind of compact car will average less than 22 miles per gallon, from 22 to 25 miles per gallon, or more than 25 miles per gallon. Find the probability that among ten such cars tested, three will average less than 22 miles per gallon, five will average from 22 to 25 miles per gallon, and two will average more than 25 miles per gallon.
    (b) Suppose that the probabilities are 0.50, 0.20, 0.20, and 0.10 that a state income tax form will be filled out correctly, that it will contain only errors favoring the taxpayer, that it will contain only errors favoring the government, or that it will contain both kinds of errors. What is the probability that among 12 such tax forms (randomly chosen for audit) five will be filled out correctly, three will contain only errors favoring the taxpayer, three will contain only errors favoring the government, and one will contain both kinds of errors?

7.5 The Poisson Distribution

171

7.6

The Mean of a Probability Distribution

On page 134 we said that a child in the age group from 6 to 16 can be expected to go to the dentist 1.9 times a year, and we pointed out that this figure is a mathematical expectation, which was obtained by multiplying 0, 1, 2, 3, 4, ..., by the respective probabilities that a child in this age group will visit a dentist that many times a year. On the same page, we also showed that similar calculations lead to the result that a certain airline's Sky Harbor Airport office can expect 2.75 complaints per day about the handling of luggage.

EXAMPLE If we apply this procedure to the examples of Section 7.1, we find that, between them, the two salespersons can expect to sell

$$0(0.05) + 1(0.18) + 2(0.39) + 3(0.38) = 2.10$$

of the three Ford Mustangs, the number of points we can expect in one roll of a die is

$$1 \cdot \frac{1}{6} + 2 \cdot \frac{1}{6} + 3 \cdot \frac{1}{6} + 4 \cdot \frac{1}{6} + 5 \cdot \frac{1}{6} + 6 \cdot \frac{1}{6} = 3\frac{1}{2}$$

and the number of heads we can expect in four flips of a balanced coin is

$$0 \cdot \frac{1}{16} + 1 \cdot \frac{4}{16} + 2 \cdot \frac{6}{16} + 3 \cdot \frac{4}{16} + 4 \cdot \frac{1}{16} = 2$$

Of course, they cannot actually sell 2.1 cars and we cannot actually roll a $3\frac{1}{2}$ with a die; like all mathematical expectations, these figures must be looked upon as averages.

In general, if a random variable takes on the values $x_1, x_2, x_3, \ldots,$ or x_k, with the probabilities $f(x_1), f(x_2), f(x_3), \ldots,$ and $f(x_k)$, its expected value (or its mathematical expectation) is given by

$$x_1 \cdot f(x_1) + x_2 \cdot f(x_2) + x_3 \cdot f(x_3) + \cdots + x_k \cdot f(x_k)$$

and it is customary to refer to this quantity as the **mean of the random variable,** or the **mean of its distribution.** Using the $\sum$ notation, we can write

Mean of probability distribution

$$\mu = \sum x \cdot f(x)$$

where the mean of the distribution of a random variable, like the mean of a population, is denoted by the Greek letter μ (*mu*). The notation is the same, for as we pointed out in connection with the binomial distribution, when we observe a value of a random variable, we refer to its distribution as the population we are sampling. For instance, Figure 7.2 on page 159 pictures the population we are sampling when we observe the number of successes in six trials and $p = 0.30$, and Figure 7.3 on page 166 pictures the population we are sampling when we observe a value of a random variable having the Poisson distribution with $\lambda = 3.1$.

EXAMPLE   Using the probabilities obtained on page 159, we can now say that if the custom official's claim that 30 percent of all persons returning from Europe fail to declare one or more purchases on which there is duty is correct, we can expect that

$$\mu = 0(0.118) + 1(0.303) + 2(0.324) + 3(0.185) + 4(0.060)$$
$$+ 5(0.010) + 6(0.001)$$
$$= 1.802$$

of six persons returning from Europe will fail to declare one or more purchases on which there is duty.

When a random variable can take on many different values, the calculation of μ usually becomes very laborious. For instance, if we want to know how many among 800 customers entering a store can be expected to make a purchase, and the probability that any one of them will make a purchase is 0.40, we will first have to calculate the 801 probabilities corresponding to 0, 1, 2, . . . , or 800 of them making a purchase. However, if we think for a moment, we might argue that in the long run 40 percent of the customers will make a purchase, 40 percent of 800 is 320, and, hence, we can expect that 320 of the 800 customers will make a purchase. Similarly, if a balanced coin is flipped 1,000 times, we can argue that in the long run heads will come up 50 percent of the time, and, hence, that we can expect $1,000(0.50) = 500$ heads. These two values are, indeed, correct; both problems deal with binomial distributions, and it can be shown that in general

Mean of binomial distribution

$$\mu = n \cdot p$$

for the mean of a binomial distribution. In words, the mean of a binomial distribution is simply the product of the number of trials and the probability of success on an individual trial.

EXAMPLE   We can now use this formula to verify the result which we obtained for the example dealing with the persons who fail to declare purchases made in Europe on which there is duty. Since we had $n = 6$

and $p = 0.30$, we can now argue that $\mu = 6(0.30) = 1.80$, and it should be apparent that the small difference of 0.002 between this exact value and the one obtained before is due to our rounding the probabilities to three decimals. Also, for the number of heads we can expect in four flips of a coin, we now get $\mu = 4 \cdot \frac{1}{2} = 2$, and this agrees with the result obtained in the beginning of this section.

It is important to remember that the formula $\mu = n \cdot p$ applies only to binomial distributions. There are other formulas for other distributions; for instance, for the hypergeometric distribution the formula for the mean is

Mean of hypergeometric distribution

$$\mu = \frac{n \cdot a}{a + b}$$

EXAMPLE With reference to the example on page 163 where an inspector randomly chooses for examination eight of 16 trucks, of which five emit excessive amounts of pollutants and 11 do not, we have $a = 5, b = 11$, and $n = 8$. Hence, the inspector can expect to catch

$$\mu = \frac{8 \cdot 5}{5 + 11} = 2.5$$

of the faulty trucks. This should not come as a surprise—he examines half of the trucks and he can expect to catch half of the faulty ones.

Also, the mean of the Poisson distribution is simply $\mu = \lambda$. Formal proofs of all these special formulas may be found in any textbook on mathematical statistics.

7.7

The Standard Deviation of a Probability Distribution

In Chapter 3 we saw that there are many situations in which we must describe, in addition to the mean or some other measure of location, the variability (spread, or dispersion) of a set of data. As we indicated in that chapter, the most widely used statistical measures of variation are the variance and its square root, the standard deviation, which both measure variability by averaging the squared deviations from the mean. For probability distributions, we measure variability in almost the same way, but instead of averaging the squared deviations from the mean, we calculate their expected value. If x is a value of some random variable whose probability distribution has the mean μ, the deviation from the mean is $x - \mu$ and we define the **variance of the probability distribution** as the expected value of the squared deviation from the mean, namely, as

Variance of probability distribution

$$\sigma^2 = \sum (x - \mu)^2 \cdot f(x)$$

where the summation extends over all values taken on by the random variable. As in the preceding section, and for the same reason, we denote descriptions of probability distributions with the same symbols as descriptions of populations. The square root of the variance defines the **standard deviation of a probability distribution,** and we write

Standard deviation of probability distribution

$$\sigma = \sqrt{\sum (x - \mu)^2 \cdot f(x)}$$

EXAMPLE   To illustrate the calculation of the standard deviation of a probability distribution, let us refer again to the example dealing with the number of persons who fail to declare purchases made in Europe. Values of this random variable and their probabilities are shown in the first two columns of the table below, and since the mean of this distribution was shown to be $\mu = 1.8$, we have

Number of persons	Probability	Deviation from mean	Squared deviation from mean	$(x - \mu)^2 f(x)$
0	0.118	−1.8	3.24	0.38232
1	0.303	−0.8	0.64	0.19392
2	0.324	0.2	0.04	0.01296
3	0.185	1.2	1.44	0.26640
4	0.060	2.2	4.84	0.29040
5	0.010	3.2	10.24	0.10240
6	0.001	4.2	17.64	0.01764
				$\sigma^2 = 1.26604$

The values in the column on the right were obtained by multiplying each squared deviation from the mean by its probability, and their sum is the variance of the distribution. Also, $\sigma = \sqrt{1.26604} = 1.13$ rounded to two decimals.

In this example, the calculations were quite easy, owing mainly to the fact that the deviations from the mean were small numbers given to one decimal. If the deviations from the mean are large numbers, or if they are given to several decimals, it is usually worthwhile to simplify the calculations by using the computing formula for σ^2 given in Exercise 3 on page 178.

As in the case of the mean, the calculation of the variance or the standard deviation can generally be simplified when we deal with special kinds of probability distributions. For instance, for the binomial distribution we have the formula

Standard deviation of binomial distribution

$$\sigma = \sqrt{np(1 - p)}$$

EXAMPLE For the example dealing with the number of persons who fail to declare purchases made in Europe, we have $n = 6$ and $p = 0.30$, so that

$$\sigma = \sqrt{6(0.30)(0.70)} = \sqrt{1.26} = 1.12$$

and this value differs from the result obtained before by the small rounding error of $1.13 - 1.12 = 0.01$.

EXAMPLE For the number of heads in four flips of a balanced coin, where $n = 4$ and $p = \frac{1}{2}$, the formula yields

$$\sigma = \sqrt{4 \cdot \frac{1}{2} \cdot \frac{1}{2}} = 1$$

In Exercise 7 on page 179 the reader will be asked to verify this result by the long method.

There also exist special formulas for the standard deviations of other special distributions, and they may be found in more advanced texts.

7.8
Chebyshev's Theorem

Intuitively speaking, the variance and the standard deviation of a probability distribution measure its spread or its dispersion: When σ is small, the probability is high that we will get a value close to the mean, and when σ is large, we are more likely to get a value far away from the mean. This important idea is expressed rigorously in Chebyshev's theorem, which we introduced in Section 3.7 as it pertains to frequency distributions. For probability distributions, the theorem can be stated as follows:

Chebyshev's
theorem

> *The probability that a random variable will take on a value within k*
> *standard deviations of the mean is at least*
>
> $$1 - \frac{1}{k^2}$$

Thus, the probability of getting a value within two standard deviations of the mean (a value between $\mu - 2\sigma$ and $\mu + 2\sigma$) is at least $\frac{3}{4}$, the probability of getting a value within five standard deviations of the mean (a value between $\mu - 5\sigma$ and $\mu + 5\sigma$) is at least $\frac{24}{25}$, and the probability of getting a value within ten standard deviations of the mean (a value between $\mu - 10\sigma$ and $\mu + 10\sigma$) is at least $\frac{99}{100}$. The quantity k in Chebyshev's theorem can be any positive number, although the theorem becomes trivial when k is 1 or less.

EXAMPLE Suppose that the number of telephone calls which a theatrical agent receives each day from her clients is a random variable whose distribution has the mean $\mu = 14$ and the standard deviation $\sigma = 3.50$. Using Chebyshev's theorem with $k = 2$, we can say that the probability is at least 0.75 (or that the odds are at least 3 to 1) that on any one day she will receive between $14 - 2(3.50) = 7$ and $14 + 2(3.50) = 21$ calls from her clients.

EXAMPLE Suppose that, betting on tails, we have won only 128 times in 400 flips of a coin, and we are beginning to wonder whether the game we are playing is fair. If the game is fair, we have a fifty–fifty chance of winning on each flip of the coin, and the number of tails in 400 flips of the coin is a random variable having the binomial distribution with $n = 400$ and $p = \frac{1}{2}$, so that

$$\mu = np = 400 \cdot \frac{1}{2} = 200$$

and

$$\sigma = \sqrt{np(1 - p)} = \sqrt{400 \cdot \frac{1}{2} \cdot \frac{1}{2}} = 10$$

Since the 128 tails we got differs from $\mu = 200$ by 72, which is more than seven standard deviations, we can argue that the probability is at least $1 - \dfrac{1}{7^2} = \dfrac{48}{49} = 0.9796$ that we should get a value within seven standard deviations of the mean (or get between 130 and 270 tails) and, hence, it may well be justified to say that the game is not fair.

EXERCISES

1 Suppose the probabilities are 0.4, 0.3, 0.2, and 0.1 that among three recently married couples, 0, 1, 2, or 3 will be divorced within two years. Find the mean and the variance of this probability distribution.

2 The following table gives the probabilities that a probation officer will receive 0, 1, 2, 3, 4, 5, or 6 reports of probation violations on any given day:

Number of violations	0	1	2	3	4	5	6
Probability	0.15	0.22	0.31	0.18	0.09	0.04	0.01

Calculate the mean and the standard deviation of this probability distribution.

3 Using the rules for summations given in Section 3.10, we can derive the following short-cut formula for the variance (and, hence, the standard deviation):

$$\sigma^2 = \sum x^2 \cdot f(x) - \mu^2$$

The advantage of this formula is that we do not have to work with the deviations from the mean. Instead, we subtract μ^2 from the sum of the products obtained by multiplying the square of each value of the random variable by the corresponding probability.

(a) Use this formula to find the variance of the probability distribution of Exercise 1.

(b) Use this formula to calculate the standard deviation of the distribution of Exercise 2.

(c) Use this formula to recalculate σ for the example dealing with the number of persons who fail to declare purchases made in Europe, for which we obtained $\sigma = 1.13$ on page 175.

4 As can easily be verified by means of the formula for the binomial distribution (or by listing all 32 possibilities), the probabilities of getting 0, 1, 2, 3, 4, or 5 heads in five flips of a balanced coin are $\frac{1}{32}, \frac{5}{32}, \frac{10}{32}, \frac{10}{32}, \frac{5}{32}$, and $\frac{1}{32}$. Find the mean of this distribution and also the variance using

(a) the formula defining σ^2 on page 175;

(b) the short-cut formula of Exercise 3;

(c) the special formula $\sigma^2 = np(1 - p)$.

5 Use the probabilities obtained in Exercise 11 on page 169 to find the mean and the standard deviation of the distribution of the number of patients, among ten coming to a certain medical building, who have to wait in their doctor's waiting room for at least 45 minutes. Verify the results by substituting $n = 10$ and $p =$

0.60 into the special formulas for the mean and the standard deviation of a binomial distribution. (*Hint:* To calculate σ use the short-cut formula of Exercise 3.)

6   Use the probabilities in the table on page 155 and the value $\mu = 3.5$ obtained on page 172, to verify that $\sigma^2 = \frac{35}{12}$ for the distribution of the number of points rolled with a balanced die.

7   Use the probabilities in the table on page 155 and the value $\mu = 2$ obtained on page 172, to verify that $\sigma = 1$ for the distribution of the number of heads obtained in four flips of a balanced coin.

8   Find the mean and the standard deviation of the distribution of each of the following random variables (having binomial distributions):
   (a) The number of heads obtained in 676 flips of a balanced coin.
   (b) The number of 3's obtained in 720 rolls of a balanced die.
   (c) The number of persons (among 600 invited) who will attend the opening of a new branch bank, when the probability is 0.30 that any one of them will attend.
   (d) The number of defectives in a sample of 600 parts made by a machine, when the probability is 0.04 that any one of the parts is defective.
   (e) The number of students (among 800 interviewed) who do not like the food served at the university cafeteria, when the probability is 0.65 that any one of them does not like the food.

9   With reference to the example on page 163, we showed that the probabilities are, respectively, 0.359, 0.128, and 0.013 that the inspector will catch 3, 4, or 5 of the company's five trucks which emit excessive amounts of pollutants.
   (a) Show that the probabilities corresponding to his catching 0, 1, or 2 of the faulty trucks are, respectively, 0.013, 0.128, and 0.359.
   (b) Use all these probabilities to calculate the mean and the standard deviation of the distribution of the number of faulty trucks which the inspector will catch.
   (c) Use the special formula on page 174 to verify the value obtained for μ in part (b).

10   Among ten faculty members considered for promotions there are six men and four women.
   (a) If two of them are chosen at random, find the probabilities that 0, 1, or 2 women will be included.
   (b) Use the probabilities obtained in part (a) to calculate the mean and the standard deviation of this probability distribution.
   (c) Use the special formula on page 174 to verify the value obtained for μ in part (b).

11   Referring to the table on page 166, calculate the mean and the variance of the distribution of the number of policy holders who file a claim during any given year. Use these results to verify that the mean and the variance of a Poisson distribution are given by the formulas

$$\mu = \lambda \quad \text{and} \quad \sigma^2 = \lambda$$

where $\lambda = np$ when the Poisson distribution is used to approximate the binomial distribution.

7.8 Chebyshev's Theorem

179

12 If the number of gamma rays emitted per second by a certain radioactive substance is a random variable having the Poisson distribution with $\lambda = 2.5$, the probabilities that it will emit 0, 1, 2, 3, 4, 5, 6, 7, 8, or 9 gamma rays in any one second are, respectively, 0.082, 0.205, 0.256, 0.214, 0.134, 0.067, 0.028, 0.010, 0.003, and 0.001.

    (a) Use these probabilities to calculate the mean of this distribution.

    (b) Use the short-cut formula of Exercise 3 to calculate the variance of this distribution.

    (c) Use the special formulas of Exercise 11 to verify the results obtained in parts (a) and (b).

13 The daily number of customers served lunch on a weekday by a certain restaurant is a random variable with $\mu = 142$ and $\sigma = 12$. According to Chebyshev's theorem, with what probability can we assert that between 82 and 202 customers will be served lunch by the restaurant on any given weekday?

14 The number of marriage licenses issued in a certain city during the month of June averages $\mu = 134$ with a standard deviation of $\sigma = 7.5$.

    (a) What does Chebyshev's theorem with $k = 9$ tell us about the number of marriage licenses issued there during a month of June?

    (b) According to Chebyshev's theorem, with what probability can we assert that between 74 and 194 marriage licenses will be issued there during a month of June?

15 Use Chebyshev's theorem to verify that the probability is at least $\frac{35}{36} = 0.972$ that

    (a) in 900 flips of a balanced coin there will be between 360 and 540 heads, and hence the proportion of heads will be between 0.40 and 0.60;

    (b) in 10,000 flips of a balanced coin there will be between 4,700 and 5,300 heads, and hence the proportion of heads will be between 0.47 and 0.53;

    (c) in 1,000,000 flips of a balanced coin there will be between 497,000 and 503,000 heads, and hence the proportion of heads will be between 0.497 and 0.503.

This exercise serves to illustrate the law of large numbers (mentioned on page 85), according to which the proportion of successes approaches the probability of a success when the number of trials becomes larger and larger.

7.9

Technical Note (Simulation) ★

In recent years, simulation techniques have been applied to many problems in the various sciences, and if the processes being simulated involve an element of chance, we refer to these techniques as **Monte Carlo methods.** Such methods have been used, for instance, to study traffic flow on proposed freeways, to conduct war "games," to study the spread of epidemics or human behavior during times of natural disaster (say, a flood or an earthquake), and to study the scattering of neutrons or the collisions of photons with electrons. Also, in business research, such methods are used to solve inventory problems, production scheduling, or the effects of advertising campaigns, and to study many other

FIGURE 7.4
Spinner for generating random digits.

situations involving over-all planning and organization. In most cases, this will eliminate the cost of building and operating expensive equipment, and in some instances it can be used when direct experimentation is impossible.

Although Monte Carlo methods are sometimes based on actual gambling devices, it is usually expedient to use published tables of **random numbers** (or **random digits**). Such tables consist of many pages on which the digits 0, 1, 2, . . . , and 9 are set down in a "random" fashion, much as they would appear if they were generated one at a time by a chance or gambling device giving each digit the same probability of $\frac{1}{10}$. Some early tables of random numbers were copied from pages of census data or from tables of 20-place logarithms, but they were found to be deficient in various ways. Nowadays, such tables are made with the use of electronic computers, but it would be possible to generate a table with a carefully constructed spinner like that of Figure 7.4.

EXAMPLE To illustrate the use of a table of random numbers (in this case Table IX on pages 470 through 473), let us play "Heads or tails" without actually flipping a coin. Letting 0, 2, 4, 6, and 8 represent heads, and 1, 3, 5, 7, and 9 represent tails, and arbitrarily choosing the fourth column on page 470 starting at the top, we get 3, 9, 8, 1, 5, 1, 6, 3, 2, 4, . . . , and we interpret this as tails, tails, heads, tails, tails, tails, heads, tails, heads, heads, ...

EXAMPLE Repeated flips of any number of coins, say, three coins, can be simulated in the same way. If we use the first three columns of the table on page 471, starting with the sixteenth row, we read out the random numbers 659, 900, 972, 219, 411, 237, 599, 826, 838, 619, . . . , and, counting the number of even digits in each case, we interpret this as 1, 2, 1, 1, 1, 1, 0, 3, 2, 1, . . . , heads.

Since the probabilities of 0, 1, 2, or 3 heads in three flips of a balanced coin (or when flipping three balanced coins) are $\frac{1}{8}$, $\frac{3}{8}$, $\frac{3}{8}$, and $\frac{1}{8}$, as can easily be verified by listing the eight equally likely possibilities, we could also use the scheme shown in the following table:

Number of heads	Probability	Random digits
0	$\frac{1}{8}$	0
1	$\frac{3}{8}$	1, 2, 3
2	$\frac{3}{8}$	4, 5, 6
3	$\frac{1}{8}$	7

Omitting the digits 8 and 9 wherever they may occur, we would thus interpret the random digits 8, 1, 3, 0, 7, 4, 3, 6, 9, 4, 8, 3, 5, 8, 0, ... , in the twelfth row of the table on page 470 as 1, 1, 0, 3, 2, 1, 2, 2, 1, 2, 0, ... , heads.

Of the two methods used in this example, the first has the disadvantage that we still have to count how many of the digits are even and how many are odd; the second has the disadvantage that some of the digits will have to be omitted, and this may conceivably lead to situations where many random digits will have to be omitted before we find enough that can be used. To avoid such difficulties, it is generally preferable to express the probabilities as decimals (rounded, if necessary).

EXAMPLE Since the probabilities for 0, 1, 2, or 3 heads are, respectively, 0.125, 0.375, 0.375, and 0.125 when we flip three balanced coins, we could use the following scheme:

Number of heads	Probability	Random numbers
0	0.125	000–124
1	0.375	125–499
2	0.375	500–874
3	0.125	875–999

Note that we used three-digit random numbers because the probabilities are given to three decimals, and that we allocated 125 (or one-eighth) of the random numbers from 000 through 999 to 0 heads, 375 (or three-eighths) to 1 head, 375 (or three-eighths) to 2 heads, and 125 (or one-eighth) to 3 heads. If we arbitrarily use the 16th, 17th, and 18th columns of the table on page 471 starting with the sixth row, we get 974, 611, 345, 664, 041, 203, 531, 421, 031, 925, . . . , and we interpret this as 3, 2, 1, 2, 0, 1, 2, 1, 0, 3, . . . , heads.

The allocation of random numbers to the various values of a random variable can be facilitated by referring to the corresponding cumulative probabilities, as is illustrated by the following example:

EXAMPLE Suppose that the probabilities are, respectively, 0.008, 0.037, 0.089, 0.146, 0.179, 0.175, 0.143, 0.100, 0.062, 0.034, 0.016, 0.007, 0.003, and 0.001 that 0, 1, 2, 3, . . . , or 13 cars will arrive at the toll booth of a bridge in any given minute during the early afternoon. To simulate the arrival of cars at this toll booth, we might use the following scheme:

Number of cars	Probability	Cumulative probability	Random numbers
0	0.008	0.008	000–007
1	0.037	0.045	008–044
2	0.089	0.134	045–133
3	0.146	0.280	134–279
4	0.179	0.459	280–458
5	0.175	0.634	459–633
6	0.143	0.777	634–776
7	0.100	0.877	777–876
8	0.062	0.939	877–938
9	0.034	0.973	939–972
10	0.016	0.989	973–988
11	0.007	0.996	989–995
12	0.003	0.999	996–998
13	0.001	1.000	999

The cumulative probabilities are the respective probabilities that at most 0 cars, at most 1 car, at most 2 cars, . . . , and at most 13 cars will arrive at the toll booth in any given minute during the early afternoon, and it should be observed that *in each case the last random digit is one less than the number formed by the last three digits of the corresponding cumulative probability.* For instance, 007 is one less than 008, 044 is one less than 045, and 133 is one less than 134.

Now, if we simulate the arrival of cars at the toll booth by using the 6th, 7th, and 8th columns of the table on page 470 starting with the 21st row, we get 836, 712, 524, 762, 325, 081, 960, 594, 473, 370, 305, 178, 523, 184, 368, 864, 676, 975, 553, and 618, and this means that during 20 one-minute intervals the number of cars arriving at the toll booth is 7, 6, 5, 6, 4, 2, 9, 5, 5, 4, 4, 3, 5, 3, 4, 7, 6, 10, 5, and 5.

EXERCISES

1 Use random numbers to simulate 100 flips of a balanced coin. Although it does not really matter which five digits represent heads and which five digits represent tails, use the same scheme as in the example on page 181.

2 Using the digits 1, 2, 3, 4, 5, and 6 to represent the corresponding faces of a die (and omitting 0, 7, 8, and 9), simulate 120 rolls of a balanced die. Also construct a table showing how many times the die came up 1, 2, 3, 4, 5, and 6, and also the corresponding expected frequencies, which are all equal to 20.

3 The probabilities that a real estate broker will sell 0, 1, 2, 3, 4, 5, or 6 houses in a week are, respectively, 0.14, 0.27, 0.27, 0.18, 0.09, 0.04, and 0.01.
    (a) Distribute the two-digit random numbers from 00 through 99 among these seven possibilities so that the corresponding random numbers can be used to simulate the number of houses the real estate broker sells in a week.
    (b) Use the results of part (a) to simulate the real estate broker's weekly sales during 25 consecutive weeks.

4 Suppose that the probabilities are, respectively, 0.41, 0.37, 0.16, 0.05, and 0.01 that there will be 0, 1, 2, 3, or 4 UFO sightings in a certain region on any one day.
    (a) Distribute the two-digit random numbers from 00 through 99 among the five values of this random variable, so that the corresponding random numbers can be used to simulate the sighting of UFO's in the given region.
    (b) Use the result of part (a) to simulate the sighting of UFO's in the given region on 30 days.

5 With reference to the example on page 158 and the probabilities shown in Figure 7.2, assign three-digit random numbers to $x = 0, 1, 2, 3, 4, 5$, and 6, and simulate the number of persons, in 30 samples of six each, who fail to declare one or more purchases made in Europe. [Since the sum of the probabilities is 1.001, let $f(1) = 0.302$.]

6 With reference to the example on page 165 and the probabilities shown in Figure 7.3, assign three-digit random numbers to $x = 0, 1, 2, 3, 4, 5, 6, 7, 8, 9$, and 10, and simulate the insurance company's "experience" over a period of twelve years.

7 Suppose the probabilities are 0.2466, 0.3452, 0.2417, 0.1128, 0.0395, 0.0111, 0.0026, and 0.0005 that there will be 0, 1, 2, 3, 4, 5, 6, or 7 polluting spills in the Great Lakes on any one day.
    (a) Distribute the four-digit random numbers from 0000 through 9999 to the eight values of this random variable, so that the corresponding random numbers can be used to simulate polluting spills in the Great Lakes.
    (b) Use the results of part (a) to simulate the number of polluting spills in the Great Lakes on 40 consecutive days.

8 Depending upon the availability of parts, a company can manufacture 3, 4, 5, or 6 units of a certain item per week with corresponding probabilities of 0.10, 0.40, 0.30, and 0.20. The probabilities that there will be a weekly demand for 0, 1, 2, 3, ..., or 8 units are, respectively, 0.05, 0.10, 0.30, 0.30, 0.10, 0.05, 0.05, 0.04, and 0.01. If a unit is sold during the week that it is made, it will yield a profit of $100; this profit is reduced by $20 for each week that a unit has to be stored. Use random numbers to simulate the operations of this company for 50 consecutive weeks and estimate their expected weekly profit.

BIBLIOGRAPHY   More detailed tables of binomial probabilities may be found in

ROMIG, H. G., *50–100 Binomial Tables*. New York: John Wiley & Sons, Inc., 1953.

Tables of the Binomial Probability Distribution, National Bureau of Standards Applied Mathematics Series No. 6. Washington, D.C.: U.S. Government Printing Office, 1950.

and a detailed table of Poisson probabilities is given in

MOLINA, E. C., *Poisson's Exponential Binomial Limit*. Princeton, N.J.: D. Van Nostrand Company, Inc., 1947.

Among the many published tables of random numbers, one of the most widely used is

RAND CORPORATION, *A Million Random Digits with 100,000 Normal Deviates*. New York: Macmillan Publishing Co., Inc., third printing 1966.

8

The Normal
Distribution

Continuous sample spaces and continuous random variables arise whenever we deal with quantities that are measured on a continuous scale—for instance, when we measure the speed of a car, the amount of alcohol in a person's blood, the net weight of a package of frozen food, or the amount of tar in a cigarette. In actual practice, we always round measurements to the nearest whole unit or to a few decimals, but this does not take away from the fact that there exists a continuum of possibilities, and we could ask for probabilities associated with individual numbers or points. However, if we did this, we would find that (for all practical purposes) such probabilities are always zero—surely, we should be willing to give any odds that a car will not be traveling exactly $20\pi = 62.83185\ldots$ miles per hour, and we should be willing to give any odds that the weight of a package of frozen food will not be exactly $\sqrt{64.1} = 8.00624756\ldots$ ounces. This does not matter, though, for in the continuous case we are not really interested in probabilities associated with individual numbers or points, but in probabilities associated with intervals or regions. For instance, we might ask for the probability that (at a certain checkpoint) a car will be traveling between 55 and 65 miles per hour, or that a package of frozen food will weigh between 7.95 and 8.05 ounces. In this chapter we shall learn how to determine, and work with, probabilities relating to continuous sample spaces and continuous random variables.

8.1
Continuous Distributions

When we first discussed histograms in Chapter 2, we pointed out that the frequencies, percentages, or proportions associated with the various classes are given by the areas of the rectangles; as is illustrated by Figures 7.2 and 7.3, this is true also for probabilities associated with the values of random variables such as the ones discussed in Chapter 7. In the continuous case, we also represent probabilities by means of areas, as is illustrated in Figure 8.1, but instead of areas of rectangles we consider areas under continuous curves. The first diagram of Figure 8.1 represents the probability distribution of a random variable which takes on only the values 0, 1, 2, ..., 9, and 10, and the probability of getting a 3, for example, is given by the area of the white rectangle. The second diagram refers to a continuous random variable which can take on any value on the interval from 0 to 10, and the probability of getting a value on the interval from 2.5 to 3.5 is given by the area of the white region under the curve. Similarly, the area of the dark region under the curve gives the probability of getting a value greater than 8.

Continuous curves such as the one shown in the second diagram of Figure 8.1 are the graphs of functions which we refer to as **probability densities,** or more informally as **continuous distributions.** A probability density is characterized by the fact that **the area under the curve between any two values** a **and** b **(see Figure 8.2) gives the probability that a random variable having this continuous distribution will take on a value on the interval from** a **to** b**.** It follows from this that the values of a probability density cannot be negative, and that the total area under the curve (representing the certainty

FIGURE 8.1

Histogram of probability distribution and graph of continuous distribution.

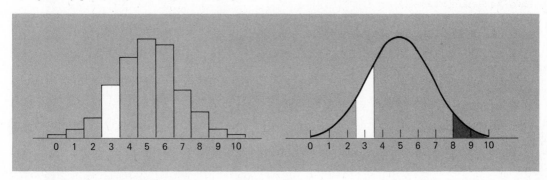

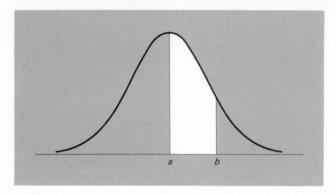

FIGURE 8.2
Continuous distribution.

that a random variable must take on one of its values) is always equal to 1. These rules correspond to the ones given for probability distributions at the end of Section 7.2.

EXAMPLE If we approximate a family income distribution with a smooth curve, as in Figure 8.3, we can determine what proportion of the incomes falls into any given interval (or the probability that the income of a family, chosen at random, will fall into the interval) by looking at the corresponding area under the curve. By comparing the area of the white region on the right of Figure 8.3 with the total area under the curve (representing 100 percent), we can judge by eye that roughly 10 to 12 percent of the families have incomes of $24,000 or more. Similarly, it can be seen that about 40 to 45 percent of the families have incomes of $12,000 or less.

Statistical descriptions of continuous distributions are as important as descriptions of probability distributions or distributions of observed data, but most of them, including the mean and the standard deviation, cannot be defined without the use of calculus. Nevertheless, we can always picture a continuous distribution as being approximated by a histogram of a proba-

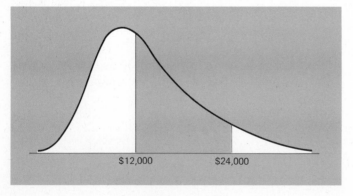

FIGURE 8.3
Curve approximating family income distribution.

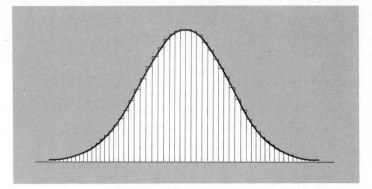

FIGURE 8.4

Continuous distribution approximated by histogram of probability distribution.

bility distribution (see Figure 8.4) whose mean and standard deviation can be calculated. Then, if we choose histograms with narrower and narrower classes, the means and the standard deviations of the corresponding probability distributions will approach the mean and the standard deviation of the continuous distribution. Actually, the mean and the standard deviation of a continuous distribution measure the same properties as the mean and the standard deviation of a probability distribution—the expected value of a random variable having the given distribution, and the expected value of its squared deviations from the mean. More intuitively, the mean μ of a continuous distribution is a measure of its "center" or "middle," and the standard deviation σ of a continuous distribution measures its "dispersion" or "spread."

8.2
The Normal Distribution

Among the many continuous distributions used in statistics, the **normal distribution** is by far the most important. Its study dates back to eighteenth-century investigations into the nature of experimental errors. It was observed that discrepancies among repeated measurements of the same physical quantity displayed a surprising degree of regularity; their patterns (distribution), it was found, could be closely approximated by a certain kind of continuous distribution curve, referred to as the "normal curve of errors" and attributed to the laws of chance. The mathematical properties of this kind of continuous distribution curve and its theoretical basis were first investigated by Pierre Laplace (1749–1827), Abraham de Moivre (1667–1745), and Karl Friedrich Gauss (1777–1855).

The graph of a normal distribution is a bell-shaped curve that extends indefinitely in both directions. Although this may not be apparent from a small drawing like that of Figure 8.5, the curve comes closer and closer to the horizontal axis without ever reaching it, no matter how far we go in either direction away from the mean. Fortunately, it is seldom necessary to extend

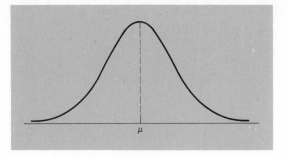

FIGURE 8.5
Normal distribution.

the tails of a normal distribution very far because the area under that part of the curve lying more than 4 or 5 standard deviations away from the mean is for most practical purposes negligible. This can be seen from the values given at the bottom of Table I on page 455.

An important feature of a normal distribution is that its mathematical equation is completely determined if we know its mean and its standard deviation; in other words, there is one and only one normal distribution with a given mean μ and a given standard deviation σ. In practice, we find areas under the graph of a normal distribution, or simply **areas under a normal curve,** in special tables, such as Table I at the end of the book. To use this table, we must understand what is meant by the normal distribution in its **standard form,** or as it is usually called, the **standard normal distribution.**

Since the equation of the normal distribution depends on μ and σ, we get different curves and, hence, different areas for different values of μ and σ.

EXAMPLE   Figure 8.6 shows the superimposed graphs of two normal distributions, one having $\mu = 10$ and $\sigma = 5$ and the other having $\mu = 20$ and $\sigma = 10$. The area under the curve, say, between 12 and 15, is obviously not the same for the two distributions.

FIGURE 8.6
Two normal distributions.

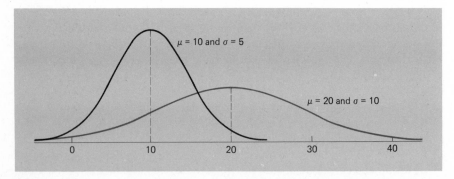

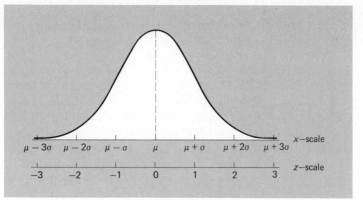

FIGURE 8.7

Change of scale.

It is physically impossible, but also unnecessary, to construct separate tables of normal-curve areas for all conceivable pairs of values of μ and σ; instead, we tabulate these areas only for a normal distribution having $\mu = 0$ and $\sigma = 1$, the so-called standard normal distribution. Then, we obtain areas under any normal curve by performing a simple change of scale (see Figure 8.7), in which we convert the units of measurement into **standard units, standard scores,** or **z-scores,** by means of the formula

Standard units

$$z = \frac{x - \mu}{\sigma}$$

In this new scale, z simply tells us how many standard deviations the corresponding x-value lies above or below the mean.

EXAMPLE   To find the area between 12 and 15 under the normal curve with $\mu = 10$ and $\sigma = 5$ in Figure 8.6, we determine the area under the standard normal distribution between

$$z = \frac{12 - 10}{5} = 0.40 \quad \text{and} \quad z = \frac{15 - 10}{5} = 1.00$$

Similarly, to find the area between 12 and 15 under the normal curve with $\mu = 20$ and $\sigma = 10$ in Figure 8.6, we determine the area under the standard normal distribution between

$$z = \frac{12 - 20}{10} = -0.80 \quad \text{and} \quad z = \frac{15 - 20}{10} = -0.50$$

8.2   The Normal Distribution

191

In the first case, the z-values are both positive because 12 and 15 both exceed $\mu = 10$, and in the second case they are both negative because 12 and 15 are both less than $\mu = 20$. (See also Figure 8.10.)

The table we use in problems like this is Table I at the end of the book, whose entries are the areas under the standard normal distribution between the mean ($z = 0$) and $z = 0.00, 0.01, 0.02, 0.03, \ldots, 3.08$, and 3.09, and also $z = 4.0$, $z = 5.0$, and $z = 6.0$. In other words, the entries in Table I are areas under the standard normal distribution like the white area in Figure 8.8. Table I has no entries corresponding to negative values of z, for these are not needed by virtue of the symmetry of any normal distribution curve about its mean.

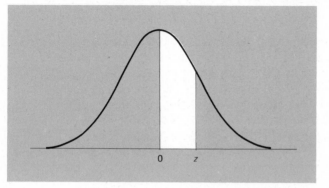

0   z

FIGURE 8.8

Tabulated areas under the graph of the standard normal distribution.

EXAMPLE To find the area under the standard normal distribution between $z = -1.26$ and $z = 0$, we look up the area between $z = 0$ and $z = 1.26$. In Table I this area is 0.3962.

Questions concerning areas under normal distributions arise in various ways, and the ability to find any desired area quickly can be a big help. Although the table gives only areas between $z = 0$ and selected positive values of z, we often have to find areas to the left or to the right of given positive or negative values of z, or areas between two given values of z. This is easy, provided we remember exactly what areas are represented by the entries in Table I, and also that the standard normal distribution is symmetrical about $z = 0$, so that the area to the left of $z = 0$ and the area to the right of $z = 0$ are both equal to 0.5000.

EXAMPLE The probability of getting a z less than 0.94 (the area under the curve to the left of $z = 0.94$) is $0.5000 + 0.3264 = 0.8264$, and the probability of getting a z greater than -0.65 (the area under the curve to the right of $z = -0.65$) is $0.5000 + 0.2422 = 0.7422$ (see Figure 8.9). Similarly, the probability of getting a z greater than 1.76 is $0.5000 - 0.4608 = 0.0392$, and the probability of getting a z less than -0.85 is $0.5000 - 0.3023 = 0.1977$ (see Figure 8.9). The

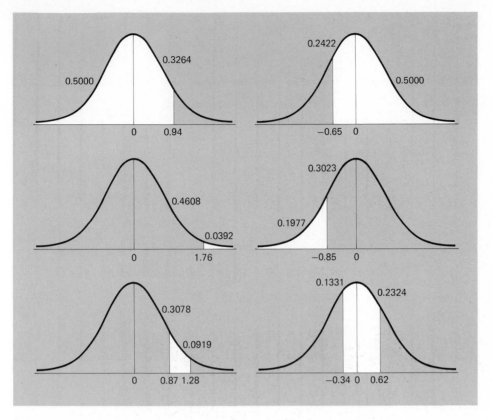

FIGURE 8.9

Areas under normal distributions.

probability of getting a z between 0.87 and 1.28 is $0.3997 - 0.3078 = 0.0919$, and the probability of getting a z between -0.34 and 0.62 is $0.1331 + 0.2324 = 0.3655$ (see Figure 8.9).

EXAMPLE For the normal distribution with $\mu = 10$ and $\sigma = 5$, the probability of getting a value between 12 and 15 is given by the area of the white region of the first diagram of Figure 8.10. Using the result obtained on page 191 and looking up the areas corresponding to $z = 0.40$ and $z = 1.00$, we find by subtraction that the area under the curve between $z = 0.40$ and $z = 1.00$ is $0.3413 - 0.1554 = 0.1859$. Similarly, for the normal distribution with $\mu = 20$ and $\sigma = 10$, the probability of getting a value between 12 and 15 is given by the area of the white region of the second diagram of Figure 8.10. Looking up the areas corresponding to $z = 0.50$ and $z = 0.80$, we find by subtraction that the area under the curve between $z = -0.80$ and $z = -0.50$ (the values we obtained on page 191) is $0.2881 - 0.1915 = 0.0966$.

8.2 The Normal Distribution

193

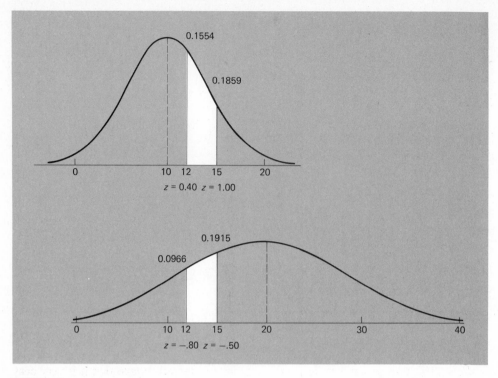

FIGURE 8.10

Areas under normal distributions.

There are also problems in which we are given areas under normal curves and are asked to find the corresponding values of *z*, as in the following example.

EXAMPLE To find a *z* which is such that the area under the curve to its right is 0.1500, we observe from Figure 8.11 that this *z* corresponds to an entry of $0.5000 - 0.1500 = 0.3500$ in Table I. Referring to this

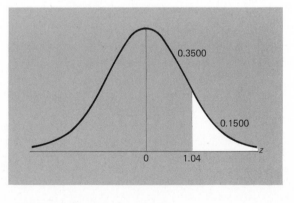

FIGURE 8.11

Normal distribution.

table, we find that the entry closest to 0.3500 is 0.3508, and that the corresponding z is 1.04.

Table I also enables us to verify the remark often heard that **for reasonably symmetrical bell-shaped distributions, about 68 percent of the values will fall within one standard deviation of the mean, about 95 percent will fall within two standard deviations of the mean, and over 99 percent will fall within three standard deviations of the mean.** These figures apply to normal distributions, and in parts (a), (b), and (c) of Exercise 6 on page 198 the reader will be asked to show that 0.6826 of the area under the standard normal distribution falls between $z = -1$ and $z = 1$, that 0.9544 of the area falls between $z = -2$ and $z = 2$, and that 0.9974 of the area falls between $z = -3$ and $z = 3$. The results of parts (d) and (e) of that exercise also provide a basis for our earlier remark that (although the "tails" extend indefinitely in both directions) the area under a normal curve beyond 4 or 5 standard deviations from the mean is negligible.

8.3

A Check for "Normality" ★

There are various ways in which we can test whether an observed frequency distribution fits the over-all pattern of a normal curve. The one we shall discuss here is not the best; it is largely subjective, but it has the decided advantage that it is very easy to perform.

EXAMPLE   To illustrate this technique, let us refer again to the sulfur oxides emission data, which we used as an example in Chapters 2 and 3. If we divide each of the cumulative frequencies in the table on page 16 by 80, the total frequency, and then multiply by 100 to express the figures as percentages, we obtain the following cumulative "less than" percentage distribution:

Tons of sulfur oxides	Cumulative percentage
Less than 4.95	0.00
Less than 8.95	3.75
Less than 12.95	16.25
Less than 16.95	33.75
Less than 20.95	65.00
Less than 24.95	86.25
Less than 28.95	97.50
Less than 32.95	100.00

To eliminate the gaps between successive classes, we used the class boundaries here instead of the class limits.

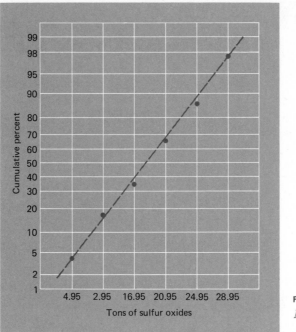

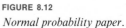

FIGURE 8.12

Normal probability paper.

Before we plot this cumulative percentage distribution on the special graph paper shown in Figure 8.12, let us briefly examine its scales. As can be seen from Figure 8.12, the cumulative percentage scale is already printed on the graph paper in the special way which makes it suitable for our purpose. The other scale consists of equal subdivisions. This kind of graph paper is called **normal probability paper,** or **arithmetic probability paper,** and it can be obtained in the bookstores of most colleges and universities.

If we plot on this kind of graph paper the cumulative "less than" percentages which correspond to the class boundaries of a distribution and the points thus obtained follow the general pattern of a straight line, we can consider this as evidence that the distribution has roughly the shape of a normal distribution.

EXAMPLE   Returning to our example and plotting the cumulative percentages as in Figure 8.12, we find that the points are all close to the dashed line, and hence conclude that the distribution of the sulfur oxides emission data has roughly the shape of a normal distribution. Observe that in Figure 8.12 we did not plot the cumulative percentages which correspond to 4.95 and 32.95; as we have pointed out earlier, we never quite reach 0 or 100 percent of the area under the curve, no matter how far we go in either direction away from the mean.

It must be understood that the method described here is only a crude (and highly subjective) way of checking whether a distribution follows the pattern of a normal curve; indeed, only large and obvious departures from a straight line constitute real evidence that a distribution does not follow the pattern of a normal curve. A more objective method of checking for "normality" will be given in Chapter 13.

EXERCISES

1 Suppose that a continuous random variable takes on values on the interval from 2 to 10 and that the graph of its distribution, called a **uniform density**, is given by the horizontal line of Figure 8.13.

   (a) What probability is represented by the white region of the diagram and what is its value?

   (b) What is the probability that the random variable will take on a value less than 7? Would the answer be the same if we asked for the probability that the random variable will take on a value less than or equal to 7?

   (c) What is the probability that the random variable will take on a value between 2.7 and 8.8?

2 Suppose that a continuous random variable takes on values on the interval from 0 to 4 and that the graph of its distribution, called a **triangular density**, is given by the line of Figure 8.14.

   (a) Verify that the total area under the curve is equal to 1.

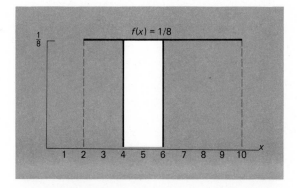

FIGURE 8.13
Uniform density.

FIGURE 8.14
Triangular density.

(b) What is the probability that the random variable will take on a value less than 1?

(c) What is the probability that the random variable will take on a value greater than 2?

(d) What is the probability that the random variable will take on a value between 1.5 and 2.5?

3 Find the area under the standard normal distribution which lies
   (a) between $z = 0$ and $z = 0.87$;
   (b) between $z = -1.66$ and $z = 0$;
   (c) to the right of $z = 0.48$;
   (d) to the right of $z = -0.27$;
   (e) to the left of $z = 1.30$;
   (f) to the left of $z = -0.79$;
   (g) between $z = 0.55$ and $z = 1.12$;
   (h) between $z = -1.05$ and $z = -1.75$;
   (i) between $z = -1.95$ and $z = 0.44$.

4 Find the area under the standard normal distribution which lies
   (a) between $z = -0.72$ and $z = 0.75$;
   (b) to the right of $z = -2.20$;
   (c) to the left of $z = -0.15$;
   (d) between $z = 2.15$ and $z = 2.35$;
   (e) to the right of $z = 1.40$;
   (f) between $z = -0.36$ and $z = -0.24$;
   (g) to the left of $z = 0.93$.

5 Find z if
   (a) the normal-curve area between 0 and z is 0.4726;
   (b) the normal-curve area to the left of z is 0.9868;
   (c) the normal-curve area to the right of z is 0.7704;
   (d) the normal-curve area to the left of z is 0.3085;
   (e) the normal-curve area to the right of z is 0.1314;
   (f) the normal-curve area between $-z$ and z is 0.8502;
   (g) the normal-curve area between $-z$ and z is 0.9700.

6 Find the normal-curve area between $-z$ and z if
   (a) $z = 1.00$;   (e) $z = 5.00$;
   (b) $z = 2.00$;   (f) $z = 1.96$;
   (c) $z = 3.00$;   (g) $z = 2.33$;
   (d) $z = 4.00$;   (h) $z = 2.58$.

7 In later chapters we shall let z_α denote the value of z for which the area under the normal curve to its right is equal to α (Greek lowercase letter *alpha*). Find
   (a) $z_{0.10}$;   (d) $z_{0.02}$;
   (b) $z_{0.05}$;   (e) $z_{0.01}$;
   (c) $z_{0.025}$;   (f) $z_{0.005}$.

8 A random variable has a normal distribution with the mean $\mu = 80.0$ and the standard deviation $\sigma = 4.8$. What are the probabilities that this random variable will take on a value
   (a) less than 87.2?   (c) between 81.2 and 86.0?
   (b) greater than 76.4?   (d) between 71.6 and 88.4?

9 A normal distribution has the mean $\mu = 62.4$. Find its standard deviation if 20 percent of the area under the curve lies to the right of 79.2.

10　A random variable has a normal distribution with the standard deviation $\sigma = 5$. Find its mean if the probability is 0.8264 that the random variable will take on a value less than 52.5.

11　Convert the distribution of Exercise 3 on page 22 into a cumulative "less than" percentage distribution, and use normal probability paper to judge whether the distribution has roughly the shape of a normal curve.

12　Convert the distribution of part (b) of Exercise 9 on page 57 into a cumulative "less than" percentage distribution, and use normal probability paper to judge whether the distribution has roughly the shape of a normal curve.

13　Plot the cumulative "less than" percentage distribution of whichever data you grouped among those of Exercises 8, 9, and 10 on page 23 on normal probability paper, and judge whether the distribution has roughly the shape of a normal curve.

14　Normal probability paper can be used to obtain crude estimates of the mean and the standard deviation of a distribution which has roughly the shape of a normal curve. To estimate the mean, we have only to observe that since the normal distribution is symmetrical about the mean, the area under the curve to the left of the mean is 0.5000. Hence, if we check the 50 percent mark on the vertical scale and go horizontally to the line we fit to the points (for instance, the dashed line of Figure 8.12), then the corresponding value on the horizontal scale provides an estimate of the mean of the distribution. To estimate the standard deviation, we observe that the areas under the curve to the left of $z = -1$ and $z = +1$ are roughly 0.16 and 0.84. Hence, if we check 16 percent and 84 percent on the vertical scale, we can judge by the straight line we have fitted to the points what values on the horizontal scale correspond to $z = -1$ and $z = +1$; their difference divided by 2 provides an estimate of the standard deviation of the distribution.

    (a) Use this method to estimate the mean and the standard deviation of the distribution of the sulfur oxides emission data from Figure 8.12. Compare the results with the exact values previously obtained in the text.

    (b) Use this method to estimate the mean and the standard deviation of the distribution of part (b) of Exercise 9 on page 57.

15　Another continuous distribution, called the **exponential distribution,** has many important applications. If a random variable has an exponential distribution with the mean μ, the probability that it will take on a value between 0 and any given positive value x is $1 - e^{-x/\mu}$ (see Figure 8.15). Here e is the constant which

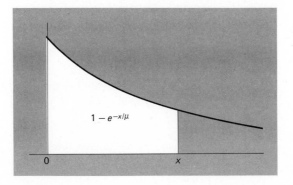

FIGURE 8.15

Exponential distribution.

appears also in the formula for the Poisson distribution, and values of $e^{-x/\mu}$ can be obtained directly from Table XII.

  (a) Find the probabilities that a random variable having an exponential distribution with $\mu = 10$ takes on a value less than 3, a value between 4 and 6, a value greater than 15.

  (b) The lifetime of a certain electronic component is a random variable which has an exponential distribution with a mean of 2,000 hours. What is the probability that such a component will last at most 1,800 hours? What is the probability that such a component will last anywhere from 4,000 hours to 5,000 hours?

  (c) According to medical research, the time between successive reports of a rare tropical disease is a random variable having an exponential distribution with a mean of 120 days. What is the probability that the time between successive reports of the disease will exceed 48 days?

8.4

Applications of the Normal Distribution

Let us now consider some applied problems in which it will be assumed that the distributions of the data can be approximated closely with normal curves.

EXAMPLE Suppose that the amount of cosmic radiation to which a person is exposed while flying by jet across the United States is a random variable having a normal distribution with a mean of 4.35 mrem and a standard deviation of 0.59 mrem.[†] What is the probability that a person will be exposed to more than 5.00 mrem of cosmic radiation on such a flight? The answer to this question is given by the area of the white region of the first diagram of Figure 8.16; that is, the area under the curve to the right of

$$z = \frac{5.00 - 4.35}{0.59} = 1.10$$

Since the entry in Table I corresponding to $z = 1.10$ is 0.3643, we find that the probability is $0.5000 - 0.3643 = 0.1357$, or approximately 0.14, that a person will be exposed to more than 5.00 mrem of cosmic radiation on such a flight.

Continuing with this example, let us also determine the probability that a passenger on a jet flight across the United States will be exposed to anywhere from 3.00 to 4.00 mrem of cosmic radiation.

[†] This unit of radiation stands for "milli (1/1,000) roentgen equivalent man."

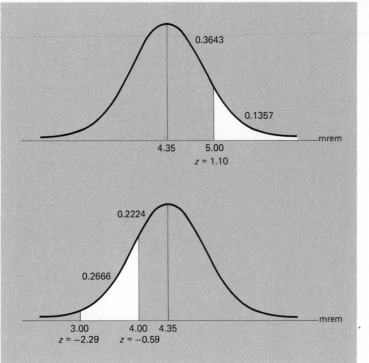

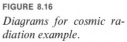

FIGURE 8.16
Diagrams for cosmic radiation example.

This probability is given by the area of the white region of the second diagram of Figure 8.16, namely, the area between

$$z = \frac{3.00 - 4.35}{0.59} = -2.29 \quad \text{and} \quad z = \frac{4.00 - 4.35}{0.59} = -0.59$$

Since the corresponding entries in Table I are 0.4890 for $z = 2.29$ and 0.2224 for $z = 0.59$, we find that the probability is 0.4890 − 0.2224 = 0.2666, or approximately 0.27.

EXAMPLE   Suppose that the actual amount of instant coffee which a filling machine puts into "6-ounce" jars varies from jar to jar, and it may be looked upon as a random variable having a normal distribution with a standard deviation of 0.04 ounce. If only 2 percent of the jars are to contain less than 6 ounces of coffee, what must be the mean fill of these jars? In this example we are given $\sigma = 0.04$, a normal curve area (that of the white region of Figure 8.17), and we are asked to find μ. Since the value of z for which the entry in

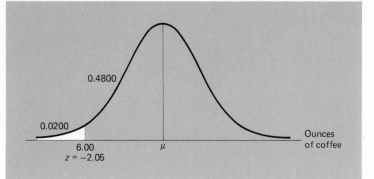

0.4800

0.0200

6.00
z = −2.05

μ

Ounces
of coffee

FIGURE 8.17

*Diagram for instant coffee
filling example.*

Table I comes closest to 0.5000 − 0.0200 = 0.4800 is 2.05, we have

$$-2.05 = \frac{6.00 - \mu}{0.04}$$

and, solving for μ, we get

$$6.00 - \mu = -2.05(0.04) = -0.082$$

and then

$$\mu = 6.00 + 0.082 = 6.082 \text{ ounces}$$

or 6.08 ounces to the nearest hundredth of an ounce. This "giveaway" may not be satisfactory so far as the coffee processor is concerned, but it can be reduced to $\mu = 6.05$ ounces if the variability of the filling machine can be reduced to $\sigma = 0.025$ ounce, or to $\mu = 6.02$ ounces if the variability of the filling machine can be reduced to $\sigma = 0.01$ ounce (see Exercise 6 on page 207).

Although, strictly speaking, the normal distribution applies to continuous random variables, it is often used to approximate distributions of **discrete random variables,** which can take on only a finite number of values or as many values as there are positive integers. In many situations this yields satisfactory results, provided that we make the **continuity correction** illustrated in the following example.

EXAMPLE   In a study of aggressive behavior, male white mice, returned to the group in which they live after five weeks of isolation, averaged 18.6 fights in the first five minutes with a standard deviation of 3.3 fights. If it can be assumed that the distribution of this random variable (the number of fights into which such a mouse gets under the

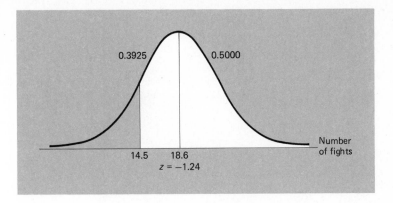

FIGURE 8.18
Diagram for example dealing with aggressive behavior of mice.

stated conditions) can be approximated closely with a normal distribution, what is the probability that such a mouse will get into at least 15 fights in the first five minutes? The answer is given by the area of the white region of Figure 8.18; the area to the right of 14.5, not 15. The reason for this is that the number of fights in which such a mouse gets involved is a whole number. Hence, if we want to approximate the distribution of this random variable with a normal curve, we must "spread" its values over a continuous scale, and we do this by representing each whole number k by the interval from $k - \frac{1}{2}$ to $k + \frac{1}{2}$. For instance, 5 is represented by the interval from 4.5 to 5.5, 10 is represented by the interval from 9.5 to 10.5, 20 is represented by the interval from 19.5 to 20.5, and the probability of 15 or more is given by the area under the curve to the right of 14.5. Accordingly, we get

$$ z = \frac{14.5 - 18.6}{3.3} = -1.24 $$

and it follows from Table I that the area of the white region of Figure 8.18—the probability that such a mouse will get into at least 15 fights in the first five minutes—is $0.5000 + 0.3925 = 0.8925$, or approximately 0.89.

All the examples of this section dealt with random variables having normal distributions, or distributions which can be approximated closely with normal curves. Whenever we observe a value (or values) of a random variable having a normal distribution, we say that we are sampling a **normal population**; this is consistent with the terminology introduced at the end of Section 7.3.

8.4 **Applications of the Normal Distribution**

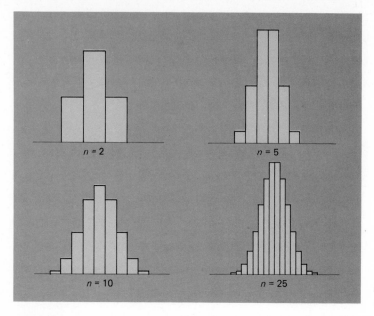

Binomial distributions with
$p = \frac{1}{2}$.

8.5

The Normal Approximation to the Binomial Distribution

The normal distribution is sometimes introduced as a continuous distribution which provides a very close approximation to the binomial distribution when *n*, the number of trials, is very large and *p*, the probability of a success on an individual trial, is close to $\frac{1}{2}$. Figure 8.19 shows the histograms of binomial distributions having $p = \frac{1}{2}$ and $n = 2, 5, 10$, and 25, and it can be seen that with increasing *n* these distributions approach the symmetrical bell-shaped pattern of the normal distribution. In fact, a normal curve with the mean $\mu = np$ and the standard deviation $\sigma = \sqrt{np(1 - p)}$ can often be used to approximate a binomial distribution even when *n* is fairly small and *p* differs from $\frac{1}{2}$, but is not too close to either 0 or 1. A good rule of thumb is to use this approximation only when np and $n(1 - p)$ are both greater than 5.

EXAMPLE To illustrate the normal-curve approximation to the binomial distribution, let us consider the probability of getting 6 heads and 10 tails in 16 flips of a balanced coin. To find the exact value, we substitute $n = 16$, $x = 6$, and $p = \frac{1}{2}$ into the formula for the binomial distribution, getting

$$f(6) = \binom{16}{6} \left(\frac{1}{2} \right)^6 \left(1 - \frac{1}{2} \right)^{10} = 8{,}008 \left(\frac{1}{2} \right)^{16} = \frac{8{,}008}{65{,}536}$$

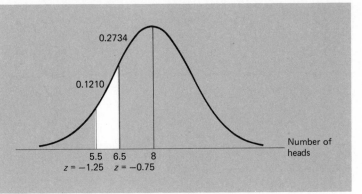

FIGURE 8.20

Normal-curve approximation to binomial distribution.

or 0.1222 rounded to four decimals. To find the normal-curve approximation to this probability, we use the continuity correction and represent 6 heads by the interval from 5.5 to 6.5 (see Figure 8.20). Since $\mu = 16 \cdot \frac{1}{2} = 8$ and $\sigma = \sqrt{16 \cdot \frac{1}{2} \cdot \frac{1}{2}} = 2$, we have in standard units $z = \dfrac{5.5 - 8}{2} = -1.25$ for 5.5 and $z = \dfrac{6.5 - 8}{2} = -0.75$ for 6.5. The corresponding entries in Table I are 0.3944 and 0.2734, and the approximate probability of 6 heads and 10 tails in 16 flips of a balanced coin is $0.3944 - 0.2734 = 0.1210$, only 0.0012 less than the exact value rounded to four decimals.

The normal-curve approximation to the binomial distribution is particularly useful in problems where we would otherwise have to use the formula for the binomial distribution repeatedly to obtain the values of many different terms.

EXAMPLE Suppose we want to know the probability that at least 70 of 100 mosquitos will be killed by a new insect spray, when the probability is 0.75 that any one of them will be killed by the spray. In other words, we want to know the probability of getting at least 70 successes in 100 trials when the probability of a success on an individual trial is 0.75. If we tried to solve this problem by using the formula for the binomial distribution, we would have to find the sum of the probabilities corresponding to 70, 71, 72, . . . , 99, and 100 successes. This would obviously involve a tremendous amount of work, but using the normal-curve approximation, we need only find the area of the white region of Figure 8.21, the area to the right of 69.5. We are again using the continuity correction according to which 70 is represented by the interval from 69.5 to 70.5, 71 is represented by the interval from 70.5 to 71.5, and so on.

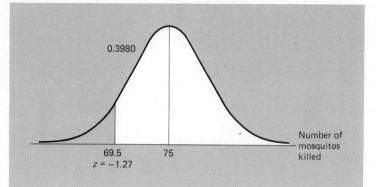

0.3980

69.5   75

z = −1.27

Number of mosquitos killed

FIGURE 8.21

Normal-curve approxima-tion to binomial distribu-tion.

Since $\mu = 100(0.75) = 75$ and $\sigma = \sqrt{100(0.75)(0.25)} = 4.33$, we find that in standard units 69.5 becomes

$$z = \frac{69.5 - 75}{4.33} = -1.27$$

and that the probability is $0.3980 + 0.5000 = 0.8980$. The actual value of the probability, looked up in the tables by Romig listed in the Bibliography at the end of Chapter 7, is 0.8962. So, the error of the approximation is only 0.0018.

EXERCISES

1 With reference to the example on page 200, find the probabilities that a person flying by jet across the United States will be exposed to
(a) less than 3.50 mrem of cosmic radiation;
(b) anywhere from 4.10 to 4.50 mrem of cosmic radiation.

2 Suppose that during periods of transcendental meditation the reduction of a person's oxygen consumption may be looked upon as a random variable having a normal distribution with $\mu = 38.6$ cc per minute and $\sigma = 4.3$ cc per minute. Find the probabilities that during a period of transcendental meditation a person's oxygen consumption will be reduced by
(a) at least 45.5 cc per minute;
(b) at most 36.0 cc per minute;
(c) anywhere from 32.0 to 42.0 cc per minute;
(d) anywhere from 40.0 to 45.0 cc per minute.

3 The burning time of an experimental rocket is a random variable which has a normal distribution with $\mu = 4.36$ seconds and $\sigma = 0.04$ second. What are the probabilities that this kind of rocket will burn for
(a) less than 4.25 seconds;
(b) more than 4.40 seconds;
(c) 4.30 to 4.42 seconds?

4 The lengths of the sardines received by a cannery have a mean of 4.54 inches and a standard deviation of 0.25 inch. If the distribution of these lengths has roughly

the shape of a normal distribution, what percentage of all these sardines are
  (a) longer than 5.00 inches?
  (b) shorter than 4.00 inches?
  (c) from 4.40 to 4.60 inches long?

5 In a photographic process, the developing time of prints may be looked upon
   as a random variable having a normal distribution with a mean of 15.28 seconds
   and a standard deviation of 0.12 second. Find the probabilities that it will take
   (a) at least 15.50 seconds to develop one of the prints;
   (b) at most 15.00 seconds to develop one of the prints;
   (c) from 15.10 to 15.40 seconds to develop one of the prints;
   (d) from 15.05 to 15.15 seconds to develop one of the prints.

6 With reference to the filling-machine example on page 201, verify that about
   2 percent of the jars will contain less than 6 ounces of coffee if
   (a) $\mu = 6.05$ ounces and $\sigma = 0.025$ ounce;
   (b) $\mu = 6.02$ ounces and $\sigma = 0.01$ ounce.

7 With reference to the example on page 202, find the probabilities that (under the
   given conditions) a mouse will get into
   (a) at most 20 fights;
   (b) anywhere from 16 to 22 fights.

8 If the yearly number of major earthquakes, the world over, is a random variable
   whose distribution can be closely approximated with a normal distribution
   having $\mu = 20.8$ and $\sigma = 4.5$, find the probabilities that there will be
   (a) exactly 18 major earthquakes in any given year;
   (b) at least 22 major earthquakes in any given year;
   (c) from 20 to 25 earthquakes, inclusive, in any given year.

9 A taxicab driver knows from experience that the number of fares he will pick
   up in an evening is a random variable with $\mu = 23.7$ and the standard deviation
   $\sigma = 4.2$. Assuming that the distribution of this random variable can be approximated with a normal distribution, find the probabilities that in an evening the
   driver will pick up
   (a) exactly 20 fares;
   (b) at least 18 fares;
   (c) at most 25 fares;
   (d) from 15 to 21 fares.

10 Use the normal-curve approximation to find the probability of getting 7 heads
   and 7 tails in 14 flips of a balanced coin, and compare the result with the exact
   value (rounded to three decimals) in Table VI.

11 A student answers each of the 48 questions on a multiple-choice test, each with
   four possible answers, by randomly drawing a card from an ordinary deck of
   52 playing cards and checking the first, second, third, or fourth answer depending
   on whether the card drawn is a spade, heart, diamond, or club. Use the normal-
   curve approximation to find the probabilities that the student will get
   (a) exactly 15 correct answers;
   (b) at least 15 correct answers.
   Compare these results with the corresponding exact values, which are 0.0767 and
   0.1999 (rounded to four decimals) according to the National Bureau of Standards
   Tables listed in the Bibliography at the end of Chapter 7.

8.5 The Normal Approximation to the Binomial Distribution

12 If 70 percent of all persons flying across the Atlantic Ocean feel the effect of the time difference for at least 24 hours, what is the probability that among 150 persons flying across the Atlantic Ocean, at least 100 will feel the effect of the time difference for at least 24 hours?

13 If 22 percent of all patients with high blood pressure have bad side effects from a certain kind of medicine, what is the probability that among 120 patients with high blood pressure who are treated with this medicine more than 30 will have bad side effects?

14 If 62 percent of all clouds seeded with silver iodide show spectacular growth, what is the probability that among 40 clouds seeded with silver iodide at most 20 will show spectacular growth?

15 To illustrate the law of large numbers which we mentioned in Section 4.3, find the probabilities that the proportion of heads will be anywhere from 0.49 and 0.51 when a balanced coin is flipped
   (a) 100 times;
   (b) 1,000 times;
   (c) 10,000 times.

8.6

Technical Note (Simulation) ★

There are many ways in which we can simulate observations of continuous random variables, but the theory on which they are based are beyond the scope of this text. Limiting ourselves to random variables having normal distributions, we shall illustrate here only the use of published tables of **random normal numbers,** also called **random normal deviates.** Such tables consist of many pages on which numbers (rounded to three decimals in Table X at the end of the book) are set down in a random fashion, much as they would appear if they were generated one at a time by a chance or gambling device which "produces" values of a random variable having the standard normal distribution.

To simulate values of a random variable having a normal distribution with a given mean μ and a given standard deviation σ, we look up values of z in a table of random normal numbers, and then change them into values of x, the random variable under consideration, by using the formula

Changing from standard units

$$x = \mu + \sigma z$$

This formula may be obtained by solving for x the equation $z = \dfrac{x - \mu}{\sigma}$.

As in the use of tables of ordinary random numbers, we should always be fairly "random" in choosing the page and the place from which we start.

EXAMPLE Suppose we want to simulate five values of a random variable having a normal distribution with $\mu = 41.3$ and $\sigma = 12.5$. If we arbitrarily use the eighth column of the table on page 475 starting with the sixth row, we get -0.745, 0.655, -1.115, 0.027, and -2.520, and the corresponding values of the random variable are $41.3 + 12.5(-0.745) = 32.0$, $41.3 + 12.5(0.655) = 49.5$, $41.3 + 12.5(-1.115) = 27.4$, $41.3 + 12.5(0.027) = 41.6$, and $41.3 + 12.5(-2.520) = 9.8$.

EXERCISES

1 The amount of time it takes a person to learn how to operate a certain machine is a random variable having a normal distribution with $\mu = 5.6$ hours and $\sigma = 1.2$ hours. Simulate the amount of time it takes eight persons to learn how to operate the machine.

2 The distribution of the grades which college-bound high school seniors get on a certain standardized test can be approximated closely by a normal distribution with $\mu = 54.3$ and $\sigma = 6.2$. Rounding the results to the nearest whole numbers, simulate the grades which 12 college-bound high school seniors get on this test.

3 Suppose that the increase in the pulse rate of a person performing a certain task is a random variable whose distribution can be closely approximated by a normal distribution with $\mu = 28.40$ and $\sigma = 4.17$. Simulate the increase in the pulse rate of 20 persons performing this task. Round the results to the nearest whole numbers.

4 The distribution of the weights of the grapefruits shipped by a large orchard can be closely approximated by a normal distribution with $\mu = 19.6$ ounces and $\sigma = 2.2$ ounces. Simulate the weight of 24 grapefruits shipped by this orchard.

BIBLIOGRAPHY

More detailed tables of normal-curve areas may be found in many handbooks of statistical tables; for instance, in

PEARSON, E. S., and HARTLEY, H. O., *Biometrika Tables for Statisticians*, *3rd ed.* Cambridge: Cambridge University Press, 1966.

Extensive tables of random normal numbers are given in the RAND Corporation tables listed on page 185. The construction of such tables is discussed in the introduction to these tables and also in Chapter 4 of

MILLER, I., and FREUND, J. E., *Probability and Statistics for Engineers*, *2nd ed.* Englewood Cliffs, N.J.: Prentice-Hall, Inc., 1977.

9

Sampling and Sampling Distributions

In most statistical investigations, the main purpose is to make sound generalizations based on samples about the parameters of the populations from which the samples came. Note the word "sound," because the question of when and under what conditions samples permit such generalizations is not easily answered. For instance, if we want to estimate the average amount of money that persons spend on their vacations, would we take as our sample the amounts spent by the deluxe-class passengers on a 30-day ocean cruise; or would we attempt to estimate, or predict, the wholesale prices of all farm products on the basis of the prices of fresh asparagus alone? Obviously not, but just which vacationers and which farm products we should include in our samples, and how many of them, is neither intuitively clear nor self-evident.

In most of the methods which we shall study in the remainder of this book, it will be assumed that we are dealing with a particular kind of sample called a **random sample**. This attention to random samples is due to their permitting valid, or logical, generalizations, and hence they are widely used in practice. As we shall see, however, random sampling is not always practical, feasible, or even desirable, and some other sampling procedures will be mentioned briefly in Sections 9.3 through 9.6.

9.1

Finite and Infinite Populations

In the beginning of Chapter 3 we distinguished between populations and samples, stating that a population consists of all conceivably or hypothetically possible observations (instances, or occurrences) of a given phenomenon, while a sample is simply part of a population. For the work which follows, let us now distinguish between two kinds of populations—**finite populations** and **infinite populations.**

A finite population is one which consists of a finite number, or fixed number, of elements (items, objects, measurements, or observations). Examples of finite populations are the net weights of the 10,000 cans of paint in a production lot, the SAT scores of all the freshmen admitted to a certain university in 1979, and the daily high temperatures recorded at a weather station during the years 1975–1978.

In contrast to finite populations, a population is said to be infinite if there is, hypothetically at least, no limit to the number of elements it can contain. The population which consists of the results of all hypothetically possible rolls of a pair of dice is an infinite population, and so is the population which consists of all conceivable possible measurements of the weight of a piece of rock.

9.2

Random Sampling

To introduce the idea of a random sample from a finite population, let us ask the following three questions: **1. How many distinct samples of size n can be drawn from a finite population of size N? 2. How is a random sample to be defined? 3. How can a random sample be drawn in actual practice?**

To answer the first question, we refer to the rule for combinations according to which r objects can be selected from a set of n distinct objects in $\binom{n}{r}$ ways. With a change of letters, we can say that the number of different samples of size n which can be drawn from a finite population of size N is $\binom{N}{n}$.

EXAMPLE For instance, $\binom{12}{2} = 66$ different samples of size $n = 2$ can be drawn from a finite population of size $N = 12$, and $\binom{100}{3} = 161{,}700$ different samples of size $n = 3$ can be drawn from a finite population of size $N = 100$.

To answer the second question, we make use of the answer to the first one and define a **simple random sample** (or more briefly, a **random sample**) from a finite population as **a sample which is chosen in such a way that each of the** $\binom{N}{n}$ **possible samples has the same probability,** $1 \big/ \binom{N}{n}$, **of being selected.**

EXAMPLE If a finite population consists of the $N = 5$ elements a, b, c, d, and e (which might be the incomes of five persons, the weights of five guinea pigs, the prices of five commodities, and so on), there are $\binom{5}{3} = 10$ possible distinct samples of size $n = 3$; they consist of the elements abc, abd, abe, acd, ace, ade, bcd, bce, bde, and cde. If we choose one of these samples in such a way that each has the probability $\frac{1}{10}$ of being chosen, we call this sample a random sample.

With regard to the third question of how to take a random sample in actual practice, we could, in simple cases like the one above, write each of the $\binom{N}{n}$ possible samples on a slip of paper, put these slips into a hat, shuffle them thoroughly, and then draw one without looking. Such a procedure is obviously impractical, if not impossible, in more realistically complex problems of sampling; we mentioned it here only to make the point that the selection of a random sample must depend entirely on chance.

Fortunately, we can take a random sample without actually resorting to the tedious process of listing all possible samples. We can list instead the N individual elements of a finite population, and then take a random sample by choosing the elements to be included in the sample one at a time, making sure that in each of the successive drawings each of the remaining elements of the population has the same chance of being selected. In part (b) of Exercise 9 on page 215 the reader will be asked to show that this leads to the same probability, $1 \big/ \binom{N}{n}$, for each possible sample.

EXAMPLE To take a random sample of 12 of a city's 247 drugstores, we could write each store's name (address, or some other business identification number) on a slip of paper, put the slips of paper into a box or a bag and mix them thoroughly, and then draw (without looking) 12 of the slips one after the other without replacement.

Even this relatively easy procedure can be simplified in actual practice; usually, the simplest way to take a random sample from a finite population is to refer to a table of random numbers. As we pointed out on page 181, published tables of random numbers (such as the one from which Table IX of this book is excerpted) consist of pages on which the digits 0, 1, 2, . . . , and 9 are set down in much the same fashion as they might appear if they had been generated by a chance or gambling device giving each digit the same probability of $\frac{1}{10}$ of appearing at any given place in the table.

EXAMPLE   To illustrate the use of random numbers in random sampling, let us refer again to the problem of choosing 12 of a city's 247 drugstores. Numbering the stores 001, 002, 003, . . . , 246, and 247 (say, in the order in which they are listed in the telephone directory), we arbitrarily enter the table on page 471 and read out the digits in the 26th, 27th, and 28th columns, starting with the sixth row and going down the page. This gives us a random sample consisting of the 12 drugstores whose numbers are

046   230   079   022   119   150   056   064   193   232   040   146

In selecting these numbers, we ignored the tabled numbers greater than 247; and had any number recurred, we would have ignored it too.

When lists are available and items are readily numbered, it is easy to draw random samples from finite populations with the aid of random number tables. Unfortunately, however, it is often impossible to proceed in the way we have just described. For instance, if we want to use a sample to estimate the mean outside diameter of thousands of ball bearings packed in a large crate, or if we want to estimate the mean height of the trees in a forest, it would be impossible to number the ball bearings or the trees, choose random numbers, and then locate and measure the corresponding ball bearings or trees. In these and in many similar situations, all we can do is proceed according to the dictionary definition of the word "random," namely, "haphazardly, without aim or purpose." That is, we must not select or reject any element of a population because of its seeming typicalness or lack of it, nor must we favor or ignore any part of a population because of its accessibility or lack of it, and so forth. Hopefully, such haphazard procedures will nevertheless lead to samples which may be treated as if they were, in fact, random samples.

To this point we have discussed random sampling only in connection with finite populations. The concept of a random sample from an infinite population is more difficult to explain, but a few simple illustrations will show the basic characteristics of such a sample.

EXAMPLE   Suppose we consider the results of 15 tosses of a coin as a sample from the hypothetically infinite population which consists of the results of all possible tosses of the coin. If the probability of getting heads is the same for each toss and the 15 tosses are independent, we say that the sample is random.

EXAMPLE   We would also be sampling from an infinite population if we sample with replacement from a finite population, and our sample would be random if in each draw all elements of the population have the same probability of being selected, and successive draws are independent.

Generally speaking, we require that the selection of each item in a random sample from an infinite population is controlled by the same probabilities, and that successive selections are independent of one another.

EXERCISES

1   How many different samples of size $n = 2$ can be selected from a finite population of size
    (a) $N = 6$?   (b) $N = 10$?   (c) $N = 25$?

2   How many different samples of size $n = 3$ can be selected from a finite population of size
    (a) $N = 10$?   (b) $N = 25$?   (c) $N = 50$?

3   What is the probability of each possible sample if
    (a) a random sample of size 4 is to be drawn from a finite population of size 12?
    (b) a random sample of size 5 is to be drawn from a finite population of size 22?

4   Referring to the example on page 212, where a random sample of size 3 is drawn from the finite population which consists of the elements a, b, c, d, and e, find
    (a) the probability that any specific element (say, the element b) will be contained in the sample;
    (b) the probability that any specific pair of elements (say, the elements b and d) will be contained in the sample.

5   List all possible choices of four of the following six corporations: General Motors, IBM, Shell Oil, Coca-Cola, Polaroid, and American Air Lines. If a person randomly selects four of these corporations to invest in their stock, find
    (a) the probability of each possible sample;
    (b) the probability that any particular one of these corporations (say, Shell Oil) will be included in the sample;
    (c) the probability that any particular pair of these corporations (say, IBM and Polaroid) will be included in the sample.

6   A newspaper reporter wants to interview 15 of the 823 persons who signed a petition for the recall of a local government official. If these persons are numbered

serially from 1 through 823 on the petition, which ones (by number) would the reporter select for an interview, if she selects her sample using the first three columns of the table on page 470, going down the page beginning with the sixth row?

7   The employees of a company have badges numbered serially from 1 through 568. Use the 11th, 12th, and 13th columns of the table on page 471, going down the page starting with the 10th row, to select (by number) a random sample of eight of the company's employees to serve on a grievance committee.

8   Explain why each of the following samples does not qualify as a random sample from the required population or might otherwise fail to give the desired information:

   (a) To predict a municipal election, a public opinion poll telephones persons selected haphazardly from the city's telephone directory.

   (b) To determine public sentiment about certain foreign trade agreements, an interviewer asks voters: "Do you feel that this unfair practice should be stopped?"

   (c) To determine the average annual income of its graduates 10 years after graduation, a university's alumni office sent questionnaires in 1978 to all members of the class of 1968, and the estimate is based on the questionnaires returned.

   (d) To study executives' reaction to its copying machines, the Xerox corporation hires a research organization to ask executives the question: "How do you like using Xerox copies?"

   (e) To study consumer reaction to a new convenience food, a house-to-house survey is conducted during weekday mornings, with no provisions for return visits in case no one is at home.

9   On page 212 we said that a random sample can be drawn from a finite population by choosing the elements to be included in the sample one at a time, making sure that in each of the successive drawings each of the remaining elements of the population has the same chance of being selected. To verify that this will give the correct probability to each sample, let us refer to the example on page 212, where we dealt with random samples of size 3 drawn from the finite population which consists of the elements a, b, c, d, and e. To find the probability of drawing any particular sample (say, b, c, and e), we can argue that the probability of getting one of these three letters on the first draw is $\frac{3}{5}$, the probability of then getting one of the remaining two letters on the second draw is $\frac{2}{4}$, and the probability of then getting the third letter on the third draw is $\frac{1}{3}$. Multiplying these three probabilities, we find that the probability of getting the particular sample is $\frac{3}{5} \cdot \frac{2}{4} \cdot \frac{1}{3} = \frac{1}{10}$, and this agrees with the value obtained on page 215.

   (a) Use the same kind of argument to verify that for each possible random sample of size 3, drawn one at a time from a finite population of size 100,

$$\text{the probability is } 1 \Big/ \binom{100}{3} = \frac{1}{161,700}.$$

   (b) Use the same kind of argument to verify in general that for each possible random sample of size n, drawn one at a time from a finite population

$$\text{of size } N, \text{ the probability is } 1 \Big/ \binom{N}{n}.$$

10 Making use of the fact that among the $\binom{N}{n}$ samples of size n which can be drawn from a finite population of size N there are $\binom{N-1}{n-1}$ which contain a specific element, show that the probability that any specific element of the population will be contained in a random sample of size n is $\dfrac{n}{N}$.

9.3
Sample Designs ⋆

So far we discussed only random samples, and we did not even consider the possibility that under certain conditions there may be samples which are better (say, easier to obtain, cheaper, or more informative) than random samples, and we did not go into any details about the question of what might be done when random sampling is impossible. Indeed, there are many other ways of selecting a sample from a population, and there is an extensive literature devoted to the subject of designing sampling procedures.

In statistics, a **sample design** is a definite plan, completely determined before any data are actually collected, for obtaining a sample from a given population. Thus, the plan to take a simple random sample of 12 of a city's 247 drugstores by using a table of random numbers in a prescribed way constitutes a sample design. In what follows, we shall discuss briefly some of the most important kinds of sample designs.

9.4
Systematic Sampling ⋆

In some instances, the most practical way of sampling is to select, say, every 20th name on a list, every 12th house on one side of a street, every 50th piece coming off an assembly line, and so on. Sampling of this sort is called **systematic sampling,** and an element of randomness is usually introduced into this kind of sampling by using random numbers to pick the unit with which to start. Although a systematic sample may not be a random sample in accordance with the definition, it is often reasonable to treat systematic samples as if they were random samples; indeed, in some instances, systematic samples actually provide an improvement over simple random samples inasmuch as the samples are spread more evenly over the entire populations.

The real danger in systematic sampling lies in the possible presence of hidden periodicities. For instance, if we inspect every 40th piece made by a particular machine, the results would be very misleading if, because of a regularly recurring failure, every 10th piece produced by the machine is blemished. Also, a systematic sample might yield biased results if we interview the residents of every 10th house along a certain route and it so happens that each house thus chosen is a corner house on a double lot.

9.5

Stratified Sampling ★

When we know something about the makeup of a population (that is, if we know something about its composition) and this is of relevance to our investigation, we may be able to improve on random sampling by **stratification.** This is a procedure which consists of stratifying (or dividing) the population into a number of non-overlapping subpopulations, or **strata,** and then taking a sample from each stratum. If the items selected from each stratum constitute a simple random sample, the entire procedure (first stratification and then simple random sampling) is called **stratified (simple) random sampling.**

EXAMPLE Suppose we want to estimate the mean weight of four persons on the basis of a sample of size 2; the (unknown) weights of the four persons are 115, 135, 185, and 205 pounds, so that μ, the mean weight we want to estimate, is 160 pounds. If we take an ordinary random sample of size 2 from this population, the $\binom{4}{2} = 6$ possible samples are 115 and 135, 115 and 185, 115 and 205, 135 and 185, 135 and 205, and 185 and 205, and the corresponding means are 125, 150, 160, 160, 170, and 195. Observe that since each of these samples has the probability $\frac{1}{6}$, the probability is $\frac{2}{6}$ that the mean of such a sample will differ from $\mu = 160$ by as much as 35 and, hence, provide a poor estimate of the mean weight of the four persons.

Now suppose we know that two of the four persons are men and two are women, and we decide to stratify our sample (by sex) by randomly choosing one of the two men and one of the two women. Assuming that the two smaller weights are those of the two women, we find that there are only the four stratified samples 115 and 185, 115 and 205, 135 and 185, and 135 and 205, and that the corresponding means are 150, 160, 160, and 170. In this case, none of the means differs from $\mu = 160$ by more than 10 and our chances of getting a good (close) estimate of the mean weight of the four persons are greatly improved.[†]

Essentially, the goal of stratification is to form strata in such a way that there is some relationship between being in a particular stratum and the

---

[†] For more details about stratification in problems of statistical inference, see Exercise 36 on page 245 and Exercise 10 on page 258.

answer sought in the statistical study, and that within the separate strata there is as much homogeneity (uniformity) as possible. In our example there is such a connection between sex and weight and there is much less variability in weight within each of the two groups than there is within the entire population.

In the above example, we used **proportional allocation**, which means that the sizes of the samples from the different strata were proportional to the sizes of the strata. In general, if we divide a population of size N into k strata of size $N_1, N_2, \ldots,$ and N_k, and take a sample of size n_1 from the first stratum, a sample of size n_2 from the second stratum, $\ldots$, and a sample of size n_k from the kth stratum, we say that the allocation is proportional if

$$\frac{n_1}{N_1} = \frac{n_2}{N_2} = \cdots = \frac{n_k}{N_k}$$

or, at least, if these ratios are as nearly equal as possible. As can easily be verified (see Exercise 5 on page 221), this leads to the formula

Sample sizes for proportional allocation

$$n_i = \frac{N_i}{N} \cdot n \qquad for\ i = 1, 2, \ldots, and\ k$$

where n is the total size of the sample, that is, $n = n_1 + n_2 + \cdots + n_k$. When necessary, we use the integers closest to the values yielded by this formula.

EXAMPLE In the example which dealt with the weights of the four persons, we had $N_1 = 2$, $N_2 = 2$, $n_1 = 1$, and $n_2 = 1$, so that $\dfrac{n_1}{N_1} = \dfrac{n_2}{N_2} = \dfrac{1}{2}$, and the allocation is, indeed, proportional.

EXAMPLE To consider another illustration, suppose that a sample of size $n = 60$ is to be drawn from a population of size $N = 4,000$, which is divided into three strata of size $N_1 = 2,000$, $N_2 = 1,200$, and $N_3 = 800$. Substitution into the formula for proportional allocation yields

$$n_1 = \frac{2,000}{4,000} \cdot 60 = 30, \qquad n_2 = \frac{1,200}{4,000} \cdot 60 = 18,$$

$$n_3 = \frac{800}{4,000} \cdot 60 = 12$$

for the sizes of the three samples. Note that we could also have argued directly that the three strata constitute 50, 30, and 20 percent of the population, and that the corresponding percentages of $n = 60$ are $60(0.50) = 30$, $60(0.30) = 18$, and $60(0.20) = 12$.

This illustrates proportional allocation, but there exist other ways of allocating portions of a sample to the different strata; one of these, called **optimum allocation,** is described in Exercise 8 on page 221.

Stratification is not restricted to a single variable of classification, or characteristic, and populations are often stratified according to several characteristics. For instance, in a system-wide survey designed to determine the attitude of its students, say, toward a new tuition plan, a state college system with 17 colleges might stratify the students with respect to class standing, sex, major, and college. So, part of the sample would be allocated to junior women majoring in engineering in college A, part to senior men majoring in English in college L, and so on. Up to a point, stratification like this, called **cross stratification,** will often increase the precision (reliability) of estimates and other generalizations, and it is widely used, particularly in opinion sampling and market surveys.

In stratified sampling, the cost of taking random samples from the individual strata is often so expensive that interviewers are simply given quotas to be filled from the different strata, with few (if any) restrictions on how they are to be filled. For instance, in determining voters' attitudes toward increased medical coverage for elderly persons, an interviewer working a certain area might be told to interview 6 male self-employed homeowners under 30 years of age, 10 female wage earners in the 45–60 age bracket who live in apartments, 3 retired males over 60 who live in trailers, and so on, with the actual selection of the individuals being left to the interviewer's discretion. This is called **quota sampling,** and it is a convenient, relatively inexpensive, and sometimes necessary procedure, but as it is often executed, the resulting samples do not have the essential features of random samples. In the absence of any controls on their choice, interviewers naturally tend to select individuals who are most readily available—persons who work in the same building, shop in the same store, or perhaps reside in the same general area. Quota samples are thus essentially **judgment samples,** and inferences based on such samples generally do not lend themselves to any sort of formal statistical evaluation.

9.6
Cluster Sampling ★

To illustrate another important kind of sampling, suppose that a large foundation wants to study the changing patterns of family expenditures in the San Diego area. In attempting to complete schedules for 1,200 families, the foundation finds that simple random sampling is practically impossible, since suitable lists are not available and the cost of contacting families scattered over a wide area (with possibly two or three callbacks for the not-at-homes) is very high.

One way in which a sample can be taken in this situation is to divide the total area of interest into a number of smaller, nonoverlapping areas, say, city blocks. A number of these blocks are then randomly selected, with the ultimate sample consisting of all (or samples of) the families residing in these blocks.

In the kind of sampling described in the above example, called **cluster sampling,** the total population is divided into a number of relatively small subdivisions, which are themselves clusters of still smaller units, and then some of these subdivisions, or clusters, are randomly selected for inclusion in the over-all sample. If the clusters are geographic subdivisions, as in our example, this kind of sampling is also called **area sampling.** To give another example of cluster sampling, suppose that the Dean of Students of a university wants to know how fraternity men at the school feel about a certain new regulation. He can take a cluster sample by interviewing some or all of the members of several randomly selected fraternities.

Although estimates based on cluster samples are usually not as reliable as estimates based on simple random samples of the same size (see Exercise 7 on page 221), they are usually more reliable per unit cost. Referring again to the survey of family expenditures in the San Diego area, it is easy to see that it may well be possible to take a cluster sample several times the size of a simple random sample for the same cost. It is much cheaper to visit and interview families living close together in clusters than families selected at random over a wide area.

In practice, several of the methods of sampling we have discussed may well be used in the same study. For instance, if government statisticians want to study the attitude of American elementary school teachers toward certain federal programs, they might first stratify the country by states or some other geographic subdivisions. To take a sample from each stratum, they might then use cluster sampling, subdividing each stratum into a number of smaller geographic subdivisions (say, school districts), and finally they might use simple random sampling or systematic sampling to select a sample of elementary school teachers within each cluster.

EXERCISES

1   The following are the numbers of commercial FM radio stations in operation in 1973 in the 50 states (listed in alphabetic order): 56, 3, 18, 44, 161, 30, 21, 5, 97, 71, 4, 7, 110, 79, 43, 31, 67, 48, 15, 35, 40, 87, 45, 42, 50, 10, 17, 11, 14, 27, 19, 100, 75, 10, 113, 37, 21, 119, 7, 43, 10, 68, 134, 10, 6, 62, 42, 27, 79, and 1. List the ten possible systematic samples of size 5 that can be taken from this list by starting with one of the first ten numbers and then taking each tenth number on the list.

2   The following are the numbers of restaurants in twenty cities: 18, 23, 12, 10, 20, 11, 7, 11, 116, 34, 9, 8, 19, 25, 9, 83, 28, 9, 13, and 15.
    (a) List the five possible systematic samples of size 4 that can be taken from this list by starting with one of the first five numbers and then taking each fifth number on the list.

(b) Calculate the mean of each of the five samples obtained in part (a) and verify that their mean equals the average (mean) number of restaurants in the given twenty cities.

3 To generalize the example on page 217, suppose we want to estimate the mean weight of six persons, whose (unknown) weights are 115, 125, 135, 185, 195, and 205 pounds.
   (a) List all possible random samples of size 2 which can be taken from this population, calculate their means, and determine the probability that the mean of such a sample will differ by more than 5 from $\mu = 160$, the mean weight of the six persons.
   (b) Supposing that the first three weights are those of women and the other three are those of men, list all possible stratified samples of size 2 which can be taken by randomly choosing one of the two women and one of the two men, calculate their means, and determine the probability that the mean of such a sample will differ by more than 5 from $\mu = 160$, the mean weight of the six persons.

4 Among the 300 persons empaneled for jury duty, 150 are whites, 100 are blacks, and 50 are Chicanos. In how many ways can we choose a stratified 2 percent sample of these 300 persons,
   (a) if one third of the sample is to be allocated to each group?
   (b) if the allocation is to be proportional?

5 If the sample sizes for the individual strata are calculated according to the formula on page 218, verify that
   (a) the allocation is, indeed, proportional;
   (b) the sum of the sample sizes is actually equal to n.

6 A stratified sample of size $n = 80$ is to be taken from a population of size $N = 2,000$, which consists of four strata for which $N_1 = 500$, $N_2 = 1,200$, $N_3 = 200$, and $N_4 = 100$. If we use proportional allocation, how large a sample must we take from each stratum?

7 With reference to Exercise 3, list all possible cluster samples of size 2 which can be taken by randomly choosing either two of the three women or two of the three men, calculate their means, and determine the probability that the mean of such a sample will differ by more than 5 from $\mu = 160$, the mean weight of the six persons. If we compare this probability with those obtained in parts (a) and (b) of Exercise 3, what does this tell us about the relative merits of simple random sampling, stratified sampling, and cluster sampling in the given situation?

8 In stratified sampling with proportional allocation, the importance of differences in stratum size is accounted for by letting the larger strata contribute relatively more items to the sample. However, strata differ not only in size but also in variability, and it would seem reasonable to take larger samples from the more variable strata and smaller samples from the less variable strata. If we let σ_1, $\sigma_2, \ldots,$ and σ_k denote the standard deviations of the k strata, we can account for both, differences in stratum size and differences in stratum variability, by requiring that

$$\frac{n_1}{N_1 \sigma_1} = \frac{n_2}{N_2 \sigma_2} = \cdots = \frac{n_k}{N_k \sigma_k}$$

9.6 Cluster Sampling

221

or that these ratios be as nearly equal as possible. This is called **optimum allocation,** and it can be shown that it leads to the formula

$$n_i = \frac{n \cdot N_i \sigma_i}{N_1 \sigma_1 + N_2 \sigma_2 + \cdots + N_k \sigma_k} \qquad for \ i = 1, 2, \ldots, and \ k$$

Referring to the example on page 218, suppose that the standard deviations of the three strata are $\sigma_1 = 8$, $\sigma_2 = 15$, and $\sigma_3 = 32$. Substituting these values together with $n = 60$, $N_1 = 2,000$, $N_2 = 1,200$, and $N_3 = 800$ into the formula, we get

$$n_1 = \frac{60 \cdot 2,000 \cdot 8}{2,000 \cdot 8 + 1,200 \cdot 15 + 800 \cdot 32}$$

which equals 16 rounded to the nearest whole number, and $n_2 = 18$ and $n_3 = 26$.
  (a) A population is divided into two strata so that $N_1 = 10,000$, $N_2 = 30,000$, $\sigma_1 = 45$, and $\sigma_2 = 60$. How should a sample of size $n = 100$ be allocated to the two strata, if we use optimum allocation?
  (b) A population is divided into three strata so that $N_1 = 5,000$, $N_2 = 2,000$, $N_3 = 3,000$, $\sigma_1 = 15$, $\sigma_2 = 18$, and $\sigma_3 = 5$. How should a sample of size $n = 84$ be allocated to the three strata, if we use optimum allocation?
9 If we want to estimate the mean of a population on the basis of a stratified sample, we first calculate the means of the k samples, $\bar{x}_1, \bar{x}_2, \ldots,$ and $\bar{x}_k$, and then we combine them using the formula

$$\bar{x}_w = \frac{N_1 \bar{x}_1 + N_2 \bar{x}_2 + \cdots + N_k \bar{x}_k}{N_1 + N_2 + \cdots + N_k}$$

This is a weighted mean (see Exercise 16 on page 38) of the individual $\bar{x}$'s, and the weights are the sizes of the strata.
    The records of a casualty insurance company show that among 3,800 claims filed against the company over a period of time, 2,600 were minor claims (under \$200), while the other 1,200 were major claims (\$200 or more). To estimate the average size of these claims, the company takes a 1 percent sample, proportionally allocated to the two strata, with the following results (rounded to the nearest dollar):

Minor claims: 42, 115, 63, 78, 45, 148, 195, 66, 18, 73,
          55, 89, 170, 41, 92, 103, 22, 138, 49, 62,
          88, 113, 29, 71, 58, 83
Major claims: 246, 355, 872, 649, 253, 338, 491, 860, 755,
          502, 488, 311

  (a) Find the means of these two samples and then determine their weighted mean, using as weights the respective sizes of the two strata.

(b) Verify that the result of part (a) equals the ordinary mean of the 38 claims; that is, verify for this example that proportional allocation is **self-weighting.**

10 Verify symbolically that for stratified sampling with proportional allocation the weighted mean given by the formula of Exercise 9 equals the ordinary mean of all the sample values obtained for all the strata. (*Hint:* Make use of the fact that $n_i \cdot \bar{x}_i$ equals the sum of the values obtained for the ith stratum.)

9.7
Sampling Distributions

Let us now introduce the concept of the **sampling distribution** of a statistic, probably the most basic concept of statistical inference. As we shall see, this concept is closely related to the idea of chance variation, or chance fluctuations, which we mentioned earlier to emphasize the need for measuring the variability of data. In this chapter we shall concentrate mainly on the sample mean and its sampling distribution, but in some of the exercises at the end of this section and in later chapters we shall consider also the sampling distributions of other statistics.

There are two ways of approaching the study of sampling distributions. One, based on appropriate mathematical theory, leads to what is called a **theoretical sampling distribution;** the other, based (in reality or by simulation) on repeated samples from the same population, leads to what is called an **experimental sampling distribution.** The latter will prove to be very useful in our study because it provides experimental verification of theorems which cannot be derived formally at the level of this book.

EXAMPLE Let us first present the idea of a **theoretical sampling distribution of the mean** by actually listing all possible random samples of size $n = 2$ which can be drawn from the finite population of size $N = 5$, whose elements are the numbers 3, 5, 7, 9, and 11. The mean and the standard deviation of this population (which we shall need later in this example) are

$$\mu = \frac{3 + 5 + 7 + 9 + 11}{5} = 7$$

and

$$\sigma = \sqrt{\frac{(3-7)^2 + (5-7)^2 + (7-7)^2 + (9-7)^2 + (11-7)^2}{5}}$$
$$= \sqrt{8}$$

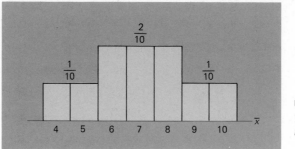

FIGURE 9.1

Theoretical sampling distribution of the mean.

in accordance with the formulas on pages 30 and 44. If we take a sample of size $n = 2$ from this population, there are $\binom{5}{2} = 10$ possibilities, and they are 3 and 5, 3 and 7, 3 and 9, 3 and 11, 5 and 7, 5 and 9, 5 and 11, 7 and 9, 7 and 11, and 9 and 11. The means of these samples are 4, 5, 6, 7, 6, 7, 8, 8, 9, and 10, and if sampling is random so that each sample has the probability $\frac{1}{10}$, we obtain the following sampling distribution of the mean:

$\bar{x}$	Probability
4	$\frac{1}{10}$
5	$\frac{1}{10}$
6	$\frac{2}{10}$
7	$\frac{2}{10}$
8	$\frac{2}{10}$
9	$\frac{1}{10}$
10	$\frac{1}{10}$

A histogram of this distribution is shown in Figure 9.1.

An examination of this sampling distribution reveals some pertinent information relative to the problem of estimating the mean of the given population on the basis of a random sample of size 2. For instance, we see that corresponding to $\bar{x} = 6, 7,$ or 8, the probability is $\frac{6}{10}$ that a sample mean will not differ from the population mean by more than 1, and that corresponding to $\bar{x} = 5, 6, 7, 8,$ or 9, the probability is $\frac{8}{10}$ that a sample mean will not differ from the population mean by more than 2.

Further useful information about this theoretical sampling distribution of the mean can be obtained by calculating its mean $\mu_{\bar{x}}$ and its standard deviation $\sigma_{\bar{x}}$. (The subscript $\bar{x}$ is used here to

distinguish these parameters from those of the original population.)
Following the definitions of the mean and the variance of a probability distribution on pages 172 and 175, we get

$$\mu_{\bar{x}} = 4 \cdot \frac{1}{10} + 5 \cdot \frac{1}{10} + 6 \cdot \frac{2}{10} + 7 \cdot \frac{2}{10} + 8 \cdot \frac{2}{10} + 9 \cdot \frac{1}{10} + 10 \cdot \frac{1}{10}$$

$$= 7$$

and

$$\sigma_{\bar{x}}^2 = (4-7)^2 \cdot \frac{1}{10} + (5-7)^2 \cdot \frac{1}{10} + (6-7)^2 \cdot \frac{2}{10} + (7-7)^2 \cdot \frac{2}{10}$$

$$+ (8-7)^2 \cdot \frac{2}{10} + (9-7)^2 \cdot \frac{1}{10} + (10-7)^2 \cdot \frac{1}{10}$$

$$= 3$$

so that $\sigma_{\bar{x}} = \sqrt{3}$. Thus, we find that the mean of the sampling distribution of the $\bar{x}$'s equals the mean of the population, and although the exact relationship is by no means obvious, the standard deviation of the sampling distribution of the $\bar{x}$'s is smaller than that of the population.

The relationships demonstrated in this example are of fundamental importance and we shall state them formally later in this section. Meanwhile, let us turn to the problem of constructing an **experimental sampling distribution of the mean,** hoping thereby to gain further insight into the "behavior" of sample estimates $\bar{x}$ of a population mean μ.

EXAMPLE   To present the idea of such an experimental sampling distribution, suppose that an operator of tow trucks wants to determine how many service calls he can expect to receive on a weekday afternoon. He needs this information in connection with the purchase of new equipment. Suppose, furthermore, that in a sample of five weekday afternoons he received 19, 15, 11, 12, and 21 service calls. The mean of this sample is

$$\bar{x} = \frac{19 + 15 + 11 + 12 + 21}{5} = 15.6$$

and in the absence of any other information he may use this figure as an estimate of μ, the number of service calls he can expect to receive on a weekday afternoon. It stands to reason, however, that if the sample had consisted of the number of service calls he received on five other weekdays, the mean would probably not have been 15.6. Indeed, if the operator of the tow trucks took several samples, each consisting of the number of service calls he received on five weekdays, he might get such discrepant estimates of μ as 13.6, 18.4, 12.2, and 17.0.

To get some idea of how the means of such samples might vary purely as the result of chance, suppose that the number of service calls the tow truck operator receives on a weekday afternoon is a random variable having the Poisson distribution of Section 7.5 with the mean $\lambda = 16$, and let us perform an experiment which consists of simulating the drawing of 50 random samples, each consisting of five observations. How this is done is explained in Section 9.10; so far as the work here is concerned, we shall merely use the results which are shown in the following table:

Sample	Number of service calls	Sample	Number of service calls
1	16, 15, 14, 12, 18	26	10, 18, 19, 13, 20
2	20, 18, 16, 19, 14	27	14, 19, 16, 13, 21
3	21, 18, 17, 26, 25	28	18, 14, 23, 23, 14
4	16, 10, 9, 19, 15	29	17, 16, 11, 17, 11
5	17, 16, 20, 17, 7	30	16, 13, 10, 14, 20
6	20, 8, 17, 16, 13	31	13, 17, 22, 19, 18
7	22, 21, 16, 15, 13	32	15, 13, 16, 14, 21
8	11, 23, 12, 20, 14	33	16, 15, 8, 12, 23
9	17, 22, 21, 16, 20	34	16, 15, 11, 20, 13
10	18, 13, 15, 11, 12	35	17, 17, 16, 21, 14
11	15, 11, 14, 14, 18	36	18, 20, 14, 26, 18
12	22, 15, 13, 19, 11	37	13, 16, 17, 11, 6
13	20, 17, 11, 19, 15	38	17, 19, 15, 19, 16
14	15, 16, 16, 15, 17	39	11, 13, 18, 23, 18
15	15, 16, 17, 17, 16	40	12, 25, 21, 18, 8
16	13, 15, 15, 13, 18	41	21, 12, 14, 17, 16
17	12, 11, 19, 17, 16	42	19, 10, 15, 16, 18
18	15, 26, 19, 20, 15	43	20, 10, 15, 15, 19
19	15, 11, 21, 8, 17	44	16, 10, 26, 14, 20
20	12, 21, 10, 15, 16	45	18, 12, 13, 19, 9
21	10, 11, 9, 11, 11	46	15, 14, 21, 17, 11
22	12, 12, 24, 11, 5	47	12, 13, 13, 14, 12
23	16, 18, 14, 9, 11	48	18, 8, 21, 14, 15
24	17, 9, 18, 16, 9	49	20, 17, 16, 18, 19
25	11, 17, 19, 20, 17	50	19, 17, 16, 13, 15

Each of these samples contains five values—the simulated number of service calls which the tow truck operator received on five week-

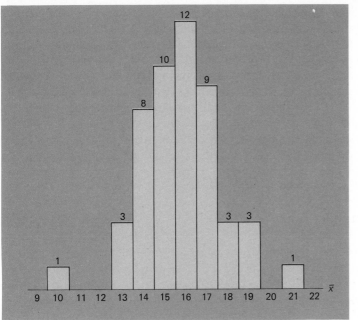

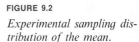

FIGURE 9.2

Experimental sampling distribution of the mean.

day afternoons—and the means of these 50 samples are

15.0	17.4	21.4	13.8	15.4	14.8	17.4	16.0	19.2	13.8
14.4	16.0	16.4	15.8	16.2	14.8	15.0	19.0	14.4	14.8
10.4	12.8	13.6	13.8	16.8	16.0	16.6	18.4	14.4	14.6
17.8	15.8	14.8	15.0	17.0	19.2	12.6	17.2	16.6	16.8
16.0	15.6	15.8	17.2	14.2	15.6	12.8	15.2	18.0	16.0

Grouped into a table having the classes 9.5–10.5, 10.5–11.5, . . . , and 20.5–21.5, we can picture the distribution of these means as in Figure 9.2[†]

Inspection of this experimental sampling distribution of $\bar{x}$ tells us a great deal about the way in which the means of random samples drawn from the given population tend to scatter among themselves due to chance. For instance, we find that the smallest mean is 10.4 and the largest is 21.4; furthermore, 31 out of 50 (or 62 percent) of the means are between 14.5 and 17.5, and 48 out of 50 (or 96 percent) of the means are between 12.5 and 19.5. Since the mean of the Poisson population from which the samples came is known to be $\mu = \lambda = 16$, we find that 62 percent of the sample means are "off" (differ from the population mean) by less than 1.5, and that 96 percent of the sample means are "off" by less than 3.5.

[†] There is no risk of ambiguity here, since the division of integers by 5 will always leave us with 0, 2, 4, 6, or 8 tenths, and never with 5 tenths.

The two examples we have given in this section are really only learning devices designed to introduce the concept of a sampling distribution. They do not reflect what we do in actual practice, where we ordinarily base an inference on one sample, not on 50, and cannot enumerate all possible samples. What should be understood at this point is that (1) a single mean is only one of many (perhaps, infinitely many) possibilities, (2) the distribution of the (actual or hypothetical) totality of these means is called a sampling distribution of the mean, and (3) the many means which form such a distribution are scattered more or less widely (and, as we shall see, predictably) about the population mean μ.

9.8

The Standard Error of the Mean

In most practical situations we can determine how close a sample mean might be to the mean of the population from which it came, by referring to two theorems which express essential facts about sampling distributions of the mean. The first of these theorems expresses formally what we discovered in connection with the example on page 225, namely, that the mean of the sampling distribution of $\bar{x}$ equals the mean of the population, and that its standard deviation is smaller than the standard deviation of the population. It may be phrased as follows: **For random samples of size n taken from a population having the mean μ and the standard deviation σ, the theoretical sampling distribution of $\bar{x}$ has the mean $\mu_{\bar{x}} = \mu$, and its standard deviation is**

Standard error of the mean

$$\sigma_{\bar{x}} = \frac{\sigma}{\sqrt{n}} \quad or \quad \sigma_{\bar{x}} = \frac{\sigma}{\sqrt{n}} \cdot \sqrt{\frac{N-n}{N-1}}$$

depending on whether the population is infinite or finite of size N.

It is customary to refer to $\sigma_{\bar{x}}$, the standard deviation of the sampling distribution of the mean, as the **standard error of the mean.** Its role in statistics is fundamental, since it measures the extent to which sample means can be expected to fluctuate, or vary, due to chance. Clearly, some knowledge of this variability is essential in determining how well an $\bar{x}$ estimates the mean of the population from which the sample came. As intuition suggests correctly, the smaller $\sigma_{\bar{x}}$ is (the less the $\bar{x}$'s are spread out) the better are our chances that the estimate will be close. What determines the size of $\sigma_{\bar{x}}$, and hence the goodness of an estimate, can be seen from the formulas above. Both formulas show that the standard error of the mean *increases* as the variability of the population *increases* (in fact, it is directly proportional to σ), and that it *decreases* as the number of items in the sample *increases*. With respect to the latter, note that substitution into either formula yields $\sigma_{\bar{x}} = \sigma$ for $n = 1$, and that $\sigma_{\bar{x}} = 0$ only for $n = N$; in other words, $\sigma_{\bar{x}}$ takes on values between 0 and σ, and is 0 only when the sample includes the entire population.

The factor $\sqrt{\dfrac{N - n}{N - 1}}$ in the second formula for $\sigma_{\bar{x}}$ is called the **finite population correction factor,** for without it the two formulas for $\sigma_{\bar{x}}$ (for finite and infinite populations) are the same. In practice, it is omitted unless the sample constitutes at least 5 percent of the population, for otherwise it is so close to 1 that it has little effect on the value of $\sigma_{\bar{x}}$.

EXAMPLE When $n = 100$ and $N = 10,000$, the sample constitutes but 1 percent of the population, and

$$\sqrt{\frac{N - n}{N - 1}} = \sqrt{\frac{10,000 - 100}{10,000 - 1}} = 0.995$$

This is so close to 1 that the correction factor can be omitted in most practical applications.

To get a feeling for the two formulas for $\sigma_{\bar{x}}$, let us return for a moment to the two sampling distributions constructed in the preceding section.

EXAMPLE In the first of the two examples we constructed the theoretical sampling distribution of the mean for random samples of size $n = 2$ drawn without replacement from a finite population of size $N = 5$, for which $\mu = 7$ and $\sigma = \sqrt{8}$. Substituting these values into the second formula for $\sigma_{\bar{x}}$, we get

$$\sigma_{\bar{x}} = \frac{\sqrt{8}}{\sqrt{2}} \cdot \sqrt{\frac{5 - 2}{5 - 1}} = \sqrt{3}$$

This agrees with the result obtained on page 225, where we showed directly that the standard deviation of the given theoretical sampling distribution of the mean is $\sqrt{3}$.

EXAMPLE In the tow-truck-service-calls example we took 50 random samples of size $n = 5$ from a population having the Poisson distribution with the mean $\lambda = 16$ and, hence, the standard deviation $\sigma = \sqrt{\lambda} = \sqrt{16} = 4$ (see Exercise 11 on page 179). Substituting these values into the first formula for $\sigma_{\bar{x}}$, we get

$$\sigma_{\bar{x}} = \frac{4}{\sqrt{5}} = 1.79$$

and it will be of interest to see how close the standard deviation of the 50 sample means in our experiment comes to this theoretical value. Actually calculating the mean and the standard deviation of the 50 sample means on page 226, we get 15.75 and 1.94, which are fairly close to $\mu_{\bar{x}} = 16$ and $\sigma_{\bar{x}} = 1.79$. They are sufficiently close, at least, to say that they provide experimental support for the theorem (on page 228) about the mean and the standard deviation of the sampling distribution of the mean.

9.9

The Central Limit Theorem

When we estimate the mean of a population, we usually attach a probability to a measure of the error of our estimate. For instance, if we use Chebyshev's theorem, we can assert with a probability of at least $1 - \dfrac{1}{k^2}$ that the mean of a random sample will differ from the mean of the population from which it came by less than $k \cdot \sigma_{\bar{x}}$. In other words, when we use the mean of a random sample to estimate the mean of a population, we can assert with a probability of at least $1 - \dfrac{1}{k^2}$ that our error will be less than $k \cdot \sigma_{\bar{x}}$.

EXAMPLE If we take a random sample of size $n = 64$ from an infinite population with $\sigma = 20$, we can assert with a probability of at least $1 - \dfrac{1}{2^2} = 0.75$ that the sample mean will be "off" (that is, differ from the population mean either way) by less than $2 \cdot \dfrac{20}{\sqrt{64}} = 5$.

Similarly, if we use the mean of a random sample of size $n = 100$ to estimate the mean of another infinite population with $\sigma = 4$, we can assert with a probability of at least $1 - \dfrac{1}{5^2} = 0.96$ that our estimate will be "off" by less than $5 \cdot \dfrac{4}{\sqrt{100}} = 2$.

Observe how we can thus make probability statements about errors of estimates without having to go through the tedious process of constructing the corresponding theoretical sampling distributions.

Since Chebyshev's theorem applies to any distribution, it can always be used as in the preceding example, but there exists another theorem, the **central limit theorem,** which often enables us to make much stronger probability statements about the possible size of the error of an estimate of a population mean. In words, it can be stated as follows:

Central limit theorem

> *If n (the sample size) is large, the sampling distribution of the mean can be approximated closely with a normal distribution.*

This theorem is of fundamental importance in statistics, as it justifies the use of normal-curve methods in a wide range of problems; it applies to infinite populations, and also to finite populations when n, though large, constitutes but a small portion of the population. It is difficult to say precisely how large n must be so that the central limit theorem applies, but unless the population distribution has a very unusual shape, $n = 30$ is usually regarded as sufficiently large. When the population we are sampling has, itself, roughly the shape of a normal curve, the same will be true also for the sampling distribution of the mean *regardless of the size of n*. This explains, perhaps, why the distribution of Figure 9.2 is fairly symmetrical and bell-shaped (like a normal curve) even though the sample size is only $n = 5$.

To illustrate the importance of the central limit theorem, let us re-examine the last example and also the experimental sampling distribution of Section 9.7.

EXAMPLE Using Chebyshev's theorem with $k = 2$ we could assert with a probability of at least $1 - \dfrac{1}{2^2} = 0.75$ that the mean of a random sample of size $n = 64$ drawn from an infinite population with $\sigma = 20$ will differ from the population mean by less than 5. Using the central limit theorem we must determine the normal-curve area between $z = -2$ and $z = 2$, and Table I shows that the answer is $0.4772 + 0.4772 = 0.9544$ (see also Figure 9.3). Compared to "at least 0.75," we can thus make the much stronger assertion that the probability is 0.9544 that the mean of a random

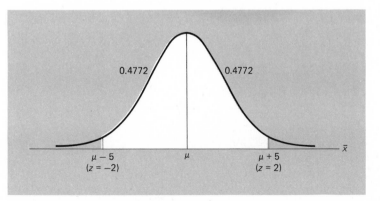

0.4772   0.4772

$\mu - 5$ μ $\mu + 5$
$(z = -2)$ $(z = 2)$

FIGURE 9.3

Sampling distribution of the mean.

9.9   The Central Limit Theorem

sample of size $n = 64$ from the given population will differ from the population mean by less than 5.

In the second part of the example, Chebyshev's theorem with $k = 5$ enabled us to assert with a probability of at least $1 - \dfrac{1}{5^2} = 0.96$ that the mean of a random sample of size $n = 100$ drawn from an infinite population with $\sigma = 4$ will differ from the population mean by less than 2. Using the central limit theorem we must determine the normal-curve area between $z = -5$ and $z = 5$, and Table I shows that the answer is $0.4999997 + 0.4999997 = 0.9999994$. Compared to "at least 0.96," we can thus make the much stronger assertion that the probability is 0.9999994 that the mean of a random sample of size $n = 100$ from the given population will differ from the population mean by less than 2.

EXAMPLE For the sampling distribution which we constructed from the 50 sample means in the tow-truck-service-calls example (see Figure 9.2), the agreement between the experimental results and the results which we would expect according to the central limit theorem (or on the basis of the assumption that the population, itself, has roughly the shape of a normal distribution) is remarkably good. In our example, 31 of the 50 means (or 62 percent) fell between 14.5 and 17.5, or within 1.5 of the population mean $\mu = 16$. If we apply normal-curve theory, the probability of getting a sample mean between 14.5 and 17.5 (with $\sigma_{\bar{x}} = 1.79$ in this example) is given by the area of the white region of Figure 9.4, namely, by the area under the standard normal distribution between

$$z = \frac{14.5 - 16}{1.79} = -0.84 \quad \text{and} \quad z = \frac{17.5 - 16}{1.79} = 0.84$$

It follows from Table I that this area is $0.2995 + 0.2995 = 0.5990$. In other words, we can expect about 60 percent of the sample means

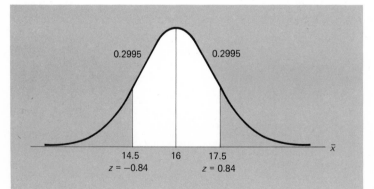

FIGURE 9.4

Sampling distribution of the mean.

to fall on the interval from 14.5 to 17.5, and this is remarkably close to the 62 percent which actually fell on this interval. Using the same kind of argument, we can also show that we can expect about 95 percent of the sample means to fall on the interval from 12.5 to 19.5 (see Exercise 17 on page 236), and this is very close to the 96 percent which actually fell on this interval.

EXERCISES

1   Random samples of size 2 are drawn from the finite population which consists of the numbers 5, 6, 7, 8, 9, and 10.
   (a) Show that the mean of this population is $\mu = 7.5$ and that its standard deviation is $\sigma = \sqrt{\frac{35}{12}}$.
   (b) List the 15 possible random samples of size 2 that can be drawn from this finite population and calculate their means.
   (c) Using the results of part (b) and assigning each possible sample the probability $\frac{1}{15}$, construct the sampling distribution of the mean for random samples of size 2 from the given finite population.
   (d) Calculate the mean and the standard deviation of the probability distribution obtained in part (c) and verify the results with the use of the theorem on page 228.

2   Repeat parts (b), (c), and (d) of Exercise 1 for random samples of size 3 from the given finite population.

3   The finite population of Exercise 1 can be converted into an infinite population if we sample with replacement, that is, if each value is replaced before the next one is drawn.
   (a) List the 36 possible ordered samples of size 2 that can be drawn with replacement from the given population and calculate their means. (By "ordered" we mean that the samples 7 and 10 and 10 and 7, for example, are considered different samples.)
   (b) Using the results of part (a) and assigning each possible sample the probability $\frac{1}{36}$, construct the sampling distribution of the mean for random samples of size 2 drawn with replacement from the given population.
   (c) Calculate the mean and the standard deviation of the probability distribution obtained in part (b), and verify the results with the use of the theorem on page 228.

4   Convert the 50 samples on page 226 into 25 samples of size $n = 10$ by combining samples 1 and 26, samples 2 and 27, . . . , and samples 25 and 50. Calculate the mean of each of these samples of size 10 and determine their mean and their standard deviation. Compare this mean and this standard deviation with the corresponding values expected in accordance with the theorem on page 228.

5   When we sample from an infinite population, what happens to the standard error of the mean (and, hence, to the size of the error we are exposed to when we use $\bar{x}$ to estimate μ) if the sample size is
   (a) increased from 60 to 240?
   (b) increased from 200 to 450?
   (c) increased from 25 to 225?
   (d) decreased from 640 to 40?

9.9   **The Central Limit Theorem**

6 What is the value of the finite population correction factor when
  (a) $n = 5$ and $N = 200$?
  (b) $n = 10$ and $N = 300$?
  (c) $n = 100$ and $N = 5,000$?

7 Show that if the mean of a random sample of size n is used to estimate the mean of an infinite population with the standard deviation σ, there is a fifty–fifty chance that the error (that is, the difference between $\bar{x}$ and μ) will be less than $0.6745 \cdot \dfrac{\sigma}{\sqrt{n}}$. It has been the custom to refer to this quantity as the **probable error of the mean**; nowadays, this term is used mainly in military applications.
  (a) If a random sample of size $n = 64$ is taken from an infinite population with $\sigma = 20.6$, what is the size of the probable error of the mean?
  (b) If a random sample of size $n = 100$ is drawn from a very large population (consisting of the fines paid for speeding violations in a certain county in 1978) which has the standard deviation $\sigma = \$8.25$, determine the probable error of the mean and explain its significance.

8 With reference to Exercise 2 on page 220, assign each of the systematic samples the probability $\frac{1}{5}$, determine the standard deviation of the corresponding sampling distribution of the mean, and compare it with $\sigma_{\bar{x}} = 12.24$, the standard error of the mean for random samples of size $n = 4$ from the given finite population.

9 In the example on page 217, we compared stratified samples from a population which consists of four weights with ordinary random samples of the same size.
  (a) Assigning each of the random samples on page 217 the probability $\frac{1}{6}$, show that the mean and the standard deviation of this sampling distribution of the mean are $\mu_{\bar{x}} = 160$ and $\sigma_{\bar{x}} = 21.0$.
  (b) Assigning each of the stratified samples on page 217, the probability $\frac{1}{4}$, show that the mean and the standard deviation of this sampling distribution of the mean are $\mu_{\bar{x}} = 160$ and $\sigma_{\bar{x}} = 7.1$.
  Discuss these results and see also part (a) of Exercise 36 on page 245.

10 In the example on page 217, the two cluster samples which consist of choosing, and weighing, either the two women or the two men have the means 125 and 195. Assigning the probability $\frac{1}{2}$ to each of these means, calculate the standard deviation of this sampling distribution of the mean, and compare it with the corresponding values for simple random sampling and stratified sampling obtained in parts (a) and (b) of Exercise 9.

11 Find the medians of the 50 samples on page 226, group them as we grouped the corresponding means, and construct a histogram such as that of Figure 9.2. Comparing the standard deviation of this **experimental sampling distribution of the median** with that of the corresponding experimental sampling distribution of the mean (which was 1.94), what can we say about the relative reliability of the median and the mean in estimating the mean of the given population?

12 For random samples of size n from a normal population, the **standard error of the median** (namely, the standard deviation of the sampling distribution of the median) is $\sqrt{\dfrac{\pi}{2}} \cdot \dfrac{\sigma}{\sqrt{n}}$, or approximately $1.25 \cdot \dfrac{\sigma}{\sqrt{n}}$, where σ is the standard deviation of the population. Compare the value obtained for the standard deviation

of the medians in Exercise 11 with the value we can expect according to this formula with $\sigma = 4$ and $n = 5$.

13 When we calculate the standard deviations of the 50 samples on page 226 and group them into a distribution with the classes 0.5–1.5, 1.5–2.5, ..., and 6.5–7.5, we obtain the following **experimental sampling distribution of** s:

Sample standard deviation	Frequency
0.5–1.5	4
1.5–2.5	6
2.5–3.5	12
3.5–4.5	16
4.5–5.5	8
5.5–6.5	2
6.5–7.5	2

There are no ambiguities here since none of the values is exactly 1.5, 2.5, ..., or 6.5.

(a) Calculate the mean and the standard deviation of this sampling distribution.

(b) For large samples, the formula $\dfrac{\sigma}{\sqrt{2n}}$ is sometimes used for the **standard error of a sample standard deviation.** Substituting $\sigma = 4$ and $n = 5$ into this formula, calculate the value of this standard error for the given example and compare it with the value of the standard deviation obtained for the experimental sampling distribution in part (a). Since $n = 5$ is not exactly a "large" sample, we should not be surprised if the results are not too close.

14 The mean of a random sample of size $n = 36$ is used to estimate the mean of an infinite population having the standard deviation $\sigma = 9$. What can we assert about the probability that the error will be less than 4.5, if we use
(a) Chebyshev's theorem?
(b) the central limit theorem?

15 The mean of a random sample of size $n = 25$ is used to estimate the mean of a very large population (consisting of the attention spans of persons over 65), which has a standard deviation of $\sigma = 2.4$ minutes. What can we assert about the probability that the error will be less than 1.2 minutes, if we use
(a) Chebyshev's theorem?
(b) the central limit theorem?

16 The mean of a random sample of size $n = 144$ is used to estimate the mean of a very large population (consisting of the weights of certain animals), which has a standard deviation of 3.6 ounces. If we use the central limit theorem, what can we assert about the probability that the error will be
(a) less than 0.50 ounce?
(b) less than 1.20 ounces?
(c) greater than 0.20 ounce?
(d) greater than 0.40 ounce?

9.9 **The Central Limit Theorem**

17 Verify the statement made on page 233 that for the given example the central limit theorem would lead us to expect that 95 percent of the means should fall on the interval from 12.5 to 19.5.

18 If measurements of the specific gravity of a metal can be looked upon as a sample from a normal population having the standard deviation $\sigma = 0.025$, what is the probability that the mean of a random sample of size $n = 16$ will be "off" by at most 0.01?

19 If the distribution of the weights of all men traveling by air between Dallas and El Paso has a mean of 163 pounds and a standard deviation of 18 pounds, what is the probability that the combined weight of 36 men traveling on such a flight is more than 6,012 pounds?

9.10

Technical Note (Simulation) ★

The experimental sampling distribution of Section 9.7 was obtained by simulating values of a random variable having the Poisson distribution with the parameter $\lambda = 16$. The technique we used was described in Section 7.9, and it was performed by using the following scheme:

Number of service calls	Probability	Random numbers
5	0.001	000
6	0.003	001–003
7	0.006	004–009
8	0.012	010–021
9	0.021	022–042
10	0.034	043–076
11	0.050	077–126
12	0.066	127–192
13	0.082	193–274
14	0.093	275–367
15	0.099	368–466
16	0.099	467–565
17	0.093	566–658
18	0.083	659–741
19	0.070	742–811
20	0.056	812–867
21	0.043	868–910
22	0.031	911–941
23	0.022	942–963
24	0.014	964–977
25	0.009	978–986
26	0.006	987–992
27	0.003	993–995
28	0.002	996–997
29	0.001	998
30	0.001	999

The probabilities were copied from a table of Poisson probabilities, and the three-digit random numbers were assigned to the values of the random variable by working with the cumulative probabilities, as suggested in the example on page 183. In our experiment, we randomly chose a page from a table of random numbers, three adjacent columns and a place from which to start, and read off sets of five three-digit random numbers.

EXAMPLE For the first of the 50 samples on page 226 we got the random numbers 522, 412, 289, 150, and 707, and, correspondingly, recorded the sample values 16, 15, 14, 12, and 18.

Simulations like this are extremely useful in the study of sampling distributions; particularly, when the theory is complicated and no tables (such as that of the normal distribution) are available.

EXERCISES

1 Use the above scheme and Table IX to simulate 25 random samples of size $n = 4$ from the given Poisson population with $\mu = 16$ and $\sigma = 4$, calculate the mean and the standard deviation of their means, and compare the results with those expected in accordance with the theorem on page 228.

2 Use the scheme of the example on page 183 and Table IX to simulate 20 random samples, each consisting of the numbers of cars that arrive at the toll booth of the bridge in $n = 6$ one-minute intervals, and calculate their means. Also calculate the standard deviation of these means and compare it with the value expected in accordance with the theorem on page 228, given that the population standard deviation is $\sigma = 2.2$.

3 Use Table X to simulate 40 random samples of size $n = 3$ from a standard normal population and calculate their means. Also calculate the standard deviation of these means and compare it with the value expected in accordance with the theorem on page 228.

4 Determine the medians of the 40 samples of Exercise 3, calculate their standard deviation, and compare it with the value expected in accordance with the standard error formula given in Exercise 12 on page 234.

5 For random samples of size n from a normal population, the **sampling distribution of** s^2 has the mean σ^2 and the standard deviation $\sigma^2 \sqrt{\dfrac{2}{n-1}}$. To verify this theory experimentally, use Table X to simulate ten random samples of size $n = 9$ from a standard normal population, calculate s^2 for each of the ten samples, and then determine their mean and their standard deviation.

6 Based on fairly advanced theory, it can be shown that for random samples of size $n = 5$ from a normal population, the mean of the **sampling distribution of the sample range** is approximately 2.3σ, where σ is the standard deviation of the population. Thus, for a random sample of size $n = 5$ we can use the sample range divided by 2.3 as an estimate of σ; for other sample sizes we divide by different constants, as is explained in Exercise 8 on page 289. Verify this theory experimentally by determining the ranges of the 40 samples of Exercise 3, calculating their mean, and comparing it with $2.3\sigma = 2.3 \cdot 1 = 2.3$.

9.10 Technical Note (Simulation)

237

9.11
In Review

A point worth repeating is that the two examples of Section 9.7 were meant to be teaching aids, designed to convey the idea of a sampling distribution. They do not reflect what we do in actual practice, where we ordinarily base an inference on one sample and not 50, and we do not enumerate all possible samples. In Chapter 10 and in subsequent chapters we shall go further into the problem of translating theory concerning sampling distributions into methods of evaluating the merits and disadvantages of statistical procedures.

Another fact worth noting concerns the $\sqrt{n}$ appearing in the denominator of the formula for the standard error of the mean. As n becomes larger and larger and we gain more and more information, our generalizations should be subject to smaller errors, and in general, our results (inferences) should be more reliable and more precise. However, the $\sqrt{n}$ in the formulas for $\sigma_{\bar{x}}$ illustrates the fact that gains in precision or reliability are not proportional to increases in the size of the sample. That is, doubling the size of the sample does not double the reliability of $\bar{x}$ as an estimate of the mean of a population, and so on. As is apparent from the formula $\sigma_{\bar{x}} = \dfrac{\sigma}{\sqrt{n}}$ for samples from infinite populations, we must take four times as large a sample to cut the standard error in half, and nine times as large a sample to triple the reliability; namely, to divide the standard error by 3. This clearly suggests that it will seldom pay to take excessively large samples.

EXAMPLE   If we increase the sample size from 100 to 10,000 (probably at considerable expense), the size of the error we are exposed to is reduced only by a factor of 10. Similarly, if we increase the sample size from 50 to, say, 20,000, the chance fluctuations we are exposed to are reduced only by a factor of 20, and this may in no way be worth the cost of taking the 19,950 additional observations.

BIBLIOGRAPHY   Derivations of the formulas for the standard error of the mean and more general formulations (and proofs) of the central limit theorem, may be found in most textbooks on mathematical statistics; for instance, in

FREUND, J. E., *Mathematical Statistics, 2nd ed.* Englewood Cliffs, N.J.: Prentice-Hall, Inc., 1971.

Further Exercises for Chapters

7,8, and 9

1   A random variable has a normal distribution with the standard deviation $\sigma = 10$. If the probability that the random variable will take on a value less than 71.3 is 0.8264, what is the probability that it will take on a value greater than 46.9?

2   The size of animal populations is often estimated by the **capture recapture method.** In this method, n_1 of the animals are captured, marked, and released. Later, n_2 of the animals are captured, x of them are found to be marked, and all this information is used to estimate N, the size of the population.

   (a) Find a formula which expresses N in terms of n_1, n_2, and x, by equating the proportion of marked animals in the population to the proportion of marked animals in the second sample.

   (b) To determine the number of rare owls living in a certain valley, $n_1 = 3$ are captured, tagged, and released; later, $n_2 = 4$ of the owls are captured and only $x = 1$ is found to be tagged. Use the formula obtained in part (a) to estimate how many of these rare owls live in the given valley.

   (c) Another way of estimating N is to choose the value for which the result of the second sample (x "successes" in n_2 trials) is most likely. With reference to part (b), find the probabilities of getting $x = 1$ and $n_2 = 4$ for $N = 9, 10, 11, 12, 13$, and 14, and then use the **maximum likelihood method** suggested here to estimate N.

3   Find the probability that two of the four members of an Antarctic expedition will get frostbite, when the probability is 0.25 that any one of them will get frostbite and it is reasonable to assume that the "trials" are independent.

4   If a student answers the 100 questions of a true–false test by flipping a fair coin, marking "true" when the coin comes up heads and "false" when it comes up tails, what does Chebyshev's theorem with $k = 4$ tell us about the number of correct answers he will get?

5 Random samples of size $n = 2$ are drawn from the population which consists of the numbers 1, 3, 5, and 7.

    (a) Verify that the mean of this finite population is $\mu = 4$ and that its standard deviation is $\sigma = \sqrt{5}$.

    (b) List the six possible samples of size $n = 2$ that can be drawn from this finite population, calculate their means, and, assigning each of these values the probability $\frac{1}{6}$, construct the theoretical sampling distribution of the mean for random samples of size $n = 2$ from the given population.

    (c) Calculate the mean and the standard deviation of the sampling distribution obtained in part (b), and verify the results using the theorem on page 228.

6 Suppose that in Exercise 5 sampling is with replacement, so that the population is (hypothetically, at least) infinite; that is, there is no limit to the number of observations we could make.

    (a) List the 16 possible samples of size $n = 2$ that can thus be drawn from the given population (counting 3 and 5, for example, and 5 and 3 as different samples), calculate their means, and, assigning each of these values the probability $\frac{1}{16}$, construct the theoretical sampling distribution of the mean for random samples of size $n = 2$ drawn with replacement from the given population.

    (b) Calculate the mean and the standard deviation of the sampling distribution obtained in part (a), and verify the results using part (a) of Exercise 5 and the theorem on page 228.

7 A population is divided into four strata so that $N_1 = 2,000$, $N_2 = 6,000$, $N_3 = 10,000$, $N_4 = 4,000$, $\sigma_1 = 25$, $\sigma_2 = 20$, $\sigma_3 = 15$, and $\sigma_4 = 30$. How should a sample of size $n = 440$ be allocated to the four strata, if we use

    (a) proportional allocation;

    (b) optimum allocation (see Exercise 8 on page 221)?

8 Find the constant probability that a census enumerator will find any one family on her list at home in the afternoon, if the probability that all four of the families she plans to visit on an afternoon are out equals $\frac{81}{256}$.

9 A manufacturer must buy coil springs which will stand a load of at least 24 pounds. If supplier A can provide springs that can stand an average load of 28.5 pounds with a standard deviation of 2.1 pounds and supplier B can provide springs that can stand an average load of 27.3 pounds with a standard deviation of 1.6 pounds, which of the two suppliers would provide the manufacturer with the smaller percentage of unsatisfactory springs? Assume that the distributions of the loads can be approximated with normal distributions.

10 An experiment consists of rolling a pair of balanced dice, one green and one red.

    (a) Letting $(2, 4)$ represent the outcome where the green die comes up 2 and the red die comes up 4, letting $(5, 1)$ represent the outcome where the green die comes up 5 and the red die comes up 1, and so forth, draw a diagram similar to that of Figure 7.1 which shows the 36 possible outcomes.

    (b) On the diagram obtained in part (a), write next to each point the corresponding total rolled with the pair of dice, and, assuming that each point has the probability $\frac{1}{36}$, construct the probability distribution for the total number of points rolled with a pair of "honest" dice; that is, construct a table showing the probabilities of rolling a total of 2, 3, 4, ..., 11, or 12.

(c) Verify that the equation of the probability distribution of part (b) can be written as

$$f(x) = \frac{6 - |x - 7|}{36} \qquad \text{for } x = 2, 3, \ldots, 11, \text{ or } 12$$

where the absolute value $|x - 7|$ equals $x - 7$ or $7 - x$, whichever is positive or zero.

11 The mean of a random sample of size $n = 225$ is used to estimate the mean of a population having the standard deviation $\sigma = 3$. What can we assert about the probability that the error will be less than 0.30, using
   (a) Chebyshev's theorem;
   (b) the central limit theorem?

12 When we calculate the sample variances of the 50 samples on page 226 and group them into a distribution having the classes 0–4.9, 5.0–9.9, ..., 40.0–44.9, and 45.0–49.9, we obtain the following experimental sampling distribution of s^2:

s^2	Frequency
0–4.9	7
5.0–9.9	9
10.0–14.9	11
15.0–19.9	10
20.0–24.9	7
25.0–29.9	2
30.0–34.9	1
35.0–39.9	1
40.0–44.9	0
45.0–49.9	2

(a) Convert this distribution into a cumulative "less than" percentage distribution and plot it on normal probability paper. What can we judge from this about the "normality" of this sampling distribution?

(b) Calculate the mean and the standard deviation of this experimental sampling distribution and compare the results with those expected in accordance with the formulas of Exercise 5 on page 237 with $\sigma = 4$ and $n = 5$.

13 With reference to Exercise 3 on page 221, show that $\sigma_{\bar{x}} = 22.7$ for the sampling distribution of the mean for the random samples of part (a) and $\sigma_{\bar{x}} = 5.8$ for the sampling distribution of the mean for the stratified samples of part (b). What does this tell us about the effectiveness of stratification in this example? [See also part (b) of Exercise 36 on page 245.]

14 With reference to Exercise 7 on page 221, show that $\sigma_{\bar{x}} = 35.2$ for the sampling distribution of the mean for the cluster samples. If we compare this value with the two standard errors of the preceding exercise, what does this tell us about the effectiveness of using cluster sampling in this example?

15 If the time it takes to fill in a new income tax form is a random variable having a normal distribution with $\mu = 32.6$ minutes and $\sigma = 3.5$ minutes, use Table X to simulate the time it takes 12 persons to fill in the new income tax form.

16 In Figure 7.1 on page 154 we associated with each point of the sample space the corresponding total number of cars sold by the two salesmen. Draw a similar diagram showing for each point the difference 0, 1, 2, or 3 between the numbers of cars sold by the two salesmen, and, using the probabilities shown in Figure 7.1 on page 154, construct the probability distribution of this random variable. This illustrates how more than one random variable may be defined over the same sample space.

17 Among 600 plants exposed to excessive radiation, 90 show abnormal growth. If a scientist collects the seed of three of the plants chosen at random, find the probability that he will get the seed from one plant with abnormal growth and two plants with normal growth by using
    (a) the hypergeometric distribution;
    (b) the binomial distribution as an approximation.

18 A finite population of size $N = 50,000$ consists of five strata for which $N_1 = 20,000$, $N_2 = 15,000$, $N_3 = 5,000$, $N_4 = 8,000$, and $N_5 = 2,000$. If a stratified random sample of size $n = 200$ is to be taken from this population using proportional allocation, how large a sample should be taken from each stratum?

19 Suppose that the Dean of Students of a college wants to have lunch with 15 entering freshmen of a class of 588 (who are listed in the registrar's office in alphabetical order). Which ones (by numbers) will she invite, if she numbers the freshmen from 001 through 588 and chooses a random sample by means of random numbers, using the last three columns of the table on page 470 starting with the first row?

20 If we have to determine all the values of a binomial distribution, it is sometimes helpful to calculate $f(0)$ using the formula for the binomial distribution, and then calculate the other values, one after the other, using the formula

$$\frac{f(x + 1)}{f(x)} = \frac{n - x}{x + 1} \cdot \frac{p}{1 - p}$$

    (a) Verify this formula by substituting for $f(x)$ and $f(x + 1)$ the corresponding expressions given by the formula for the binomial distribution.
    (b) Use this method to find all the values of the binomial distribution with $n = 6$ and $p = \frac{1}{4}$, writing them as common fractions with the denominator 4,096.

21 Among the applicants for four teaching assistantships there are seven married men, four single men, six married women, and three single women. If the selection is random, what is the probability that the assistantships will be given to a married man, a single man, and two married women? (Hint: Divide the number of ways in which a married man, a single man, and two married women can be

selected by the number of ways in which four of the twenty applicants can be selected.)

22 In a torture test, a watch is dropped from a tall building until it breaks. If the probability that it will break any time that it is dropped from the building is 0.15 (and we assume independence), what is the probability that the watch will break the fifth time it is dropped? (*Hint:* Use the formula of Exercise 12 on page 169.)

23 Mr. Brown and Ms. Jones intend to go pheasant hunting, using Mr. Brown's dog. According to Mr. Brown the probability that his dog will find a downed bird is 0.90 and according to Ms. Jones this probability is only 0.70. In the absence of any further information, we might judge that, since it is Mr. Brown's dog, he is four times as likely to be right as Ms. Jones. Assuming that one or the other must be right, how would this judgment be affected if on the hunting trip the dog actually finds only six of the nine pheasants they down? (*Hint:* Use Bayes' theorem and look up the required probabilities in Table VI.)

24 A certain supermarket carries four grades of ground beef, and the probabilities that a shopper will buy the poorest, third best, second best, or best grade are, respectively, 0.10, 0.20, 0.40, and 0.30. Find the probability that among 12 (randomly chosen) shoppers buying ground beef at this supermarket, one will choose the poorest kind, two will choose the third best kind, six will buy the second best kind, and three will buy the best kind. (*Hint:* Use the formula of Exercise 26 on page 171.)

25 The following is the probability distribution of the number of fires that break out in a national forest on an August afternoon:

Number of fires	0	1	2	3	4	5	6
Probability	0.120	0.207	0.358	0.162	0.096	0.043	0.014

(a) Distribute the three-digit random numbers from 000 through 999 to the seven values of this random variable, so that the corresponding random numbers can be used to simulate the numbers of fires that break out in the forest on August afternoons.

(b) Use the results of part (a) to simulate the numbers of fires that break out in the forest on twenty August afternoons.

26 A panel of 300 persons chosen for jury duty includes 30 under 25 years of age. Since the jury of 12 persons chosen from this panel to judge a narcotics violation does not include anyone under 25 years of age, the youthful defendant's attorney complains that this jury is not really representative. Indeed, he argues, if the selection were random, the probability of having one of the 12 jurors under 25 years of age should be *many times* the probability of having none of them under 25 years of age. Actually, what is the ratio of these two probabilities?

27 The mean of a random sample of size $n = 64$ is used to estimate the mean lifetime of certain light bulbs. If the standard deviation of these lifetimes is $\sigma = 60$ hours, what is the probability that our estimate will be "off" either way by
(a) less than 16.5 hours?
(b) less than 6.0 hours?

28 A collection of 18 mediaeval gold coins contains four counterfeits. If two of these coins are randomly selected to be sold at auction, what are the probabilities that

    (a) neither coin is a counterfeit?

    (b) one of the coins is a counterfeit?

    (c) both coins are counterfeits?

Also calculate the mean of this probability distribution and verify the result using the formula on page 174.

29 The probabilities of 0, 1, 2, or 3 armed robberies in a Western city in any given week are 0.33, 0.41, 0.17, and 0.09. Find the mean and the variance of this probability distribution.

30 The average time required to perform job A is 88 minutes with a standard deviation of 16 minutes, and the average time required to perform job B is 113 minutes with a standard deviation of 11 minutes. Assuming normal distributions, what proportion of the time will job A take longer than the average job B, and what proportion of the time will job B take less time than the average job A?

31 Explain why each of the following samples does not qualify as a random sample from the required population or might otherwise fail to give the desired information:

    (a) To ascertain facts about their bathing habits, a sample of the citizens of a European country are asked how many times on the average they bathe each week.

    (b) To predict an election, a poll taker interviews a random sample of persons coming out of the building which houses the national headquarters of a political party.

    (c) To determine the proportion of improperly sealed cans of beets, a quality control engineer inspects every 20th can coming off an assembly line.

    (d) To estimate the average length of logs fed into a mill by a constant-speed conveyor belt, measurements are made of the length of the logs which pass a given point exactly every five minutes.

32 If 5 percent of all persons who drive cars are not properly licensed, use the Poisson approximation to determine the probability that among 100 drivers stopped at a road block, three will not be properly licensed.

33 A stratified sample of size $n = 90$ is to be taken from a group of persons including 600 who are not college graduates, 800 who hold only B.A. degrees, 400 who hold also M.A. degrees, and 200 who hold also Ph.D. degrees. What part of the sample should be allocated to each of these strata, if the allocation is to be proportional?

34 The mean of a random sample of size $n = 36$ is used to estimate the mean time it take clerks to locate a file from a standard file cabinet. If $\sigma = 0.12$ minute, what can we assert about the probability that the error of this estimate will be less than 0.05 minute, using

    (a) Chebyshev's theorem?

    (b) the central limit theorem?

35 A classical example of the use of simulation techniques in the solution of a problem of mathematics is the determination of π (the ratio of the circumference of a circle to its diameter) by probabilistic means. Early in the eighteenth century George de Buffon, a French naturalist, proved that if a very fine needle of length

a is thrown at random on a board ruled with parallel lines, the probability that the needle will intersect one of the lines is $\dfrac{2a}{\pi b}$, where b is the distance between the parallel lines. Actually performing this experiment (with a toothpick or a pin thrown $n = 100$ times on a sheet of paper ruled with parallel lines an inch apart), equate the proportion of the time the needle (toothpick or pin) crosses one of the lines to $\dfrac{2a}{\pi b}$, and solve for π.

36 If $\bar{x}$ is the mean of a stratified random sample of size n obtained by proportional allocation from a finite population of size N, which consists of k strata of size $N_1, N_2, \ldots,$ and N_k, then

$$\sigma_{\bar{x}}^2 = \sum_{i=1}^{k} \frac{(N - n)N_i^2}{nN^2(N_i - 1)} \cdot \sigma_i^2$$

where $\sigma_1^2, \sigma_2^2, \ldots,$ and σ_k^2 are the corresponding variances for the individual strata.

    (a) Use this formula to verify the value given for $\sigma_{\bar{x}}$ in part (b) of Exercise 9 on page 234, actually $\sigma_{\bar{x}} = \sqrt{50}$.

    (b) Use this formula to verify the value $\sigma_{\bar{x}} = 5.8$, actually $\sigma_{\bar{x}} = \sqrt{\frac{100}{3}}$, given in Exercise 13 on page 241.

This standard error formula, by itself, does not enable us to judge the effectiveness of a stratification; this depends on whether the variances $\sigma_1^2, \sigma_2^2, \ldots,$ and σ_k^2 are appreciably smaller than the corresponding variance σ^2 for the entire population.

10

Traditionally, statistical inferences have been divided into **estimations,** in which we estimate various unknown parameters (statistical descriptions) of populations, **tests of hypotheses,** in which we either accept or reject specific assertions about populations or their parameters, and **predictions,** in which we predict future values of random variables. In this chapter, and in those that immediately follow, we shall concentrate on problems of estimation and tests of hypotheses; methods of prediction will be taken up in Chapters 14 and 15.

Inferences About Means

10.1
Problems of Estimation

Problems of estimation arise everywhere, in science, in business, and in everyday life. In science, a psychologist may want to determine the average time that it takes an adult person to react to a visual stimulus, an engineer may need to know how much variability there is in the strength of a new alloy, and a biologist may want to determine what percentage of certain insects are born physically defective. In business, a retailer may want to determine the average income of all families living within a mile of a proposed new store, a union official may have to know how much variation there is in the time it takes members to get to work, and a banker may want to find out what proportion of all installment loan payments she can expect to be late. Finally, in everyday life, we may want to determine how long it takes on the average to mend a pair of sox, we may be interested in the variation we can expect in a child's performance in school, and we may want to find out what percentage of all one-car accidents are due to driver fatigue.

In each of these three sets of examples, the first dealt with the estimation of a mean, the second with the estimation of a measure of variation, and the third with the estimation of a percentage or proportion. Since the statistical treatment of such problems of inference differs, we shall devote this chapter to inferences about means, and then take up inferences about measures of variation and inferences about proportions in Chapters 11 and 12.

10.2
The Estimation of Means

To illustrate some of the problems we face when we estimate a mean, let us return to the example which we used in Chapter 2 to illustrate the construction of a frequency distribution.

EXAMPLE In an air pollution study, data were obtained on the daily emission of sulfur oxides by an industrial plant. The mean of the values was $\bar{x} = 18.85$ tons, and if they constitute a random sample, we can use this figure as an estimate of the plant's "true" average daily emission of sulfur oxides.

An estimate like this is called a **point estimate**, since it consists of a single number, or a single point on the real number scale. Although this is the most common way in which estimates are expressed, it leaves room for many questions. For instance, it does not tell us on how much information the estimate is based, and it does not tell us anything about the possible size of the error. And of course, we must expect an error—it is possible for an estimate based on a sample to coincide with the population parameter it is intended to estimate, but this is the exception rather than the rule, as should be clear from our discussion of the sampling distribution of the mean in Chapter 9.

EXAMPLE Returning to the air pollution study, we might thus supplement the estimate $\bar{x} = 18.85$ tons with the information that this is the mean of a random sample of size $n = 80$, whose standard deviation is $s = 5.55$ tons.

Scientific reports often present sample means in this way, together with the values of n and s, but this does not supply the reader with a coherent picture unless he has had some formal training in statistics. To make the supplementary information meaningful also to the layman, let us refer to the two theorems of Chapter 9 about the sampling distribution of the mean. According to the central limit theorem, the sampling distribution of the mean can, for large samples, be approximated closely with a normal curve. Hence, we can assert with probability $1 - \alpha$, where α is the Greek letter *alpha*, that a sample mean $\bar{x}$ will differ from the population mean μ by less than $z_{\alpha/2}$ standard errors of the mean. As was explained in Exercise 7 on page 198, $z_{\alpha/2}$ is the z-value for which the area to its right under the standard normal curve is $\alpha/2$ (see also Figure 10.1). And since the standard error of the mean is $\sigma_{\bar{x}} = \dfrac{\sigma}{\sqrt{n}}$ for random samples of size n from infinite or very large populations, we can make the following assertion: **The probability is $1 - \alpha$ that $\bar{x}$ will differ from μ by less than $z_{\alpha/2} \cdot \dfrac{\sigma}{\sqrt{n}}$, and since $\bar{x} - \mu$ is the error**

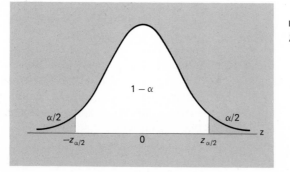

FIGURE 10.1
Standard normal distribution.

we make when we use $\bar{x}$ as an estimate of μ, **the probability is** $1 - \alpha$ **that the size of this error will be less than**

Maximum error of estimate

$$E = z_{\alpha/2} \cdot \frac{\sigma}{\sqrt{n}}$$

The two values which are most widely used for $1 - \alpha$ are 0.95 and 0.99, and from Table I we find that $z_{\alpha/2} = z_{0.025} = 1.96$ and $z_{\alpha/2} = z_{0.005} = 2.58$.

The result we have obtained involves one complication. To be able to judge the size of the error we might make when we use $\bar{x}$ as an estimate of μ, we must know σ, the population standard deviation. Since this is not the case in most practical situations, we have no choice but to replace σ with an estimate, usually the sample standard deviation s. In general, this is considered to be reasonable provided the sample size is sufficiently large, and by "sufficiently large" we mean 30 or more.

EXAMPLE In the air pollution example where we had $\bar{x} = 18.85$ tons, $n = 80$, and $s = 5.55$ tons, we can now assert with probability 0.95 that if we estimate the plant's true average daily emission of sulfur oxides as 18.85 tons, the error is less than

$$1.96 \cdot \frac{5.55}{\sqrt{80}} = 1.22 \text{ tons}$$

Of course, the error of this estimate is less than 1.22 tons or it is not, and we really do not know which, but if we had to bet, 95 to 5 (or 19 to 1) would be fair odds that the error is less than 1.22 tons. Similarly, 99 to 1 would be fair odds that the error of the estimate is less than

$$2.58 \cdot \frac{5.55}{\sqrt{80}} = 1.60 \text{ tons}$$

The formula for the maximum error to which we are exposed with a specified probability can also be used to determine the sample size needed to attain a desired degree of precision. Suppose we want to use the mean of a random sample to estimate the mean of a population, and we want to be able to assert with probability $1 - \alpha$ that the error of this estimate is less than some prescribed quantity E. We write $E = z_{\alpha/2} \cdot \frac{\sigma}{\sqrt{n}}$ and upon solving this equation for n we get

Sample size

$$n = \left[\frac{z_{\alpha/2} \cdot \sigma}{E} \right]^2$$

EXAMPLE   A college dean wants to use the mean of a random sample to estimate the average amount of time students take to get from one class to the next, and he wants this estimate to be in error by at most 0.25 minute with probability 0.95. Knowing from previous studies that $\sigma = 1.5$ minutes, he substitutes this value together with $z_{\alpha/2} = 1.96$ and $E = 0.25$ into the formula for n, and gets

$$n = \left[\frac{1.96 \cdot 1.50}{0.25} \right]^2 = 139$$

rounded up to the nearest whole number. Thus, a random sample of size $n = 139$ is required for the estimate.

As can be seen from the formula and also from the example, this method has the shortcoming that it can be used only when we know (at least approximately) the value of the standard deviation of the population whose mean we are trying to estimate.

The error we make when we use a sample mean to estimate the mean of a population is given by the difference $\bar{x} - \mu$, and the fact that, with probability $1 - \alpha$, the magnitude of this error is less than $z_{\alpha/2} \cdot \dfrac{\sigma}{\sqrt{n}}$, leads to the inequality[†]

$$-z_{\alpha/2} \cdot \frac{\sigma}{\sqrt{n}} < \bar{x} - \mu < z_{\alpha/2} \cdot \frac{\sigma}{\sqrt{n}}$$

If we now apply some relatively simple algebra, we can rewrite this inequality as

Confidence interval for μ

$$\bar{x} - z_{\alpha/2} \cdot \frac{\sigma}{\sqrt{n}} < \mu < \bar{x} + z_{\alpha/2} \cdot \frac{\sigma}{\sqrt{n}}$$

[†] In case the reader is not familiar with inequality signs, let us point out that $a < b$ means "a is less than b," while $a > b$ means that "a is greater than b." Also, $a \leqslant b$ means "a is less than or equal to b," while $a \geqslant b$ means "a is greater than or equal to b."

and we can assert with probability $1 - \alpha$ that the inequality is satisfied for any given sample. In other words, **we can assert with probability $1 - \alpha$ that the interval from $\bar{x} - z_{\alpha/2} \cdot \dfrac{\sigma}{\sqrt{n}}$ to $\bar{x} + z_{\alpha/2} \cdot \dfrac{\sigma}{\sqrt{n}}$ actually contains the population mean we are estimating.** An interval like this is called a **confidence interval,** its endpoints are called **confidence limits,** and the probability $1 - \alpha$ is called the **degree of confidence.** As before, the values most commonly used for $1 - \alpha$ are 0.95 and 0.99, and the corresponding values of $z_{\alpha/2}$ are 1.96 and 2.58. In contrast to point estimates, estimates given in the form of confidence intervals are called **interval estimates.**

When σ is unknown and n is at least 30, we replace σ by the sample standard deviation s and call the resulting confidence interval a **large-sample confidence interval for μ.**

EXAMPLE   In the air pollution study, where $\bar{x} = 18.85$ tons, $n = 80$, and $s = 5.55$ tons, we have the following 0.95 large-sample confidence interval for the plant's true average daily emission of sulfur oxides:

$$18.85 - 1.96 \cdot \frac{5.55}{\sqrt{80}} < \mu < 18.85 + 1.96 \cdot \frac{5.55}{\sqrt{80}}$$

$$17.63 < \mu < 20.07$$

Of course, the interval from 17.63 to 20.07 contains μ or it does not, and we really do not know which, but if we had to bet, 95 to 5 (or 19 to 1) would be fair odds that it does. To put it differently, the method by which the interval was obtained "works 95 percent of the time."

The corresponding 0.99 confidence interval is

$$17.25 < \mu < 20.45$$

and this illustrates the important fact that "the surer we want to be, the less we have to be sure of." In other words, **when we increase the degree of confidence, the confidence interval becomes wider and thus tells us less about the quantity we want to estimate.**

10.3

The Estimation of Means (Small Samples)

So far we have assumed not only that the sample size is large enough to treat the sampling distribution of the mean as if it were a normal distribution, but that (when necessary) σ can be replaced with s in the formula for the standard error of the mean. To develop a corresponding theory that applies also to small samples, we

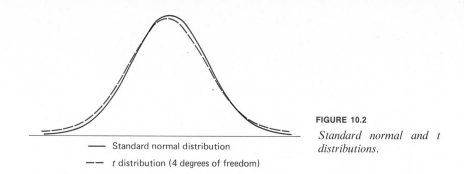

FIGURE 10.2

Standard normal and t distributions.

——— Standard normal distribution

——— *t* distribution (4 degrees of freedom)

shall now have to assume that the population we are sampling has roughly the shape of a normal distribution. We can then base our methods on the statistic

$$t = \frac{\bar{x} - \mu}{s/\sqrt{n}}$$

whose sampling distribution is called the *t* **distribution.** More specifically, it is called the **Student-*t* distribution,** or **Student's *t* distribution,** as it was first investigated by W. S. Gosset, who published his writings under the pen name "Student." As is shown in Figure 10.2, the shape of this distribution is very much like that of a normal distribution, and it is symmetrical with zero mean. The exact shape of the *t* distribution depends on the quantity $n - 1$, the sample size less one, called the **number of degrees of freedom.** [†]

For the standard normal distribution, we defined $z_{\alpha/2}$ in such a way that the area under the curve to its right equals $\alpha/2$, and, hence, the area under the curve between $-z_{\alpha/2}$ and $z_{\alpha/2}$ equals $1 - \alpha$. As is shown in Figure 10.3, the corresponding values for the *t* distribution are $-t_{\alpha/2}$ and $t_{\alpha/2}$. Since these values depend on $n - 1$, the number of degrees of freedom, they must be looked up in a special table, such as Table II at the end of this book. This table contains among others the values of $t_{0.025}$ and $t_{0.005}$ for 1 through 29 degrees of freedom, and it can be seen that $t_{0.025}$ approaches 1.96 and $t_{0.005}$ approaches 2.58 (the corresponding values for the standard normal distribution) as the number of degrees of freedom increases.

[†] It is difficult to explain at this time why one should want to assign a special name to $n - 1$. However, we shall see later in this chapter that there are other applications of the *t* distribution, where the number of degrees of freedom is defined in a different way. The reason for the term "degrees of freedom" lies in the fact that if we know $n - 1$ of the deviations from the mean, then the *n*th is automatically determined (see Exercise 9 on page 62). Since the sample standard deviation measures variation in terms of the squared deviations from the mean, we can thus say that this estimate of σ is based on $n - 1$ independent quantities or that we have $n - 1$ degrees of freedom.

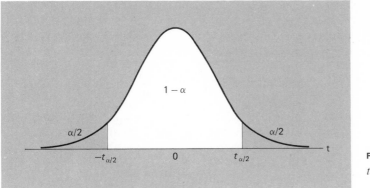

FIGURE 10.3
t distribution.

Since the *t* distribution, like the standard normal distribution, is symmetrical about its mean $\mu = 0$ (see Figure 10.3), we can now duplicate the argument on page 250 and thus arrive at the following $1 - \alpha$ **small-sample confidence interval for** μ:

Small-sample confidence interval for μ

$$\bar{x} - t_{\alpha/2} \cdot \frac{s}{\sqrt{n}} < \mu < \bar{x} + t_{\alpha/2} \cdot \frac{s}{\sqrt{n}}$$

The only difference between this confidence interval formula and the large-sample formula (with *s* substituted for σ) is that $t_{\alpha/2}$ takes the place of $z_{\alpha/2}$.

EXAMPLE   To test the durability of a new paint for white center lines, a highway department painted test strips across heavily traveled roads in eight different locations, and electronic counters showed that they deteriorated after being crossed by (to the nearest hundred) 142,600, 167,800, 136,500, 108,300, 126,400, 133,700, 162,000, and 149,400 cars. The mean and the standard deviation of these values are $\bar{x} = 140,800$ and $s = 19,200$, and since $t_{0.025}$ for $8 - 1 = 7$ degrees of freedom equals 2.365, substitution into the small-sample confidence interval formula with $1 - \alpha = 0.95$ yields

$$140,800 - 2.365 \cdot \frac{19,200}{\sqrt{8}} < \mu < 140,800 + 2.365 \cdot \frac{19,200}{\sqrt{8}}$$

$$124,700 < \mu < 156,900$$

This is an interval estimate of the average amount of traffic (car crossings) this paint can withstand before it deteriorates.

10.3   **The Estimation of Means (Small Samples)**

The method which we used earlier to determine the maximum error we make with probability $1 - \alpha$ when we use a sample mean to estimate the mean of a population can easily be adapted to small samples (provided the population we are sampling has roughly the shape of a normal distribution). All we have to do is substitute s for σ and $t_{\alpha/2}$ for $z_{\alpha/2}$ in the formula for the maximum error E given on page 249.

EXAMPLE While performing a certain task under simulated weightlessness, the pulse rate of twelve astronauts increased on the average by 27.33 beats per minute with a standard deviation of 4.28 beats per minute. Since $t_{0.005} = 3.106$ for $12 - 1 = 11$ degrees of freedom, substitution into the formula for E yields

$$E = t_{\alpha/2} \cdot \frac{s}{\sqrt{n}} = 3.106 \cdot \frac{4.28}{\sqrt{12}} = 3.84$$

Thus, if we use the mean $\bar{x} = 27.33$ beats per minute as an estimate of the true average increase in pulse rate under the given conditions, we can assert with probability 0.99 that the error of this estimate is less than 3.84 beats per minute.

10.4

The Estimation of Means (A Bayesian Method) ★

In recent years there has been mounting interest in methods of inference in which parameters (for example, the population mean μ or the population standard deviation σ) are looked upon as random variables having **prior distributions** which reflect how a person feels about the different values that a parameter can take on. Such prior considerations are then combined with direct sample evidence to obtain **posterior distributions** of the parameters, on which subsequent inferences are based. Since the method used to combine the prior considerations with the direct sample evidence is based on a generalization of Bayes' theorem of Section 5.7, we refer to such inferences as **Bayesian.**

In this section we shall present a Bayesian method of estimating the mean of a population. As we said, our prior feelings about the possible values of μ are expressed in the form of a prior distribution, and like any distribution, this kind of distribution has a mean and a standard deviation. We shall designate these values μ_0 and σ_0 and call them the **prior mean** and the **prior standard deviation.**

If we are sampling a population having the mean μ (which we want to estimate) and the standard deviation σ, if the sample is large enough to apply the central limit theorem (or if the population is normal), and if the prior distribution of μ has roughly the shape of a normal distribution, it can be shown that the posterior distribution of μ is also a normal distribution with the mean

Posterior mean

$$\mu_1 = \frac{\dfrac{n}{\sigma^2} \cdot \bar{x} + \dfrac{1}{\sigma_0^2} \cdot \mu_0}{\dfrac{n}{\sigma^2} + \dfrac{1}{\sigma_0^2}}$$

and the standard deviation σ_1, given by the formula

Posterior standard deviation

$$\frac{1}{\sigma_1^2} = \frac{n}{\sigma^2} + \frac{1}{\sigma_0^2}$$

Since μ_1, the **posterior mean,** may be used as an estimate of the mean of the population, let us examine some of its most important features. First, we note that μ_1 is a weighted mean (see Exercise 16 on page 38) of $\bar{x}$ and μ_0, and that the weights are $\dfrac{n}{\sigma^2}$ and $\dfrac{1}{\sigma_0^2}$, the reciprocals of the variances of the distribution of $\bar{x}$ and the prior distribution of μ. We see also that when no direct information is available and $n = 0$, the weight assigned $\bar{x}$ is 0, the formula reduces to $\mu_1 = \mu_0$, and the estimate is based entirely on the prior distribution. However, as more and more direct evidence becomes available (that is, as n becomes larger and larger), the weight shifts more and more toward the direct sample evidence, the sample mean $\bar{x}$. Finally, we see that when the prior feelings about the possible values of μ are vague (that is, when σ_0 is relatively large), the estimate will be based to a greater extent on $\bar{x}$. However, when there is a great deal of variability in the population which yields the direct sample evidence (that is, when σ is relatively large), the estimate will be based to a greater extent on μ_0.

EXAMPLE   Referring again to the air pollution study, suppose that before the sample data were actually obtained, the chief engineer of the plant said that his feelings about the true average daily emission of sulfur oxides can best be described by a normal distribution with the mean $\mu_0 = 17.50$ tons and the standard deviation $\sigma_0 = 2.50$ tons. Combining this prior information with the direct sample evidence, $\bar{x} = 18.85$, $n = 80$, and $s = 5.55$, we find that substitution into the formulas for μ_1 and σ_1 (with σ replaced by s) yields

$$\mu_1 = \frac{\dfrac{80}{5.55^2} \cdot 18.85 + \dfrac{1}{2.50^2} \cdot 17.50}{\dfrac{80}{5.55^2} + \dfrac{1}{2.50^2}} = 18.77$$

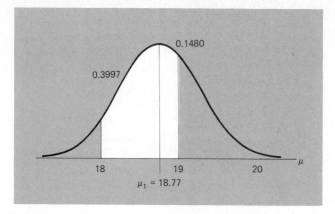

FIGURE 10.4

Posterior distribution of μ.

and

$$\frac{1}{\sigma_1^2} = \frac{80}{5.55^2} + \frac{1}{2.50^2} = 2.757 \quad \text{so that} \quad \sigma_1 = 0.60$$

To continue, we might ask for the probability that the mean of the population (that is, the plant's actual average daily emission of sulfur oxides) lies on the interval from 18.00 to 19.00 tons. The answer to this question is given by the area of the white region of Figure 10.4, namely, the area under the standard normal distribution between $z = \dfrac{18.00 - 18.77}{0.60} = -1.28$ and $z = \dfrac{19.00 - 18.77}{0.60} = 0.38$. Since the entries corresponding to 1.28 and 0.38 in Table I are 0.3997 and 0.1480, we get $0.3997 + 0.1480 = 0.5477$, or approximately 0.55, for the probability that the plant's average daily emission of sulfur oxides falls between 18.00 and 19.00 tons.

This brief introduction to Bayesian inference serves to bring out the following points: (1) **In Bayesian statistics the parameter about which an inference is made is looked upon as a random variable having a distribution of its own,** and (2) **this kind of inference permits the use of direct as well as collateral information.** To clarify the second point, let us add that in the air pollution example the prior distribution of the chief engineer may have been based on a subjective evaluation of conditions in the plant in general, or perhaps on indirect objective information about the "performance" of similar industrial plants.

EXERCISES

1 To estimate the average time required for certain repairs, an efficiency expert timed a random sample of 40 automobile mechanics in the performance of this task, getting a mean of 13.62 minutes and a standard deviation of 1.45 minutes.

With probability 0.95, what can the efficiency expert say about the maximum size of his error in estimating as 13.62 minutes the true average time it takes a mechanic to perform the repairs?

2 Use the data of Exercise 1 to construct a 0.99 confidence interval for the true average time it takes an automobile mechanic to perform the repairs.

3 A study of the annual growth of saguaro cacti showed that 50 of them, presumably a random sample, grew on the average 44.8 mm with a standard deviation of 4.7 mm during twelve months' time. Construct a 0.95 confidence interval for the true average annual growth of this kind of cactus.

4 With reference to Exercise 3, what can be said with probability 0.99 about the maximum size of the error, if the sample mean $\bar{x} = 44.8$ mm is used as an estimate of the true average annual growth of this kind of cactus?

5 A sample survey conducted in a large city in 1977 showed that 300 families spent on the average $3,943 on food with a standard deviation of $415.
   (a) Construct a 0.99 confidence interval for the mean of the population (of family food expenditures) sampled.
   (b) What can be said with probability 0.95 about the maximum size of the error, if $\bar{x} = \$3,943$ is used as an estimate of the average 1977 food expenditures of families living in the given city?

6 In an arithmetic achievement test, a random sample of 150 sixth graders from a very large school district had a mean score of 78.1 with a standard deviation of 9.5.
   (a) Construct a 0.95 confidence interval for the mean score which all the sixth graders in this school district would get if they took the test.
   (b) What can be said with probability 0.98 about the maximum size of the error made in estimating as 78.1 the average performance on this test by all sixth graders in the school district?

7 A company which writes hospitalization insurance takes a random sample of size 200 from its extensive records of claims filed in 1978 and finds that these policyholders averaged 8.30 days in the hospital with a standard deviation of 1.70 days. If 8.30 is used as an estimate of the average number of days that the company's hospitalized policyholders spent in the hospital in 1978, with what probability can it be said that this estimate is "off" by at most 0.2 day?

8 If a sample constitutes an appreciable portion of a finite population (say, 5 percent or more), the various formulas used in Section 10.2 must be modified by applying the finite population correction factor. For instance, the formula for E on page 249 becomes

$$E = z_{\alpha/2} \cdot \frac{\sigma}{\sqrt{n}} \cdot \sqrt{\frac{N-n}{N-1}}$$

   (a) A random sample of 40 drums of a chemical, drawn from among 200 such drums whose weights have the standard deviation $\sigma = 12.2$ pounds, has a mean weight of 240.8 pounds. If we estimate the mean weight of all 200 drums as 240.8 pounds, what can we assert with probability 0.95 about the possible size of the error?
   (b) Rework part (b) of Exercise 6, given that there are 750 sixth graders in the school district.

10.4 The Estimation of Means (A Bayesian Method)

257

9 Use the finite population correction factor, $\sqrt{\dfrac{N - n}{N - 1}}$, to modify the large-sample confidence interval formula on page 250, and thus make it applicable to problems in which a sample constitutes a substantial portion of a finite population.

(a) With reference to Exercise 7, construct a 0.99 confidence interval for the average number of days that the company's hospitalized policy holders spent in the hospital in 1978, given that 1,000 of them were hospitalized that year.

(b) Rework part (a) of Exercise 6, given that there are 750 sixth graders in the school district.

10 A stratified random sample of size $n = 330$ is allocated proportionally to the four strata of a finite population for which $N_1 = 2,000, N_2 = 6,000, N_3 = 10,000, N_4 = 4,000, \sigma_1 = 20, \sigma_2 = 25, \sigma_3 = 15,$ and $\sigma_4 = 30$ cm.

(a) If the mean of this sample is used to estimate the mean of the population, what can we assert with probability 0.95 about the maximum error? (*Hint*: Use the standard error formula of Exercise 36 on page 245.)

(b) If the mean of the sample is 117.39 cm, construct a 0.99 confidence interval for the mean of the population.

11 In a study of television viewing habits, it is desired to estimate the average number of hours that teenagers spend watching per week. Assuming that it is reasonable to use a standard deviation of 3.2 hours, how large a sample would be required if one wants to be able to assert with probability 0.95 that the sample mean will be "off" by at most 24 minutes?

12 It is desired to estimate the mean lifetime of a certain kind of vacuum cleaner. Given that $\sigma = 320$ days, how large a sample is needed to be able to assert with probability 0.99 that the mean of the sample will differ from the mean of the population by less than 45 days?

13 A major truck stop has kept extensive records on various transactions with its customers. If a random sample of 18 of these records show average sales of 63.84 gallons of diesel fuel with a standard deviation of 2.75 gallons, construct a 0.99 confidence interval for the mean of the population sampled.

14 Nine bearings made by a certain process have a mean diameter of 0.505 cm and a standard deviation of 0.004 cm. What can be said with probability 0.95 about the maximum size of the error, if 0.505 cm is used as an estimate of the actual average diameter of bearings made by this process?

15 In an air pollution study, an experiment station obtained a mean of 2.26 micrograms of suspended benzene-soluble organic matter per cubic meter with a standard deviation of 0.56 from a random sample of $n = 8$ different specimens.

(a) If the mean of this sample is used to estimate the corresponding true mean, what can the station assert with probability 0.99 about the maximum size of the error?

(b) Construct a 0.98 confidence interval for the true mean of the population sampled.

16 In six test runs it took 13, 14, 12, 16, 12, and 11 minutes to assemble a certain mechanical device. Construct a 0.95 confidence interval for the actual mean time it takes to assemble the device.

17 In establishing the authenticity of an ancient coin, its weight is often of critical importance. If four experts independently weighed a Phoenician tetradrachm and obtained 14.29, 14.33, 14.27, and 14.31 grams, what can they assert with probability 0.99 about the maximum size of their error, if they use the mean of these weights to estimate the true weight of the coin?

18 A dentist finds in a routine check that six prison inmates require, respectively, 2, 3, 6, 0, 4, and 3 fillings.

    (a) If he uses the mean of this sample to estimate the true mean of the population sampled, what can he say with probability 0.95 about the maximum size of his error?

    (b) Construct a 0.90 confidence interval for the average number of fillings required by the inmates of this (presumably large) prison.

19 An actuary feels that the prior distribution of the average annual losses for a certain kind of liability coverage is a normal distribution with the mean $\mu_0 = \$103.50$ and the standard deviation $\sigma_0 = \$4.30$. She also knows that for any one policy the losses will vary from year to year with the standard deviation $\sigma = \$20.85$. If the losses of a particular policy like this average $\$185.26$ over a period of ten years, find a Bayesian estimate of its true average annual losses.

20 With reference to the example on page 255 and based only on the prior feelings of the chief engineer of the plant, what is the probability that the plant's actual average daily emission of sulfur oxides lies on the interval from 18.00 to 19.00 tons?

21 A college professor is making up a final examination in history which is to be given to a large group of students. His feelings about the average grade they should get is expressed subjectively by a normal distribution which has the mean $\mu_0 = 65.2$ and the standard deviation $\sigma_0 = 1.5$.

    (a) What probability does the professor thus assign to the true average grade being somewhere on the interval from 63.0 to 68.0?

    (b) If the examination is tried on a random sample of 40 students whose grades have a mean of 72.9 and a standard deviation of 7.4, calculate the posterior mean as a Bayesian estimate of the average grade all the students in the large group should get on this test.

    (c) How does the information given in part (b) affect the probability of part (a)?

10.5
Tests of Hypotheses

                        In each of the problems mentioned in Section 10.1, somebody was interested in determining the true value of a quantity, so they were all problems of estimation. They would have concerned **tests of hypotheses,** however, if the psychologist had wanted to decide whether the average time it takes an adult person to react to the stimulus is really 0.44 second, if the engineer had wanted to determine whether the variability of the strength of the new alloy is less than that of a known substance, if the biologist had wanted to check another biologist's claim that 2.3 percent of the given insects are born physically defective, if the retailer had wanted to find out whether the actual average family income of all families living within a mile of the proposed new store is $12,500, ... , and if a D.P.S. engineer had wanted to investigate the claim that 14.5 percent of all one-car accidents are due to driver fatigue. Now it must be decided in each case whether to accept or reject a hypothesis, namely, whether to accept or reject an assertion or claim about the parameter of a population.

10.6
Two Kinds of Errors

At this point it may seem to make little difference whether we state the problems as in Section 10.1 or as in Section 10.5, but it will soon become apparent that a number of considerations arise in connection with tests of hypotheses that are not present in problems of estimation.

EXAMPLE   To illustrate the kind of situation we face when testing a statistical hypothesis, suppose that the members of an airport's planning commission are considering the possibility of redesigning the parking facilities. In particular, they want to check the claim (made by the operator of the parking facilities) that on the average cars are parked in the short-term parking area for about 42.5 minutes. So, they instruct a member of their staff to take a random sample of 50 ticket stubs showing time of arrival and time of departure, with the intention of accepting the claim if the mean of the sample falls anywhere from 40.5 to 44.5 minutes. Otherwise, they will reject the claim, and in either case they will take whatever action is thus called for in their plans.

This provides a clear-cut criterion for accepting or rejecting the claim, but unfortunately it is not infallible. Since the decision is based on a sample, there is the possibility that the sample mean will be greater than 44.5 minutes, or less than 40.5 minutes, even though the true mean is close to 42.5 minutes. The chances of this kind of error could be reduced by changing the criterion, say, by accepting the claim when the sample mean falls anywhere from 38.5 to 46.5 minutes, but this would have the undesirable effect of increasing the chances of accepting the claim when the true mean is not even close to 42.5 minutes. Thus, before adopting the original decision criterion (or, for that matter, any decision criterion) it is wise to investigate the chances that it will lead to a wrong decision.

To go ahead with this, we shall have to be more specific about the hypothesis we want to test—"close to 42.5 minutes" is not good enough. Although it does not really matter whether the true mean is 42.5 minutes, 42.503 minutes, or perhaps 42.496 minutes, the probability of erroneously rejecting the claim cannot be calculated unless we are specific. So, let us formulate the hypothesis $\mu = 42.5$ minutes and let us see what the chances are that we will nevertheless

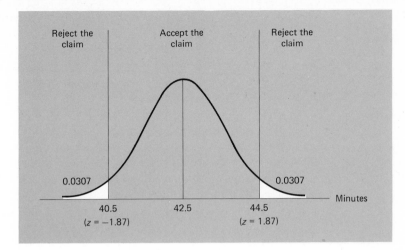

FIGURE 10.5

Test criterion.

get a sample mean less than 40.5 minutes or greater than 44.5 minutes, assuming that it is known (from similar studies) that $\sigma = 7.6$ minutes for data of this kind. This probability is given by the area of the white regions of Figure 10.5, and it can easily be determined by approximating the sampling distribution of the mean with a normal distribution. Assuming that the population sampled is large enough to be treated as infinite, we have $\sigma_{\bar{x}} = \dfrac{\sigma}{\sqrt{n}} = \dfrac{7.6}{\sqrt{50}} = 1.07$, and the dividing lines of the criterion, in standard units, are

$$z = \frac{40.5 - 42.5}{1.07} = -1.87 \quad \text{and} \quad z = \frac{44.5 - 42.5}{1.07} = 1.87$$

It follows from Table I that the area in each "tail" of the sampling distribution of Figure 10.5 is $0.5000 - 0.4693 = 0.0307$, and hence that the probability of getting a value in either tail of the sampling distribution (and erroneously rejecting the hypothesis $\mu = 42.5$ minutes) is $0.0307 + 0.0307 = 0.0614$, or approximately 0.06. Whether this is an acceptable risk is for the members of the planning commission to decide; it would have to depend on the consequences of their making such an error.

Let us now look at the other kind of situation, where the test fails to detect that μ is not equal to 42.5 minutes. Suppose, for instance, that $\mu = 45.5$ minutes, in which case the probability of nevertheless accepting the claim, $\mu = 42.5$ minutes, is given by the area of the white region of Figure 10.6, namely, the area under the

10.6 Two Kinds of Errors

261

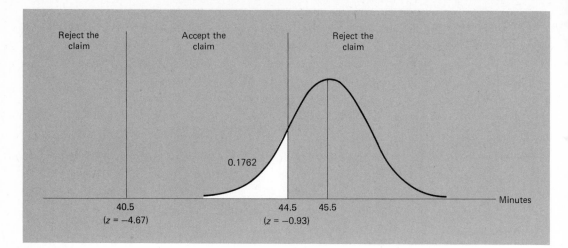

FIGURE 10.6

Test criterion.

curve between 40.5 and 44.5. The mean of the sampling distribution is now 45.5, its standard deviation is as before $\sigma_{\bar{x}} = \dfrac{7.6}{\sqrt{50}} = 1.07$, and the dividing lines of the criterion, in standard units, are

$$z = \frac{40.5 - 45.5}{1.07} = -4.67 \quad \text{and} \quad z = \frac{44.5 - 45.5}{1.07} = -0.93$$

Since the area under the curve to the left of $z = -4.67$ is negligible, it follows from Table I that the area of the white region of Figure 10.6 is $0.5000 - 0.3238 = 0.1762$, or approximately 0.18. Again, it is up to the members of the planning commission to decide whether this represents an acceptable risk.

The situation described in this example is typical of testing a statistical hypothesis, and it may be summarized in the following table, where we refer to the hypothesis being tested as hypothesis H:

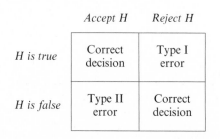

	Accept H	*Reject H*
H is true	Correct decision	Type I error
H is false	Type II error	Correct decision

If hypothesis H is true and accepted or false and rejected, the decision is in either case correct. If hypothesis H is true but rejected, it is rejected in error, and if hypothesis H is false but accepted, it is accepted in error. The first of these errors is called a **Type I error** and the probability of committing it is designated by the Greek letter α (*alpha*); the second is called a **Type II error** and the probability of committing it is designated by the Greek letter β (*beta*). Thus, in our example we showed that for the given test criterion $\alpha = 0.06$, and $\beta = 0.18$ when $\mu = 45.5$ minutes.

The scheme outlined above is reminiscent of what we did in Section 6.2. Analogous to the decision which the furniture manufacturer had to make in the example on page 137, we now have to decide whether to accept or reject hypothesis H. It is difficult to carry this analogy much further, though, for in actual practice we can seldom associate cash payoffs with the various possibilities, as we did in that example. Nevertheless, the seriousness of the consequences will generally determine whether a testing procedure presents acceptable risks.

In calculating the probability of a Type II error in our example, we arbitrarily chose the alternative value $\mu = 45.5$ minutes. However, in this problem, as in most others, there are infinitely many other alternatives, and for each one of them there is a positive probability β of erroneously accepting the hypothesis H.

EXAMPLE   Let us return to our example and calculate the probability of erroneously accepting the hypothesis $\mu = 42.5$ minutes for several other alternatives. Then we can examine these probabilities to see how well, under various conditions, the criterion controls the risks facing the planning commission.

The procedure is precisely the same as before. The probability of accepting the hypothesis $\mu = 42.5$ minutes when it is, in fact, 46.5 minutes, is just the probability that the mean of a random sample of size 50 drawn from a population with the mean 46.5 and the standard deviation 7.6 will fall anywhere from 40.5 to 44.5 minutes. By drawing a figure like Figure 10.6 with $\mu = 46.5$, and proceeding as we did above, we find that $\beta = 0.03$; namely, that the probability is 0.03 that the planning commission will erroneously decide to accept the claim that the mean parking time is 42.5 minutes, when in fact it is 46.5 minutes.

Proceeding in the same way, we obtain the other probabilities β shown in the middle column of the table given below. Since a Type II error is committed when the hypothesis H is accepted in error, the entries in the right-hand column of the table are the same as those of the middle column, except for the value which corresponds to $\mu = 42.5$ minutes. When $\mu = 42.5$ minutes, hypothesis H is true, and the probability of accepting it is the probability of not committing a Type I error; namely, $1 - \alpha = 1 - 0.06 = 0.94$.

Value of μ	Probability of type II error	Probability of accepting H
37.5	0.003	0.003
38.5	0.03	0.03
39.5	0.18	0.18
40.5	0.50	0.50
41.5	0.82	0.82
42.5	—	0.94
43.5	0.82	0.82
44.5	0.50	0.50
45.5	0.18	0.18
46.5	0.03	0.03
47.5	0.003	0.003

If we plot the probabilities of accepting H as in Figure 10.7 and fit a smooth curve, we obtain the **operating characteristic curve** of the test criterion, or simply the **OC-curve.** An operating characteristic curve provides a good overall picture of the merits of a test criterion.

EXAMPLE Examination of the curve of Figure 10.7 shows that the probability of accepting hypothesis H is greatest when it is true. For small departures from $\mu = 42.5$ minutes there is still a high probability of accepting H, and this is as it should be. Originally, the operator of the parking facilities claimed that cars are parked on the average for *about* 42.5 minutes. However, for larger and larger departures from $\mu = 42.5$ minutes in either direction, the probabilities of failing to detect them (and erroneously accepting hypothesis H) become smaller and smaller.

Of course, the operating characteristic curve of Figure 10.7 applies only to the case where the hypothesis $\mu = 42.5$ minutes is

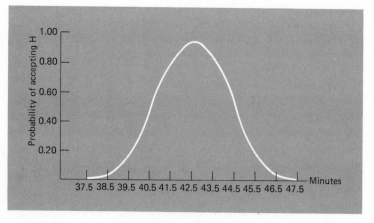

FIGURE 10.7

Operating characteristic curve.

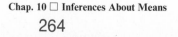

Chap. 10 □ Inferences About Means

accepted if the mean of a random sample of size 50 falls between 40.5 and 44.5 minutes and rejected otherwise, and σ is known to be 7.6 minutes. Changes in the testing procedure will automatically change the shape of the operating characteristic curve (see Exercises 7 and 8 on page 270); in fact, OC-curves can often be made to assume desired shapes by suitable choices of the sample size and/or the dividing lines of the test criteria.

If we had plotted the probabilities of rejecting H instead of those of accepting H, we would have obtained the graph of the **power function** of the test criterion rather than its operating characteristic curve. The concept of an OC-curve is used more widely in applications, especially in industrial applications, while the concept of a power function is used more widely in matters that are of theoretical interest. A detailed study of operating characteristic curves and power functions would go considerably beyond the scope of this text, and the purpose of our example was mainly to show how statistical methods can be used to measure and control the risks to which one is exposed when testing hypotheses. Of course, these methods are not limited to the particular problem concerning the average length of time that cars are parked in a given area—hypothesis H could have been the hypothesis that the average age of divorced women at the time of divorce is 35.6 years, the hypothesis that an antibiotic is 83 percent effective, the hypothesis that a computer-assisted method of instruction will on the average raise students' scores on a standard achievement test by 8.4 points, and so forth.

10.7
Null Hypotheses

In the example of the preceding section, we had less trouble with Type I errors than with Type II errors, because we formulated the hypothesis H as a **simple hypothesis**; that is, we formulated the hypothesis H so that μ took on a single value and the probability of a Type I error could be calculated. Had we formulated instead the **composite hypothesis** $\mu \neq 42.5$ minutes, the composite hypothesis $\mu < 42.5$ minutes, or the composite hypothesis $\mu > 42.5$ minutes, where in each case μ can take on more than one possible value, we could not have calculated the probability of a Type I error without specifying by how much μ differs from, is less than, or is greater than 42.5 minutes.

To be able to calculate the probability of a Type I error (that is, to know what to expect when a hypothesis is true), it is customary to formulate hypotheses to be tested as simple hypotheses; this is why we substituted $\mu = 42.5$ minutes for "the average time is about 42.5 minutes" in our example. In many instances this will require that we hypothesize the opposite of what we hope to prove. For instance, if we want to show that one method of teaching computer programming is more effective than another, we hypothesize that the two methods are equally effective. Similarly, if we want to show that one method of irrigating the soil is more expensive than another, we hypothesize that the two methods are equally expensive; and if we want to show that a new copper-bearing steel has a higher yield strength than ordinary steel, we hypothesize that the two yield strengths are the same. Since we hypothesize that there is no difference in the effectiveness of the two teaching methods, no difference in the cost of the two methods of irrigation, and no difference in the yield strength of the two kinds of steel, we call hypotheses like these **null hypotheses** and denote them H_0. Nowadays, the term "null hypothesis" is used for any hypothesis set up primarily to see whether it can be rejected, and the idea of setting up a null hypothesis is common even in nonstatistical thinking. It is precisely what we do in criminal proceedings, where an accused is assumed to be innocent unless his guilt is established beyond a reasonable doubt. The assumption that the accused is not guilty is a null hypothesis.

10.8
Significance Tests

In Section 10.6 we saw how the probability of committing a Type I error can easily be calculated when we are given a simple hypothesis, an unambiguous criterion, and enough information about the situation to apply the theory of Chapter 9. We also saw how the probabilities of Type II errors can be calculated for various alternative values of μ. The only time there is no problem in finding the probabilities of either kind of error is when we test a simple hypothesis against a simple alternative. [This would have been the case in the example on page 260 if we had been interested in testing the claim that on the average cars are parked in the given area for 42.5 minutes against the specific conflicting claim that on the average cars are parked there for, say, 45.5 minutes (see also Exercise 8 on page 280).]

Although a positive probability of committing a Type II error (accepting a false hypothesis) exists for each value of μ alternative to the value assumed under the simple hypothesis H, we can sometimes avoid Type II errors altogether.

EXAMPLE   Suppose, for instance, that a sociologist knows that in a given city licensed drivers average 1.4 traffic tickets per year with a standard deviation of 0.6. In trying to confirm her suspicion that licensed drivers over 65 average more than 1.4 traffic tickets per year, she checks the 1978 records of 40 randomly selected licensed drivers over 65 in the given city, and tests the null hypothesis that there is no difference between the performance of licensed drivers over 65 and that of all licensed drivers (that $\mu = 1.4$ applies also to licensed drivers over 65) by using the following criterion:

Reject the null hypothesis $\mu = 1.4$ (and accept the alternative $\mu > 1.4$) if the 40 licensed drivers over 65 average more than 1.6 traffic tickets; otherwise, reserve judgment (perhaps, pending further checks).

With this criterion there is no possibility of committing a Type II error; no matter what happens, the null hypothesis is never really accepted.

The procedure we have just outlined is referred to as a **significance test.** If the difference between what we expect under the null hypothesis and what we observe in a sample is too large to be reasonably attributed to chance, we reject the null hypothesis. If the difference between what we expect and what we observe is so small that it may well be attributed to chance, we say that the results are **not statistically significant.** We then accept the null hypothesis or reserve judgment, depending on whether a definite decision one way or the other is required.

EXAMPLE In the preceding example, the sociologist's suspicion that licensed drivers over 65 average more than 1.4 tickets is confirmed if the sample mean exceeds 1.6; in that case it is felt that the difference between $\mu = 1.4$ and the value obtained for $\bar{x}$ is too large to be attributed to chance (see also Exercise 6 on page 270). If the sample mean does not exceed 1.6, the sociologist's suspicion is not confirmed. Observe that we did not say that the sociologist's suspicion is wrong, or unjustified, when the sample mean does not exceed 1.6—we merely said that it is not confirmed. Thus, she may well wish to continue her investigation, perhaps, with a larger sample.

EXAMPLE Referring again to the airport parking example on page 260, we could convert the criterion into that of a significance test by writing

Reject the hypothesis $\mu = 42.5$ minutes (and accept the alternative $\mu \neq 42.5$ minutes) if the mean of the 50 sample values is less than 40.5 minutes or greater than 44.5 minutes; reserve judgment if the mean falls anywhere from 40.5 to 44.5 minutes.

So far as the rejection of the null hypothesis is concerned, the criterion has remained unchanged and the probability of a Type I error is still 0.06. However, so far as its acceptance is concerned, we are now playing it safe by reserving judgment.

Reserving judgment in a significance test is similar to what happens in court proceedings where the prosecution does not have sufficient evidence to get a conviction, but where it would be going too far to say that the defendant definitely did not commit the crime. In general, whether one can afford the luxury of reserving judgment in any given situation depends entirely on the nature of the situation. If a decision must be reached *one way or the other*, there is no way of avoiding the risk of committing a Type II error.

Since the general problem of testing hypotheses and constructing statistical decision criteria often seems confusing, at least to the beginner, it will help to proceed systematically as outlined in the following five steps: **(1) We formulate a simple (null) hypothesis so that the probability of a Type I error can be calculated. (2) We formulate an alternative hypothesis so that the rejection of the null hypothesis is equivalent to the acceptance of the alternative hypothesis.**

In the airport parking example the null hypothesis is $\mu = 42.5$ minutes, and the alternative hypothesis is $\mu \neq 42.5$ minutes (because the planning commission wants protection against the possibilities that 42.5 minutes may be too high or too low). We refer to this kind of alternative as a **two-sided alternative.** In the traffic ticket example the null hypothesis is $\mu = 1.4$, and the alternative hypothesis is $\mu > 1.4$ (because the sociologist hopes to be able to show that licensed drivers over 65 average more than 1.4 tickets per year). This is called a **one-sided alternative.** We can also write a one-sided alternative with the inequality going the other way. For instance, if we hope to be able to show that the average time required to do a certain job is less than 15 minutes, we would test the null hypothesis $\mu = 15$ minutes against the alternative $\mu < 15$ minutes.

As in the three examples of the preceding paragraph, alternative hypotheses usually specify that the population mean (or whatever other parameter may be of concern) is not equal to, greater than, or less than the value assumed under the null hypothesis. For any given problem, the choice of an appropriate alternative depends mostly on what we hope to be able to show, or better, perhaps, where we want to put the burden of proof.

EXAMPLE Suppose that a dress manufacturer, whose sewing machines average 128 dresses per work shift, is considering the purchase of new sewing machines. If he does not want to buy the new machines unless they will actually increase the output, he would test the null hypothesis $\mu = 128$ against the alternative hypothesis $\mu > 128$, and buy the new machines only if the null hypothesis can be rejected. On the other hand, if he wants to buy the new machines (which have some other nice features) unless they will actually decrease the output, he would test the null hypothesis $\mu = 128$ against the alternative hypothesis $\mu < 128$, and buy the new machines unless the null hypothesis can be rejected.

Having formulated a null hypothesis and a suitable alternative, we then proceed with the following steps[†]: **(3) We specify the probability of committing a Type I error; if possible, desired, or necessary, we may also make some specifications about the probabilities of Type II errors for specific alternatives. (4) Using suitable statistical theory, we construct a test criterion for testing the null hypothesis formulated in (1) against the alternative hypothesis formulated in (2) with α, the probability of a Type I error, as specified in (3).**

The probability of committing a Type I error is also called the **level of significance**, and it is usually set at 0.05 or 0.01. Testing a hypothesis at the level of significance $\alpha = 0.05$ simply means that we are fixing the probability of rejecting the hypothesis if it is true at 0.05. The decision to use $\alpha = 0.05$, $\alpha = 0.01$, or some other value, depends mostly on the consequences of committing a Type I error in the given situation. However, it must be kept in mind that usually we cannot make the probability of a Type I error too small either, for this will have the tendency to make the probabilities of serious Type II errors too large (see, for example, Exercise 9 on page 270).

So far as step (4) is concerned, in the airport parking example we based the criterion on the normal-curve approximation to the sampling distribution of $\bar{x}$; in general, this step depends on the statistic upon which we want to base the decision and on its sampling distribution. Looking back at our two examples, we find that the construction of a test criterion depends also on the alternative hypothesis we happen to choose. In the airport parking example, we used a **two-sided criterion (two-sided test,** or **two-tail test)** with the two-sided alternative $\mu \neq 42.5$ minutes, rejecting the null hypothesis for large or small values of $\bar{x}$; in the traffic ticket example, we used a **one-sided criterion (one-sided test,** or **one-tail test)** with the one-sided alternative $\mu > 1.4$, rejecting the null hypothesis only for large values of $\bar{x}$; and in the example on page 268 dealing with the time required to do a certain job, we would use a one-sided criterion (one-sided test, or one-tail test) with the one-sided alternative $\mu < 15$ minutes, rejecting the null hypothesis only for small values of $\bar{x}$. In general, a test is called two-sided (or two-tailed) if the null hypothesis is rejected when a value of the test statistic falls in either one or the other of the two tails of its sampling distribution, and one-sided (or one-tailed) if the null hypothesis is rejected only when a value of the test statistic falls in one specified tail of its sampling distribution.

Before we can actually perform a test, we must make one more decision. **(5) We must specify whether the alternative to rejecting the null hypothesis formulated in (1) is to accept it or to reserve judgment.** This, as we have said, depends on whether we must make a decision one way or the other on the basis of the test, or whether the circumstances permit that we delay a decision pending further study. Quite often we accept null hypotheses with the tacit hope that we are not exposed to overly high risks of committing serious

---

[†] It is here that we depart from the airport parking example, where we first specified the criterion and then calculated the probabilities of Type I and Type II errors.

Type II errors; of course, if it is necessary and we have enough information, we can calculate the probabilities needed to get an over-all picture from the operating characteristic curve of the test criterion.

Before we discuss various special tests about means in the next few sections, let us point out that the concepts we have introduced here apply equally well to hypotheses concerning proportions, standard deviations, the randomness of samples, relationships among several variables, and so on.

EXERCISES

1 Suppose that a psychological testing service is asked to check whether an executive is emotionally fit to assume the presidency of a large corporation. What type of error would it commit if it erroneously rejects the null hypothesis that the executive is fit for the job? What type of error would it commit if it erroneously accepts the null hypothesis that the executive is fit for the job?

2 Suppose we want to test the hypothesis that an antipollution device for cars is effective. Explain under what conditions we would commit a Type I error and under what conditions we would commit a Type II error.

3 Whether an error is a Type I error or a Type II error depends on how we formulate the hypothesis we want to test. To illustrate this, rephrase the hypothesis of Exercise 2 so that the Type I error becomes a Type II error, and vice versa.

4 A professor of education is concerned with the effectiveness of a programmed teaching technique.
   (a) What hypothesis is she testing, if she is committing a Type I error when she erroneously concludes that the technique is effective?
   (b) What hypothesis is she testing, if she is committing a Type II error when she erroneously concludes that the technique is effective?

5 Verify the values of the probabilities of Type II errors given in the middle column of the table on page 264.

6 Verify for the traffic ticket example on page 266 that the probability of a Type I error is 0.02.

7 Suppose that in the airport parking example on page 260 the criterion is changed so that the hypothesis $\mu = 42.5$ minutes is accepted if the sample mean falls anywhere from 41.0 to 44.0 minutes; otherwise, the hypothesis is rejected.
   (a) Find the probability of a Type I error.
   (b) Find the probabilities of Type II errors for the same ten alternative values of μ as in the table on page 264 and plot the OC-curve.

8 Suppose that in the airport parking example the sample size is increased from 50 to 60, while the criterion remains as stated on page 260.
   (a) Find the probability of a Type I error.
   (b) Find the probabilities of Type II errors for the same ten alternative values of μ as in the table on page 264 and plot the OC-curve.

9 To reduce the probability of a Type I error in the airport parking example on page 260, the criterion is modified so that the hypothesis $\mu = 42.5$ minutes is accepted if the sample mean falls anywhere from 40.0 to 45.0 minutes; otherwise, the hypothesis is rejected.
   (a) Show that this reduces the probability of a Type I error from 0.06 to 0.02.

(b) Show that for $\mu = 45.5$ the probability of a Type II error is increased from 0.18 to 0.32.

(c) Show that for $\mu = 46.5$ the probability of a Type II error is increased from 0.03 to 0.08.

10 The average drying time of a manufacturer's paint is 20 minutes. Investigating the effectiveness of a modification in the chemical composition of his paint, the manufacturer wants to test the null hypothesis $\mu = 20$ minutes against a suitable alternative, where μ is the average drying time of the paint.

(a) What alternative hypothesis should the manufacturer use if he does not want to make the modification in the chemical composition of the paint unless it is definitely superior with respect to drying time?

(b) What alternative hypothesis should the manufacturer use if the new process is actually cheaper and he wants to make the modification unless it increases the drying time of the paint?

11 With reference to Exercise 10, what simple null hypothesis and what alternative hypothesis should the manufacturer use if he does not want to make the modification unless it decreases the drying time by at least five minutes?

12 A city police department is considering replacing the tires on its cars with radial tires. If μ_1 is the average number of miles the old tires last and μ_2 is the average number of miles the new tires will last, the null hypothesis to be tested is $\mu_1 = \mu_2$.

(a) What alternative hypothesis should the department use if it does not want to buy the radial tires unless they are definitely proved to give better mileage? In other words, the burden of proof is put on the radial tires and the old tires are to be kept unless the null hypothesis can be rejected.

(b) What alternative hypothesis should the department use if it is anxious to get the new tires (which have some other good features) unless they actually give poorer mileage than the old tires? Note that now the burden of proof is on the old tires, which will be kept only if the null hypothesis can be rejected.

(c) What alternative hypothesis would the department have to use so that the rejection of the null hypothesis can lead either to keeping the old tires or to buying the new ones?

10.9

Tests Concerning Means

Having used tests concerning means to illustrate the basic principles of hypothesis testing, let us now consider how we proceed in actual practice.

EXAMPLE An oceanographer wants to check whether the average depth of the ocean in a certain area is 72.4 fathoms, as had previously been recorded. From information gathered in similar pertinent studies, she can assume that the variability of her measurements is given by a standard deviation of $\sigma = 2.1$ fathoms; also, she decides to base the test on the mean of a random sample of size $n = 35$.

Beginning with steps (1) and (2) on page 268, she formulates the null hypothesis to be tested and the alternative hypothesis as

Null hypothesis: $\mu = 72.4$ fathoms
Alternative hypothesis: $\mu \neq 72.4$ fathoms

That is, she will consider as evidence against the hypothesis $\mu = 72.4$ fathoms values of $\bar{x}$ which are either significantly less than or significantly greater than 72.4. So far as step (3) is concerned, suppose that she fixes the probability of rejecting the null hypothesis if it is true at $\alpha = 0.05$.

This illustrates the first three steps. Next, in step (4), we shall depart somewhat from the procedure used in the examples given earlier in this chapter. In both the airport parking example and the traffic ticket example, we stated the test criterion in terms of values of $\bar{x}$; now we shall base it on the statistic

Statistic for test concerning mean

$$z = \frac{\bar{x} - \mu_0}{\sigma/\sqrt{n}}$$

in which μ_0 is, by the null hypothesis, the mean of the population sampled. The reason for working with standard units, or z-values, is that it enables us to formulate criteria which are applicable to a great variety of problems, not just one.

If we approximate the sampling distribution of the mean, as before, with a normal distribution, we can now use the test criteria shown in Figure 10.8; depending on the choice of the alternative hypothesis, the dividing line of the criterion is $-z_\alpha$ or z_α for the one-sided alternatives, and they are $-z_{\alpha/2}$ and $z_{\alpha/2}$ for the two-sided alternative. As we explained earlier, z_α and $z_{\alpha/2}$ are z-values such that the area to their right under the standard normal distribution is α and $\alpha/2$. Symbolically, we can formulate these criteria as follows:

Alternative hypothesis	*Reject the null hypothesis if*	*Accept the null hypothesis or reserve judgment if*
$\mu < \mu_0$	$z < -z_\alpha$	$z \geqslant -z_\alpha$
$\mu > \mu_0$	$z > z_\alpha$	$z \leqslant z_\alpha$
$\mu \neq \mu_0$	$z < -z_{\alpha/2}$ or $z > z_{\alpha/2}$	$-z_{\alpha/2} \leqslant z \leqslant z_{\alpha/2}$

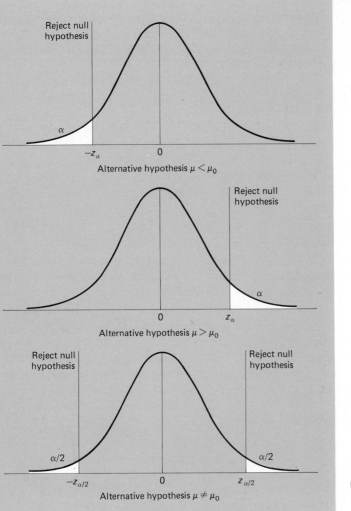

Reject null
hypothesis

α

$-z_\alpha$ 0

Alternative hypothesis $\mu < \mu_0$

Reject null
hypothesis

α

0   z_α

Alternative hypothesis $\mu > \mu_0$

Reject null
hypothesis

Reject null
hypothesis

$\alpha/2$

$\alpha/2$

$-z_{\alpha/2}$ 0 $z_{\alpha/2}$

Alternative hypothesis $\mu \neq \mu_0$

FIGURE 10.8

Test criteria.

If $\alpha = 0.05$, the dividing lines, or **critical values,** of the criteria are -1.64 or 1.64 for the one-sided alternatives, and -1.96 and 1.96 for the two-sided alternative; if $\alpha = 0.01$, the dividing lines of the criteria are -2.33 or 2.33 for the one-sided alternatives, and -2.58 and 2.58 for the two-sided alternative. All these values come directly from Table I (see Exercise 7 on page 198).

EXAMPLE   Returning now to the oceanographer, suppose that the mean of her measurements is 73.2 fathoms. Substituting this value, together with $\mu_0 = 72.4$, $\sigma = 2.1$, and $n = 35$ into the formula for z, she gets

$$z = \frac{73.2 - 72.4}{2.1/\sqrt{35}} = 2.25$$

10.9   Tests Concerning Means

Since this exceeds 1.96, she rejects the null hypothesis and concludes that the mean ocean depth in the given area does not equal 72.4 fathoms. In other words, the difference of $73.2 - 72.4 = 0.8$ between the observed mean and the hypothetical value of μ is too large to be attributed to chance.

Had the level of significance been 0.01 instead of 0.05, the hypothesis $\mu = 72.4$ could not have been rejected since $z = 2.25$ does not exceed 2.58, and the oceanographer would have to conclude that the difference between $\bar{x} = 73.2$ and $\mu_0 = 72.4$ is not statistically significant. This illustrates the important point that **the level of significance should always be specified before a significance test is actually performed. This will spare one the temptation of later choosing a level of significance which happens to suit one's particular objectives.**

In problems like this, some research workers accompany the values they obtain for z with corresponding **tail probabilities,** or **p-values,** namely, the probabilities of getting a numerical difference between $\bar{x}$ and μ_0 greater than or equal to that actually observed.

EXAMPLE In the oceanographer example, the tail probability is given by the area under the standard normal distribution to the left of -2.25 and to the right of 2.25, and it equals $2(0.5000 - 0.4878) = 0.0244$. This value falls between 0.01 and 0.05, which agrees with our earlier results, but observe that giving a tail probability does not absolve one from the responsibility of specifying the level of significance before the test is actually performed.

Since σ is unknown in many practical applications, we often have no choice but to replace it with an estimate in the formula for z. Thus, if n is large (30 or more), we substitute the sample standard deviation s for the unknown σ in the formula for z, and refer to the corresponding test as a **large-sample test for** μ. This is precisely what we shall do in the example which follows:

EXAMPLE Suppose that a trucking firm suspects that the average lifetime of 26,000 miles claimed for certain tires is too high. To check the claim at the level of significance 0.01, the firm puts 40 of these tires on its trucks and gets a mean lifetime of 25,563 miles and a standard deviation of 1,348 miles. The null hypothesis and the alternative hypothesis are

Null hypothesis: $\mu = 26{,}000$ miles
Alternative hypothesis: $\mu < 26{,}000$ miles

since the firm is interested in determining whether the tires may, perhaps, not last as long as claimed. Substituting the given values of $\bar{x}$, μ_0, n, and s (for σ) into the formula for z, the trucking firm gets

$$z = \frac{25{,}563 - 26{,}000}{1{,}348/\sqrt{40}} = -2.05$$

and since this is not less than -2.33, it cannot reject the null hypothesis. In other words, the trucking firm cannot reject the claim as being too high.

10.10

Tests Concerning Means (Small Samples)

When we do not know the value of the population standard deviation σ and the sample is small, $n < 30$, we shall assume, as on page 252, that the population we are sampling has roughly the shape of a normal distribution, and base our decision on the statistic

Statistic for small-sample test concerning mean

$$t = \frac{\bar{x} - \mu_0}{s/\sqrt{n}}$$

whose sampling distribution is the t distribution with $n - 1$ degrees of freedom. The criteria for small-sample tests concerning means based on the t statistic are those of Figure 10.8 and the table on page 272 with z replaced by t and z_α and $z_{\alpha/2}$ replaced by t_α and $t_{\alpha/2}$. As we explained on page 252, t_α and $t_{\alpha/2}$ are values for which the area to their right under the t distribution is equal to α and $\alpha/2$. All the dividing lines of such tests may be read from Table II, with the number of degrees of freedom equal to $n - 1$.

EXAMPLE   Suppose we want to decide, on the basis of a random sample of five specimens and at the level of significance $\alpha = 0.01$, whether the fat content of a certain kind of ice cream exceeds 12 percent. If the sample we get has a mean of 12.7 percent and a standard deviation of 0.38 percent, this mean is, of course, greater than 12, but it remains to be seen whether the difference between 12.7 percent and 12 percent is significant. Thus, we set up the following null and alternative hypotheses:

3. 747

Null hypothesis: $\mu = 12$ percent
Alternative hypothesis: $\mu > 12$ percent

and, substituting into the formula for t, we get

$$t = \frac{12.7 - 12}{0.38/\sqrt{5}} = 4.12$$

Since this exceeds 3.747, the value of $t_{0.01}$ for $5 - 1 = 4$ degrees of freedom, the null hypothesis must be rejected. In other words, we conclude that the average fat content of the ice cream does, indeed, exceed 12 percent. (The exact tail probability cannot be determined from Table II, but the author's *Monroe 1930* calculator yields 0.0073.)

10.11

Differences Between Means

There are many statistical problems in which we must decide whether an observed difference between two sample means can be attributed to chance. For instance, we may want to know whether there is really a difference in the average gasoline consumption of two kinds of cars, if sample data show that one kind averaged 24.6 miles per gallon while, under the same conditions, the other kind averaged 25.7 miles per gallon. Similarly, we may want to decide on the basis of samples whether men can perform a certain task faster than women, whether one kind of ceramic insulator is more brittle than another, whether the average diet in one country is more nutritious than that in another country, and so on.

The method we shall use to test whether an observed difference between two sample means can be attributed to chance, or whether it is statistically significant, is based on the following theory: If $\bar{x}_1$ and $\bar{x}_2$ are the means of two large independent random samples of size n_1 and n_2, the sampling distribution of the statistic $\bar{x}_1 - \bar{x}_2$ can be approximated closely with a normal curve having the mean $\mu_1 - \mu_2$ and the standard deviation

$$\sqrt{\frac{\sigma_1^2}{n_1} + \frac{\sigma_2^2}{n_2}}$$

where μ_1, μ_2, σ_1, and σ_2 are the means and the standard deviations of the two populations sampled. It is customary to refer to the standard deviation of this sampling distribution as the **standard error of the difference between two means.**

By "independent" samples we mean that the selection of one sample is in no way affected by the selection of the other. Thus, the theory does not apply to "before and after" kinds of comparisons, nor does it apply, say, to the comparison of the IQ's of husbands and wives. A special method for comparing the means of dependent samples is explained in Exercise 26 on page 284.

In most practical situations σ_1 and σ_2 are unknown, but if we limit ourselves to large samples (neither n_1 nor n_2 should be less than 30), we can use the sample standard deviations s_1 and s_2 as estimates of σ_1 and σ_2, and base the test of the null hypothesis $\mu_1 - \mu_2 = 0$, that the two populations have the same mean, on the statistic

Statistic for large-sample test concerning difference between two means

$$z = \frac{\bar{x}_1 - \bar{x}_2}{\sqrt{\dfrac{s_1^2}{n_1} + \dfrac{s_2^2}{n_2}}}$$

which has approximately the standard normal distribution. Indeed, this statistic is a z-value, for we calculate it by subtracting from $\bar{x}_1 - \bar{x}_2$ the mean of its sampling distribution, which under the null hypothesis is $\mu_1 - \mu_2 = 0$, and then dividing by the (estimated) standard error of the difference between two means. Depending on whether the alternative hypothesis is $\mu_1 - \mu_2 < 0$, $\mu_1 - \mu_2 > 0$, or $\mu_1 - \mu_2 \neq 0$, the criteria we use for the tests are again those shown in Figure 10.8 and also in the table on page 272, with $\mu_1 - \mu_2$ substituted for μ and 0 substituted for μ_0.

EXAMPLE   Suppose we want to see whether we can reject the null hypothesis that there is no difference in height between adult females born in two different countries. The sample data which we have at our disposal may be summarized as

$$n_1 = 120 \qquad \bar{x}_1 = 62.7 \qquad s_1 = 2.50$$
$$n_2 = 150 \qquad \bar{x}_2 = 61.8 \qquad s_2 = 2.62$$

where the measurements are in inches. Letting μ_1 and μ_2 represent the true average heights of adult females born in the two countries, the null hypothesis we shall want to test and the two-sided alternative hypothesis are

Null hypothesis: $\mu_1 - \mu_2 = 0$
Alternative hypothesis: $\mu_1 - \mu_2 \neq 0$

Setting the level of significance at $\alpha = 0.01$ and substituting into the formula for z, we get

$$z = \frac{62.7 - 61.8}{\sqrt{\dfrac{(2.50)^2}{120} + \dfrac{(2.62)^2}{150}}} = 2.88$$

10.11   Differences Between Means

277

and since this value exceeds 2.58, we conclude that the observed difference of $62.7 - 61.8 = 0.9$ inch is significant. In other words, the difference is too large to be accounted for by chance, so there must be an appreciable difference in the average height of adult females born in the two countries. Granted, the term "appreciable" is vague, but had the difference between μ_1 and μ_2 been minute, say, 0.001 inch, the chances are that the null hypothesis would not have been rejected.

10.12

Differences Between Means (Small Samples)

The significance test for the difference between two means described in the preceding section applies only to large independent random samples. However, an equivalent small-sample test is provided by the t distribution.[†] To use this test we must assume that the two independent random samples come from populations which can be approximated closely by normal distributions having the same variance. Specifically, we test the null hypothesis $\mu_1 - \mu_2 = 0$ against an appropriate one-sided or two-sided alternative with the statistic

Statistic for small-sample test concerning difference between two means

$$t = \frac{\bar{x}_1 - \bar{x}_2}{\sqrt{\dfrac{\sum (x_1 - \bar{x}_1)^2 + \sum (x_2 - \bar{x}_2)^2}{n_1 + n_2 - 2} \cdot \left(\dfrac{1}{n_1} + \dfrac{1}{n_2} \right)}}$$

where $\sum (x_1 - \bar{x}_1)^2$ is the sum of the squared deviations from the mean of the first sample while $\sum (x_2 - \bar{x}_2)^2$ is the sum of the squared deviations from the mean of the second sample. Since, by definition, $\sum (x_1 - \bar{x}_1)^2 = (n_1 - 1)s_1^2$ and $\sum (x_2 - \bar{x}_2)^2 = (n_2 - 1)s_2^2$, the above formula can be simplified somewhat when the two sample variances have already been calculated from the data.

Under the stated assumptions, it can be shown that the sampling distribution of this t statistic is the t distribution with $n_1 + n_2 - 2$ degrees of freedom. Depending on whether the alternative hypothesis is $\mu_1 - \mu_2 < 0$, $\mu_1 - \mu_2 > 0$, or $\mu_1 - \mu_2 \neq 0$, the criteria on which we base the small-sample test for the significance of the difference between two means are again those shown in Figure 10.8 and also in the table on page 272, with t substituted throughout for z, $\mu_1 - \mu_2$ substituted for μ, and 0 substituted for μ_0.

EXAMPLE Suppose we want to compare the heat-producing capacity of coal from two mines on the basis of the following data (in millions of calories per ton):

[†] For dependent samples see again Exercise 26 on page 284.

Mine 1: 8,400, 8,230, 8,380, 7,860, 7,930
Mine 2: 7,510, 7,690, 7,720, 8,070, 7,660

The means of these two random samples are 8,160 and 7,730; their difference is large, but it remains to be seen whether it is significant, say, at the level of significance $\alpha = 0.05$. Letting μ_1 and μ_2 represent the true average heat-producing capacities of coal from the two mines, the null hypothesis we shall want to test and the two-sided alternative hypothesis are

$$\text{Null hypothesis:} \qquad \mu_1 - \mu_2 = 0$$
$$\text{Alternative hypothesis:} \quad \mu_1 - \mu_2 \neq 0$$

To calculate t in accordance with the above formula, we must first determine the two quantities

$$\sum (x_1 - \bar{x}_1)^2 = (8,400 - 8,160)^2 + \cdots + (7,930 - 8,160)^2$$
$$= 253,800$$

and

$$\sum (x_2 - \bar{x}_2)^2 = (7,510 - 7,730)^2 + \cdots + (7,660 - 7,730)^2$$
$$= 170,600$$

Then, substitution of these values together with $n_1 = 5$, $n_2 = 5$, $\bar{x}_1 = 8,160$, and $\bar{x}_2 = 7,730$ into the formula for t yields

$$t = \frac{8,160 - 7,730}{\sqrt{\dfrac{253,800 + 170,600}{5 + 5 - 2} \cdot \left(\dfrac{1}{5} + \dfrac{1}{5}\right)}} = 2.95$$

Since this exceeds 2.306, the value of $t_{0.025}$ for $5 + 5 - 2 = 8$ degrees of freedom, it follows that the null hypothesis must be rejected. We conclude that the heat-producing capacity of the coal from the two mines is not the same, and it appears from the data that on the average the coal from the first mine produces more calories per ton.[†] How many more calories? That is another question, which we shall answer, at least in part, in Exercise 45 on page 357.

[†] With regard to the conclusion reached here, or that reached in any other significance test of the difference between two means, the reader would be well advised to read Section 16.2 at this time. It explains why we must be careful when we attribute statistically significant results to specific causes.

1   A law student, who wants to check a professor's claim that convicted embezzlers spend on the average 12.3 months in jail, takes a random sample of 35 such cases from court files. Using his results, namely, $\bar{x} = 11.5$ months and $s = 3.8$ months, test the null hypothesis $\mu = 12.3$ months against the alternative hypothesis $\mu \neq 12.3$ months at the level of significance $\alpha = 0.05$.

2   According to the norms established for a reading comprehension test, eighth graders should average 83.2 with a standard deviation of 8.6. If 45 randomly selected eighth graders from a certain school district averaged 86.7, test the null hypothesis $\mu = 83.2$ against the alternative hypothesis $\mu > 83.2$ at the level of significance $\alpha = 0.01$, and thus check the district superintendent's claim that her eighth graders are above average.

3   The security department of a factory wants to know whether the true average time required by the night watchman to walk his round is 25 minutes. If, in a random sample of 32 rounds, the night watchman averaged 25.8 minutes with a standard deviation of 1.5 minutes, determine at the level of significance 0.01 whether this is sufficient evidence to reject the null hypothesis $\mu = 25$ minutes.

4   Tests performed with a random sample of 40 diesel engines produced by a large manufacturer showed that they had a mean thermal efficiency of 31.8 percent with a standard deviation of 2.2 percent. Use this information and the level of significance $\alpha = 0.05$ to test the null hypothesis $\mu = 32.3$ percent against the alternative hypothesis that the true average thermal efficiency of the manufacturer's diesel engines is less than 32.3 percent.

5   In a study of new sources of food, it is reported that a pound of a certain kind of fish yields on the average 2.41 ounces of FPC (fish-protein concentrate), which is used to enrich various food products (including flour). Is this figure supported by a study in which 30 samples of this kind of fish yielded on the average 2.44 ounces of FPC (per pound of fish) with a standard deviation of 0.07 ounce, if we use

   (a) the level of significance $\alpha = 0.05$?
   (b) the level of significance $\alpha = 0.01$?

6   Find the tail probability for the tire-mileage example on page 274, namely, the probability that the test statistic will take on a value less than or equal to -2.05 when the null hypothesis is true.

7   An ambulance service claims that it takes it on the average 8.9 minutes to reach its destination in emergency calls. To check on this claim, the agency which licenses ambulance services has them timed on 50 emergency calls, getting a mean of 9.3 minutes with a standard deviation of 1.8 minutes. At the level of significance 0.05, does this constitute evidence that the figure claimed is too low?

8   If the null hypothesis $\mu = \mu_0$ is to be tested against the simple alternative hypothesis $\mu = \mu_A$, and the probabilities of Type I and Type II errors are to have the given values α and β, we must take a random sample of size

$$n = \frac{\sigma^2(z_\alpha + z_\beta)^2}{(\mu_A - \mu_0)^2}$$

where σ^2 is the variance of the population. Since this formula is based on normal-curve theory, the sample must be large or the population sampled must have roughly the shape of a normal distribution.

Suppose, for instance, that for a population with $\sigma = 6$ we want to test the null hypothesis $\mu = 200$ pounds against the alternative hypothesis $\mu = 198$ pounds. If α and β are both to be 0.05, we find that

$$n = \frac{6^2(1.64 + 1.64)^2}{(198 - 200)^2} = 97$$

rounded up to the nearest integer. Thus, a random sample of size $n = 97$ will expose us to the specified risks.

(a) Suppose we want to test the null hypothesis $\mu = \$250$ against the alternative hypothesis $\mu = \$255$ for a population whose standard deviation is $\sigma = \$12$. How large a random sample must we take, if α and β are both to be 0.01?

(b) Suppose we want to test the null hypothesis $\mu = 64$ inches against the alternative hypothesis $\mu = 61$ inches for a population whose standard deviation is $\sigma = 7.2$ inches. How large a random sample must we take, if α is to be 0.05 and β is to be 0.01? Also, for what values of $\bar{x}$ will the null hypothesis be rejected?

9 In an experiment with a new tranquilizer, the pulse rate of 16 patients was determined before they were given the tranquilizer and again five minutes later, and their pulse rate was found to be reduced on the average by 6.8 beats with a standard deviation of 1.9. Using the level of significance $\alpha = 0.05$, what can we conclude about the claim that this tranquilizer will reduce the pulse rate on the average by 7.5 beats in five minutes?

10 A random sample from a company's very extensive files shows that orders for a certain piece of machinery were filled, respectively, in 12, 10, 17, 14, 13, 18, 11, and 9 days. Use the level of significance $\alpha = 0.01$ to test the claim that on the average such orders are filled in 9.5 days. Choose the alternative hypothesis in such a way that rejection of the null hypothesis $\mu = 9.5$ days implies that it takes longer than that.

11 The yield of alfalfa from six test plots is 1.4, 1.8, 1.1, 1.9, 2.2, and 1.2 tons per acre. Test at the level of significance $\alpha = 0.05$ whether this supports the contention that the average yield for this kind of alfalfa is 1.5 tons per acre.

12 A reading teacher wants to determine whether a certain student has an average reading speed of $\mu = 750$ words per minute or whether her average reading speed is less than 750 words per minute. What can the teacher conclude at the level of significance $\alpha = 0.05$ if in six minutes the student reads, respectively, 730, 750, 740, 700, 700, and 760 words?

13 In the past, a golfer has averaged 81 on a certain course. If, with a new set of clubs, she averages 78 over five rounds with a standard deviation of 2.6, what can we conclude at the level of significance $\alpha = 0.05$ about the effect of the new clubs?

14 A random sample of 12 graduates of a secretarial school averaged 73.8 words per minute with a standard deviation of 7.9 words per minute on a typing test. Use the level of significance $\alpha = 0.01$ to test an employer's claim that the school's graduates average less than 75.0 words per minute.

10.12 **Differences Between Means (Small Samples)**

281

15 Five measurements of the tar content of a certain kind of cigarette yielded 14.5, 14.2, 14.4, 14.3, and 14.6 mg/cig (milligrams per cigarette). Show that the difference between the mean of this sample, $\bar{x} = 14.4$, and the average tar content claimed by the cigarette manufacturer, $\mu = 14.0$, is significant at $\alpha = 0.05$.

16 Suppose that in the preceding exercise the first measurement is recorded incorrectly as 16.0 instead of 14.5. Show that now the difference between the mean of the sample, $\bar{x} = 14.7$, and the average tar content claimed by the cigarette manufacturer, $\mu = 14.0$, is not significant at $\alpha = 0.05$. Explain the apparent paradox that even though the difference between $\bar{x}$ and μ has increased, it is no longer significant.

17 A sample study of the number of pieces of chalk the average professor uses was conducted at two universities. If 80 professors at one university averaged 13.6 pieces of chalk per month with a standard deviation of 2.4, while 60 professors at the other university averaged 12.7 pieces of chalk per month with a standard deviation of 3.3, test at the level of significance $\alpha = 0.01$ whether the difference between these two sample means is significant.

18 A sample study was made of the number of business lunches that executives claim as deductible expenses per month. If 40 executives in the insurance industry averaged 8.7 such deductions with a standard deviation of 1.9 in a given month, while 50 bank executives averaged 7.6 with a standard deviation of 2.1, test at the level of significance $\alpha = 0.05$ whether the difference between these two sample means is significant.

19 An investigation of two kinds of photocopying equipment showed that 60 failures of the first kind of equipment took on the average 84.2 minutes to repair with a standard deviation of 19.4 minutes, while 60 failures of the second kind of equipment took on the average 91.6 minutes to repair with a standard deviation of 18.8 minutes. Test at the level of significance $\alpha = 0.01$ whether the difference between these two sample means is significant.

20 As part of an industrial training program, some trainees are instructed by method *A*, which is straight teaching-machine instruction, and some are instructed by method *B*, which also involves the personal attention of an instructor. If random samples of size 10 are taken from large groups of trainees instructed by each of these two methods, and the scores which they obtained in an appropriate achievement test are

Method *A*: 71, 75, 65, 69, 73, 66, 68, 71, 74, 68
Method *B*: 72, 77, 84, 78, 69, 70, 77, 73, 65, 75

test the claim that method *B* is more effective. Use the level of significance $\alpha = 0.05$.

21 Six guinea pigs injected with 0.5 mg of a medication took on the average 15.4 seconds to fall asleep with a standard deviation of 2.2 seconds, while six other guinea pigs injected with 1.5 mg of the medication took on the average 11.2 seconds to fall asleep with a standard deviation of 2.6 seconds. Use the level of significance $\alpha = 0.05$ to test the null hypothesis that the difference in dosage has no effect.

22 Twelve randomly selected mature citrus trees of one variety have a mean height of 13.6 feet with a standard deviation of 1.2 feet, and fifteen randomly selected mature citrus trees of another variety have a mean height of 12.7 feet with a

standard deviation of 1.5 feet. Test at the level of significance $\alpha = 0.01$ whether the difference between the two sample means is significant.

23 The following are the numbers of sales which a random sample of nine salesmen of industrial chemicals in California and a random sample of six salesmen of industrial chemicals in Oregon made over a fixed period of time:

California: 41, 47, 62, 39, 56, 64, 37, 61, 52
Oregon: 34, 63, 45, 55, 24, 43

Use the level of significance $\alpha = 0.01$ to test whether the difference between the means of these two samples is significant.

24 Twelve measurements each of the hydrogen content (in percent number of atoms) of gases collected from the eruptions of two volcanos yielded $\bar{x}_1 = 41.2$, $\bar{x}_2 = 45.8$, $s_1 = 5.2$, and $s_2 = 6.7$. Use the level of significance $\alpha = 0.05$ to test the null hypothesis that there is no difference (with regard to hydrogen content) in the composition of the gases in the two eruptions.

25 In some problems we are interested in testing whether the difference between the means of two populations is equal to, less than, or greater than a given constant. So, we test the null hypothesis $\mu_1 - \mu_2 = \delta$ (delta), where δ is the given constant, against an appropriate alternative hypothesis. To perform this kind of test, we substitute $\bar{x}_1 - \bar{x}_2 - \delta$ for $\bar{x}_1 - \bar{x}_2$ in the numerator of the z-statistic on page 277, or the t-statistic on page 278, and otherwise proceed in the same way as before.

With reference to the illustration of Section 10.12, suppose we originally wanted to test whether the heat-producing capacity of the coal from the first mine exceeds that of the coal from the second mine by more than 100 (million calories per ton). Thus, we shall want to test the null hypothesis $\mu_1 - \mu_2 = 100$ against the alternative hypothesis $\mu_1 - \mu_2 > 100$, and the test statistic becomes

$$ t = \frac{8{,}160 - 7{,}730 - 100}{\sqrt{\dfrac{253{,}800 + 170{,}600}{5 + 5 - 2} \cdot \left(\dfrac{1}{5} + \dfrac{1}{5}\right)}} = 2.27 $$

Since this exceeds 1.860, the value of $t_{0.05}$ for $5 + 5 - 2 = 8$ degrees of freedom, it follows that the null hypothesis must be rejected. We conclude that the heat-producing capacity of the coal from the first mine exceeds that of the coal from the second mine by more than 100 (million calories per ton).

(a) Sample surveys conducted in a large county in 1950 and again in 1970 showed that in 1950 the average height of 400 ten-year-old boys was 53.2 inches with a standard deviation of 2.4 inches, while in 1970 the average height of 500 ten-year-old boys was 54.5 inches with a standard deviation of 2.5 inches. Use the level of significance $\alpha = 0.05$ to test whether the true average increase in height is at most 0.5 inch.

(b) To test the claim that the resistance of electric wire can be reduced by more than 0.050 ohm by alloying, 25 values obtained for alloyed wire yielded $\bar{x}_1 = 0.083$ ohm and $s_1 = 0.003$ ohm, and 25 values obtained for standard wire yielded $\bar{x}_2 = 0.136$ ohm and $s_2 = 0.002$ ohm. Use the level of significance $\alpha = 0.05$ to determine whether the claim has been substantiated.

10.12   Differences Between Means (Small Samples)

26 If we want to sudy the effectiveness of a new diet on the basis of weights "before and after," or if we want to study whatever differences there may be between the IQ's of husbands and wives, the methods of Sections 10.11 and 10.12 cannot be used. The samples are not independent; in fact, in each case the data are *paired*. To handle data of this kind, we work with the (signed) differences of the paired data and test whether these differences may be looked upon as a sample from a population for which $\mu = 0$. If the sample is small, we use the t test; otherwise, we use the large-sample test of Section 10.9. Apply this technique to determine the effectiveness of an industrial safety program on the basis of the following data (collected over a period of one year) on the average weekly loss of man-hours due to accidents in twelve plants "before and after" the program was put into operation: 37 and 28, 72 and 59, 26 and 24, 125 and 120, 45 and 46, 54 and 43, 13 and 15, 79 and 75, 12 and 18, 34 and 29, 39 and 35, 26 and 24. Use the level of significance $\alpha = 0.05$ to decide whether the safety program is effective.

27 The following data were obtained in an experiment designed to check whether there is a systematic difference in the weights (in grams) obtained with two different scales:

	Scale I	Scale II
Rock specimen 1	12.13	12.17
Rock specimen 2	17.56	17.61
Rock specimen 3	9.33	9.35
Rock specimen 4	11.40	11.42
Rock specimen 5	28.62	28.61
Rock specimen 6	10.25	10.27
Rock specimen 7	23.37	23.42
Rock specimen 8	16.27	16.26
Rock specimen 9	12.40	12.45
Rock specimen 10	24.78	24.75

Use the method of Exercise 26 and $\alpha = 0.01$ to test whether the difference between the means of the weights obtained with the two scales is significant.

BIBLIOGRAPHY   An informal introduction to interval estimation is given under the heading of "How to be precise though vague," in

MORONEY, M. J., *Facts from Figures*. London: Penguin Books, Inc., 1956.

A discussion of the theoretical foundation of the t distribution as well as other mathematical details omitted in this book may be found in most textbooks on mathematical statistics. Some of the theory of Bayesian estimation pertaining to Section 10.4 is discussed on pages 441–442 of

SCHLAIFER, R., *Probability and Statistics for Business Decisions*, New York: McGraw-Hill Book Co., 1959.

and in Section 5.1 of

LINDLEY, D. V., *Introduction to Probability and Statistics from a Bayesian Viewpoint*, *Part 2*. Cambridge: Cambridge University Press, 1965.

11

In Chapter 10 we learned how to judge the size of the error of an estimate of a mean, how to construct confidence intervals for means, and how to perform tests of hypotheses concerning the means of one and of two populations. As we shall see in this and some of the following chapters, very similar methods can be used for inferences about other population parameters.

In this chapter we shall concentrate on population standard deviations, or population variances, which are not only important in their own right, but which must sometimes be estimated before we can make inferences about other parameters. This was the case, for example, in Sections 10.2 and 10.9, where we made inferences about means, but had to know σ or estimate it from a sample.

Inferences About Standard Deviations

11.1

The Estimation of σ

Although there are other methods of estimating the standard deviation of a population (see, for example, Exercise 14 on page 199 and Exercise 8 on page 289), the sample standard deviation is the most widely used estimator of this parameter. Limiting our discussion here to this estimator, let us begin by constructing confidence intervals for σ based on the sample standard deviation. The theory on which these confidence intervals are based requires that the population sampled has roughly the shape of a normal distribution, in which case the statistic

Chi-square statistic

$$\chi^2 = \frac{(n-1)s^2}{\sigma^2}$$

called "chi-square," has as its sampling distribution an important continuous distribution called the **chi-square distribution.** (χ is the Greek lower case letter *chi*.) The mean of this distribution is $n - 1$ and, as with the t distribution, we call this quantity the number of degrees of freedom, or simply the **degrees of freedom.** An example of a chi-square distribution is shown in Figure 11.1; unlike the normal and t distributions, its domain is restricted to the nonnegative real numbers.

Analogous to z_α and t_α, we now define χ^2_α as the value for which the area to its right under the chi-square distribution is equal to α; like t_α, this value depends on the number of degrees of freedom. Thus, $\chi^2_{\alpha/2}$ is such that the area to its right under the curve is $\alpha/2$, while $\chi^2_{1-\alpha/2}$ is such that the area

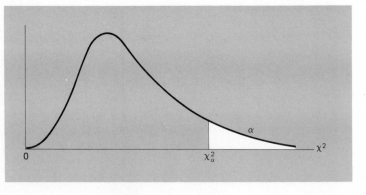

FIGURE 11.1
Chi-square distribution.

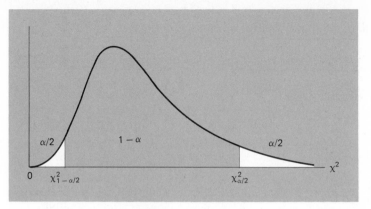

FIGURE 11.2

Chi-square distribution.

to its left under the curve is $\alpha/2$ (see also Figure 11.2). We make this distinction because the chi-square distribution, unlike the normal and t distributions, is not symmetrical. Among others, values of $\chi^2_{0.995}$, $\chi^2_{0.975}$, $\chi^2_{0.025}$, and $\chi^2_{0.005}$ are given in Table III at the end of the book for 1, 2, 3, . . . , and 30 degrees of freedom.

Referring to Figure 11.2, we find that we can assert with probability $1 - \alpha$ that a random variable having the chi-square distribution will take on a value between $\chi^2_{1-\alpha/2}$ and $\chi^2_{\alpha/2}$. Applying this result to the χ^2 statistic given on page 286, we can assert with probability $1 - \alpha$ that

$$\chi^2_{1-\alpha/2} < \frac{(n-1)s^2}{\sigma^2} < \chi^2_{\alpha/2}$$

This double inequality can be rewritten as

Confidence interval for σ^2

$$\frac{(n-1)s^2}{\chi^2_{\alpha/2}} < \sigma^2 < \frac{(n-1)s^2}{\chi^2_{1-\alpha/2}}$$

which is a $1 - \alpha$ **confidence interval for** σ^2, the population variance. Also, if we take the square root of each of the three terms in this double inequality, we get a $1 - \alpha$ confidence interval for σ, the population standard deviation.

EXAMPLE To illustrate the construction of a confidence interval for σ, let us return to the example on page 275, where five specimens of a certain kind of ice cream had a mean fat content of 12.7 percent and a standard deviation of 0.38 percent. For a 0.95 confidence interval, $\alpha = 0.05$, and we find from Table III that $\chi^2_{0.975} = 0.484$ and $\chi^2_{0.025} = 11.143$ for $5 - 1 = 4$ degrees of freedom. Hence, we get the

11.1 The Estimation of σ

following 0.95 confidence interval for σ, the actual standard deviation of the population (of fat contents) sampled:

$$\sqrt{\frac{4(0.38)^2}{11.143}} < \sigma < \sqrt{\frac{4(0.38)^2}{0.484}}$$

$$0.228 \text{ percent} < \sigma < 1.092 \text{ percent}$$

The kind of confidence interval we have just described is often referred to as a small-sample confidence interval, as it is used mainly when n is small, less than 30, and, of course, only when we can assume that the population which we are sampling has roughly the shape of a normal distribution. Otherwise, we make use of the fact that for large samples, $n \geqslant 30$, the sampling distribution of s can be approximated with a normal distribution having the mean σ and the standard deviation $\dfrac{\sigma}{\sqrt{2n}}$ (see Exercise 13 on page 235).

In this connection, $\dfrac{\sigma}{\sqrt{2n}}$ is called the **standard error of s,** and we can now assert with probability $1 - \alpha$ that

$$s - z_{\alpha/2}\frac{\sigma}{\sqrt{2n}} < \sigma < s + z_{\alpha/2}\frac{\sigma}{\sqrt{2n}}$$

Simple algebra leads to the following $1 - \alpha$ **large-sample confidence interval** for the population standard deviation σ:

Large-sample confidence interval for σ

$$\frac{s}{1 + \dfrac{z_{\alpha/2}}{\sqrt{2n}}} < \sigma < \frac{s}{1 - \dfrac{z_{\alpha/2}}{\sqrt{2n}}}$$

EXAMPLE   Referring to the illustration on page 14, which dealt with a large industrial plant's emission of sulfur oxides, we substitute $n = 80$, and $s = 5.55$ tons, and for $1 - \alpha = 0.95$ we obtain

$$\frac{5.55}{1 + \dfrac{1.96}{\sqrt{160}}} < \sigma < \frac{5.55}{1 - \dfrac{1.96}{\sqrt{160}}}$$

$$4.80 < \sigma < 6.57$$

This means that we can assert with a probability of 0.95 that the interval from 4.80 tons to 6.57 tons contains σ, the true standard deviation of the plant's daily emission of sulfur oxides.

EXERCISES

1 With reference to Exercise 14 on page 258, construct a 0.95 confidence interval for σ, the true standard deviation of the diameters of bearings made by the given process.

2 With reference to Exercise 15 on page 258, construct a 0.99 confidence interval for σ, the true standard deviation of the concentration of benzene-soluble organic matter in the air at the experiment station.

3 With reference to Exercise 16 on page 258, construct a 0.95 confidence interval for σ, the true standard deviation of the amount of time it takes to assemble the mechanical device.

4 With reference to Exercise 18 on page 259, construct a 0.99 confidence interval for σ, the true standard deviation of the number of fillings that are required by the prison inmates.

5 With reference to Exercise 3 on page 257, construct a 0.95 confidence interval for σ, the true standard deviation of the annual growth of saguaro cacti.

6 With reference to Exercise 5 on page 257, construct a 0.99 confidence interval for σ, the true standard deviation of annual family food expenditures in the given city.

7 With reference to Exercise 6 on page 257, construct a 0.98 confidence interval for σ, the true standard deviation of the arithmetic achievement scores of sixth graders in the given school district.

8 When we deal with very small samples, good estimates of the population standard deviation can often be obtained on the basis of the **sample range** (the largest sample value minus the smallest). Such quick estimates of σ are given by the sample range divided by the divisor d, which depends on the size of the sample; for samples from populations having roughly the shape of a normal distribution, its values are shown in the following table for $n = 2, 3, \ldots,$ and 12:

n	2	3	4	5	6	7	8	9	10	11	12
d	1.13	1.69	2.06	2.33	2.53	2.70	2.85	2.97	3.08	3.17	3.26

For instance, in the example on page 253, which dealt with the durability of a paint for highway center lines, we have $n = 8$ and a sample range of $167,800 - 108,300 = 59,500$ crossings. Since $d = 2.85$ for $n = 8$, we find that we can estimate σ, the true standard deviation of the population sampled, as

$$\frac{59,500}{2.85} = 20,877 \text{ crossings}$$

This is somewhat higher than the sample standard deviation $s = 19{,}200$ crossings, but not knowing the true value of σ, we cannot say which of the two estimates is actually closer.

(a) With reference to Exercise 16 on page 258, use this method to estimate the true standard deviation of the amount of time it takes to assemble the mechanical device, and compare the result with the sample standard deviation s.

(b) With reference to Exercise 17 on page 258, use this method to estimate the true standard deviation of such weights, and compare the result with the sample standard deviation s.

11.2

Tests Concerning Standard Deviations

In this section we shall consider the problem of testing the null hypothesis that a population standard deviation equals a specified constant σ_0, or that a population variance equals σ_0^2. This kind of test is required whenever we want to test the uniformity of a product, process, or operation. For instance, if we want to test whether a certain kind of glass is sufficiently homogeneous for making delicate optical equipment, whether the intelligence of a group of students is sufficiently uniform so that they can be taught in one class, whether a lack of uniformity in certain workers' performance may call for stricter supervision, and so forth.

The test of the null hypothesis $\sigma = \sigma_0$, the hypothesis that a population standard deviation equals a specified constant, is based on the same assumptions, the same statistic, and the same sampling theory as the small-sample confidence interval for σ. Assuming that our sample is random and comes from a population having roughly the shape of a normal distribution, we base our decision on the statistic

Statistic for test concerning standard deviation

$$\chi^2 = \frac{(n-1)s^2}{\sigma_0^2}$$

where n and s^2 are the sample size and the sample variance, and σ_0 is the value of the population standard deviation assumed under the null hypothesis. If the null hypothesis is true, the sampling distribution of this statistic is the chi-square distribution with $n - 1$ degrees of freedom; hence, the criteria for testing the null hypothesis $\sigma = \sigma_0$ against the alternative hypothesis $\sigma < \sigma_0$, $\sigma > \sigma_0$, or $\sigma \neq \sigma_0$ are as shown in Figure 11.3. For the one-sided alternative $\sigma < \sigma_0$, we reject the null hypothesis for values of χ^2 falling into the left-hand tail of its sampling distribution; for the one-sided alternative $\sigma > \sigma_0$, we reject the null hypothesis for values of χ^2 falling into

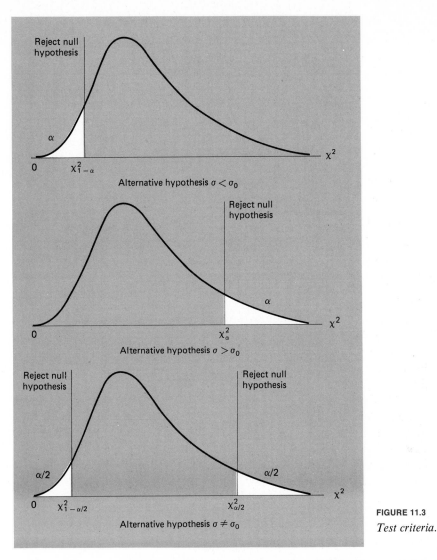

Reject null
hypothesis

α

0 $\chi^2_{1-\alpha}$

Alternative hypothesis $\sigma < \sigma_0$

Reject null
hypothesis

α

0 χ^2_{α}

Alternative hypothesis $\sigma > \sigma_0$

Reject null
hypothesis

Reject null
hypothesis

$\alpha/2$ $\alpha/2$

0 $\chi^2_{1-\alpha/2}$ $\chi^2_{\alpha/2}$

Alternative hypothesis $\sigma \neq \sigma_0$

FIGURE 11.3
Test criteria.

the right-hand tail of its sampling distribution; and for the two-sided alternative $\sigma \neq \sigma_0$, we reject the null hypothesis for values of χ^2 falling into either tail of its sampling distribution. The quantities χ^2_α, $\chi^2_{\alpha/2}$, and $\chi^2_{1-\alpha/2}$, for $n = 1, 2, 3, \ldots$, or 30 degrees of freedom, are given in Table III.

EXAMPLE   An automotive engineer, interested in certain safety features, must know whether the variability in the time it takes drivers to react to an emergency situation is less than $\sigma = 0.010$ second. To test the null hypothesis $\sigma = 0.010$ against the alternative hypothesis $\sigma < 0.010$ at the level of significance $\alpha = 0.05$, the engineer measures the reaction times of 15 persons, presumably a random sample,

11.2   Tests Concerning Standard Deviations

and gets $s = 0.006$. Substitution into the formula for χ^2 gives

$$\chi^2 = \frac{14(0.006)^2}{(0.010)^2} = 5.04$$

and since this is less than 6.571, the value of $\chi^2_{0.95}$ for $15 - 1 = 14$ degrees of freedom, the engineer rejects the null hypothesis and concludes that the variability of the reaction times is, indeed, less than $\sigma = 0.010$ second.

When n is large, $n \geqslant 30$, we can base tests of the null hypothesis $\sigma = \sigma_0$ on the same theory as the large-sample confidence intervals of the preceding section. That is, we use the statistic

<table>
<tr><td>Statistic for
large-sample
test concerning
standard deviation</td><td>$$z = \frac{s - \sigma_0}{\sigma_0/\sqrt{2n}}$$</td></tr>
</table>

whose sampling distribution is approximately the standard normal distribution, and the criteria of Figure 10.8.

EXAMPLE   The specifications for the mass production of certain springs require, among other things, that the standard deviation of their compressed lengths should not exceed 0.040 cm. To test the null hypothesis $\sigma = 0.040$ against the alternative $\sigma > 0.040$ at the level of significance $\alpha = 0.01$, a random sample of size $n = 35$ is taken from a large production lot, yielding $s = 0.053$ for the compressed lengths of the springs. Substitution into the formula for the z-statistic gives

$$z = \frac{0.053 - 0.040}{0.040/\sqrt{70}} = 2.72$$

and since this exceeds $z_{0.01} = 2.33$, the null hypothesis will have to be rejected. The springs in the production lot do not meet the specifications with regard to the variability of their compressed lengths.

11.3

Tests Concerning Two Standard Deviations

In this section we shall discuss a test concerning the equality of the standard deviations, or variances, of two populations. This test is often used in connection with the small-sample test concerning the difference between two means. For instance, in the example on page 278, dealing with the heat-producing capacity of the coal from two mines, the two sample variances were $s_1^2 = \dfrac{253,800}{4} = 63,450$ and $s_2^2 = \dfrac{170,600}{4} = 42,650$ million calories per ton and, despite what may seem to be a large difference, we assumed that the population variances were equal. We could not discuss the rationale of this assumption at that time, but without saying so, we actually tested, and were unable to reject, the null hypothesis that the populations had equal variances, before we performed the t-test for the significance of the difference between the two sample means.

Given independent random samples from two populations, we usually base tests of the equality of the population standard deviations (or variances) on the ratios s_1^2/s_2^2 or s_2^2/s_1^2, where s_1 and s_2 are the standard deviations of the two samples. Assuming that the two populations sampled have roughly the shape of normal distributions, it can be shown that the sampling distribution of such a ratio, appropriately called a **variance ratio**, is a continuous distribution called the F **distribution.** This distribution depends on the two parameters $n_1 - 1$ and $n_2 - 1$, the degrees of freedom in the sample estimates, s_1 and s_2, of the unknown population standard deviations. One difficulty with this distribution is that most tables give only values of $F_{0.05}$ (defined in the same way as $z_{0.05}$, $t_{0.05}$, and $\chi_{0.05}^2$) and $F_{0.01}$, so we can work only with the right-hand tail of the distribution. For this reason, we base our decision about the equality of two population standard deviations, σ_1 and σ_2, on the statistic

Statistic for test concerning the equality of two standard deviations

$$F = \frac{s_1^2}{s_2^2} \quad \text{or} \quad \frac{s_2^2}{s_1^2} \quad \text{whichever is larger}$$

With this statistic we reject the null hypothesis $\sigma_1 = \sigma_2$ at the level of significance α, and accept the alternative hypothesis $\sigma_1 \neq \sigma_2$, when the observed value of F exceeds $F_{\alpha/2}$ (see Figure 11.4). By using the right-hand

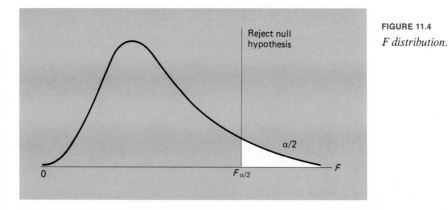

Reject null
hypothesis

$\alpha/2$

0

$F_{\alpha/2}$

F

FIGURE 11.4

F distribution.

11.3 Tests Concerning Two Standard Deviations

293

tail area of $\alpha/2$ instead of α, we compensate for the fact that we always use the larger of the two variance ratios. The necessary values of $F_{\alpha/2}$ for $\alpha = 0.10$ or 0.02, that is, $F_{0.05}$ and $F_{0.01}$, are given in Table IV at the end of the book. The number of degrees of freedom for the numerator is $n_1 - 1$ or $n_2 - 1$, depending on whether we are using the ratio s_1^2/s_2^2 or the ratio s_2^2/s_1^2; correspondingly, the number of degrees of freedom for the denominator is $n_2 - 1$ or $n_1 - 1$.

EXAMPLE   Let us return to the example on page 279, where we tested the significance of the difference between two sample means. This t-test required that the standard deviations of the populations sampled must be equal, and we can now show how we determined that this assumption was actually tenable. The sample sizes were $n_1 = 5$ and $n_2 = 5$ and the sample variances were $s_1^2 = \dfrac{253{,}800}{4} = 63{,}450$

and $s_2^2 = \dfrac{170{,}600}{4} = 42{,}650$, so that

$$F = \frac{s_1^2}{s_2^2} = \frac{63{,}450}{42{,}650} = 1.49$$

Since this falls short of 16.0, the value of $F_{0.01}$ for 4 and 4 degrees of freedom, the null hypothesis cannot be rejected at the level of significance $\alpha = 0.02$; in other words, there is no real evidence that the standard deviations of the two populations are not the same.

EXERCISES

1   In a random sample, the amounts of time which 18 women took to complete the written test for their driver's licenses had a standard deviation of 2.1 minutes. Test the null hypothesis $\sigma = 2.5$ minutes against the alternative $\sigma \neq 2.5$ minutes at the level of significance $\alpha = 0.05$.

2   If 10 determinations of the specific heat of iron had a standard deviation of 0.0086, test the null hypothesis that $\sigma = 0.0100$ for such determinations. Use the alternative hypothesis $\sigma < 0.0100$ and the level of significance $\alpha = 0.05$.

3   With reference to Exercise 11 on page 281, test the null hypothesis $\sigma = 0.1$ ton per acre against the alternative hypothesis $\sigma > 0.1$ ton per acre at the level of significance $\alpha = 0.01$.

4   With reference to Exercise 12 on page 281, test the null hypothesis $\sigma = 30$ words per minute against the alternative hypothesis $\sigma \neq 30$ words per minute at the level of significance $\alpha = 0.05$.

5   Past data indicate that the variance of measurements made on sheet metal stampings by experienced inspectors is 0.18 square inch. If a new inspector measures 100 stampings with a variance of 0.25 square inch, test at the 0.05 level of significance whether the inspector is making satisfactory measurements (against the alternative that their variability is excessive).

6   With reference to Exercise 1 on page 280, test at the level of significance $\alpha = 0.01$ whether $\sigma = 4.2$ months for the average time that convicted embezzlers spend in jail.

7   With reference to Exercise 3 on page 280, test at the level of significance $\alpha = 0.05$ whether $\sigma = 2.0$ minutes against the alternative $\sigma < 2.0$ minutes.

8   Two different lighting techniques are compared by measuring the intensity of light at selected locations in areas lighted by the two methods. If 12 measurements of the first technique have a standard deviation of 2.6 foot-candles and 16 measurements of the second technique have a standard deviation of 4.4 foot-candles, test the null hypothesis $\sigma_1 = \sigma_2$ against the alternative hypothesis $\sigma_1 \neq \sigma_2$ at the level of significance $\alpha = 0.10$.

9   With reference to Exercise 21 on page 282, test at the level of significance $\alpha = 0.02$ whether it is reasonable to assume that the two population variances are equal.

10   With reference to Exercise 22 on page 282, test at the level of significance $\alpha = 0.02$ whether it is reasonable to assume that the two population variances are equal.

11   With reference to Exercise 24 on page 283, test at the level of significance $\alpha = 0.02$ whether it is reasonable to assume that the two population variances are equal.

BIBLIOGRAPHY   Theoretical discussions of the chi-square and F distributions may be found in most textbooks on mathematical statistics; for example, in the author's *Mathematical Statistics*, referred to on page 238.

The work of this chapter is very similar to that of Chapters 10 and 11. In problems of estimation we shall again construct confidence intervals, or determine the possible size of our error. In tests of hypotheses we shall again formulate null hypotheses and alternative hypotheses, decide between one-tail tests and two-tail tests, choose a level of significance, and so forth. The main difference is that we will be concerned with other parameters—instead of population means or standard deviations, we will deal with population proportions, percentages, or probabilities.

Inferences About Proportions

12.1

The Estimation of Proportions

The information that is usually available to estimate a proportion (percentage, or probability) is the **relative frequency** of the occurrence of an event. If an event occurs x times in n trials, the relative frequency of its occurrence is $\frac{x}{n}$, and we can use this **sample proportion** to estimate the corresponding true proportion p. (As they are used here, the terms "relative frequency" and "sample proportion" are synonymous.)

EXAMPLE If a study shows that 54 of 120 cheerleaders suffered what auditory experts called "moderate to severe damage" to their voices, then $\frac{x}{n} = \frac{54}{120} = 0.45$, and we can use this figure as a point estimate of the true proportion of cheerleaders who are afflicted in this way. Similarly, a supermarket chain might estimate the proportion of its shoppers who regularly use cents-off coupons as 0.68, if a sample of 300 shoppers included 204 who regularly use cents-off coupons.

Throughout this section it will be assumed that the situations satisfy (at least approximately) the conditions underlying the binomial distribution; that is, our information will consist of the number of successes in a given number of independent trials, and it will be assumed that on each trial the probability of a success—the parameter we want to estimate—has the constant value p. Thus, the sampling distribution of the counts on which our methods will be based is the binomial distribution, whose mean and standard deviation are

$$\mu = np \quad \text{and} \quad \sigma = \sqrt{np(1 - p)}$$

The formula for σ involves p, the quantity we want to estimate, and this causes some trouble, but we can avoid it, at least for the moment, by constructing confidence intervals for p with the use of special tables. Tables V(a) and V(b) at the end of the book provide 0.95 and 0.99 confidence intervals for proportions; that is, 0.95 and 0.99 confidence intervals for the parameter p of binomial "populations."

If a sample proportion is less than or equal to 0.50, we begin by marking the value of $\frac{x}{n}$ on the bottom scale; then we go up vertically until we reach the two contour lines (curves) which correspond to the size of the sample,

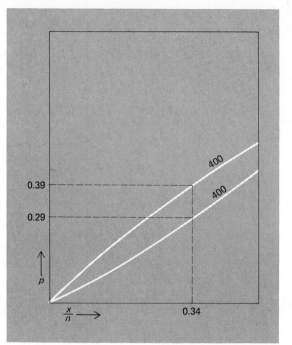

FIGURE 12.1
Confidence limits for p.

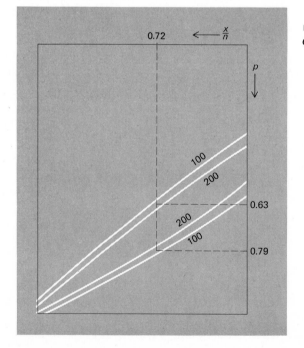

FIGURE 12.2
Confidence limits for p.

and read the confidence limits for p off the left-hand scale, as indicated in Figure 12.1. If a sample proportion is greater than 0.50, we mark the value of $\dfrac{x}{n}$ on the top scale, go down vertically until we reach the two contour lines which correspond to the size of the sample, and read the confidence limits for p off the right-hand scale, as indicated in Figure 12.2.

EXAMPLE Suppose that in a random sample of 400 persons given a flu vaccine only 136 experienced any discomfort. Marking $\dfrac{x}{n} = \dfrac{136}{400} = 0.34$ on the bottom scale of Table V(a) and proceeding as in Figure 12.1, we get the 0.95 confidence interval

$$0.29 < p < 0.39$$

for the true proportion of persons given the flu vaccine who will experience any discomfort. This means that we can assert with probability 0.95 that the interval from 0.29 to 0.39 contains the actual (population) proportion of persons who will experience any discomfort from the vaccine. A corresponding 0.99 confidence interval can be read from Table V(b) in the same way. It is

$$0.28 < p < 0.40$$

and here again, as in the estimation of means and standard deviations, we find that the interval becomes wider as the degree of confidence is increased.

EXAMPLE Suppose that in a random sample of 150 vacationers interviewed at a resort, 108 said that they chose the resort mainly because of its warm climate. Marking $\dfrac{108}{150} = 0.72$ on the top scale of Table V(a), and reading between the lines (for $n = 100$ and $n = 200$) as in Figure 12.2, we get the 0.95 confidence interval

$$0.63 < p < 0.79$$

for the actual (population) proportion of vacationers who choose the resort mainly because of its warm climate. Similarly, using Table V(b), we find that a 0.99 confidence interval for this population proportion is

$$0.61 < p < 0.81$$

Tables V(a) and V(b) have contour lines for samples of size 8, 10, 12, 16, 20, 24, 30, 40, 60, 100, 200, 400, and 1,000; for other values of n we must read between the lines as in the last example.

12.1   **The Estimation of Proportions**

We can also construct approximate large-sample confidence intervals for p on the basis of the normal-curve approximation to the binomial distribution, which is reasonable, as we said on page 204, so long as np and $n(1 - p)$ are both greater than 5. If it can be assumed that this is the case, we can assert with probability $1 - \alpha$ that x, the observed number of successes, will take on a value between $np - z_{\alpha/2}\sqrt{np(1 - p)}$ and $np + z_{\alpha/2}\sqrt{np(1 - p)}$. This can be written as

$$np - z_{\alpha/2}\sqrt{np(1 - p)} < x < np + z_{\alpha/2}\sqrt{np(1 - p)}$$

or as

$$\frac{x}{n} - z_{\alpha/2}\sqrt{\frac{p(1 - p)}{n}} < p < \frac{x}{n} + z_{\alpha/2}\sqrt{\frac{p(1 - p)}{n}}$$

after some algebraic manipulation (see Exercise 16 on page 307). This result cannot be used as is, since the unknown p itself appears in $\sqrt{\dfrac{p(1 - p)}{n}}$ on both sides of the double inequality. This quantity, $\sqrt{\dfrac{p(1 - p)}{n}}$, is called the **standard error of a proportion**, as it is, in fact, the standard deviation of the sampling distribution of a sample proportion (see Exercise 17 on page 307). To get around this, we substitute the sample proportion $\dfrac{x}{n}$ for p in $\sqrt{\dfrac{p(1 - p)}{n}}$, and thus arrive at the following $1 - \alpha$ **large-sample confidence interval for** p:

Large-sample confidence interval for p

$$\frac{x}{n} - z_{\alpha/2}\sqrt{\frac{\dfrac{x}{n}\left(1 - \dfrac{x}{n}\right)}{n}} < p < \frac{x}{n} + z_{\alpha/2}\sqrt{\frac{\dfrac{x}{n}\left(1 - \dfrac{x}{n}\right)}{n}}$$

EXAMPLE If, as on page 299, $x = 136$ persons experienced some discomfort among $n = 400$ who were given the flu vaccine, and we substitute $\dfrac{x}{n} = \dfrac{136}{400} = 0.34$ and $z_{0.025} = 1.96$ into the formula directly above, we get the 0.95 large-sample confidence interval

$$0.34 - 1.96\sqrt{\frac{(0.34)(0.66)}{400}} < p < 0.34 + 1.96\sqrt{\frac{(0.34)(0.66)}{400}}$$

$$0.294 < p < 0.386$$

Rounded to two decimals, the limits of this interval are identical with those of the interval $0.29 < p < 0.39$ which we read from Table V(a).

The large-sample theory presented here can also be used to judge the size of the error we may be making when we use a sample proportion as a point estimate of a population proportion p. In this connection, we can assert with probability $1 - \alpha$ that the size (or magnitude) of our error is less than

Maximum error of estimate

$$E = z_{\alpha/2} \sqrt{\frac{p(1 - p)}{n}} \quad or \ approximately \quad E = z_{\alpha/2} \sqrt{\frac{\frac{x}{n}\left(1 - \frac{x}{n}\right)}{n}}$$

The first of these two formulas cannot be used in practice since p is the unknown quantity we are trying to estimate, but the second formula can be used provided that n is large enough to justify the normal-curve approximation to the binomial distribution.

EXAMPLE With $x = 108$ and $n = 150$ in the study designed to determine why people visit a certain resort, $\dfrac{x}{n} = \dfrac{108}{150} = 0.72$ is an estimate of the true proportion of vacationers who go there mainly because of the warm climate, and we can assert with probability 0.99 that this estimate is in error by less than

$$E = 2.58 \sqrt{\frac{(0.72)(0.28)}{150}} = 0.095$$

rounded to three decimals.

As in the estimation of means, we can use the expression for the maximum error to determine how large a sample is needed to attain a desired degree of precision. If we want to assert with probability $1 - \alpha$ that a sample proportion will differ from the true proportion p by less than E, we can solve the equation

$$E = z_{\alpha/2} \sqrt{\frac{p(1 - p)}{n}}$$

for *n* and get

Sample size

$$n = p(1 - p)\left[\frac{z_{\alpha/2}}{E}\right]^2$$

Since the unknown population proportion *p* appears in this formula, it cannot be used unless we have some information about the possible values *p* might assume. If we have no such information, we make use of the fact that $p(1 - p)$ is $\frac{1}{4}$ when $p = \frac{1}{2}$, and less than $\frac{1}{4}$ for all other values of *p*.[†] Hence, if we use the formula

Sample size

$$n = \frac{1}{4}\left[\frac{z_{\alpha/2}}{E}\right]^2$$

we may be making the sample larger than necessary, but we can account for this by asserting with a probability of *at least* $1 - \alpha$ that the error will be less than *E*.

In case we do have some information about the possible values *p* might assume, we can take this into account by substituting into the first formula for *n* whichever of the possible values is closest to $\frac{1}{2}$. For instance, if it is reasonable to suppose that the proportion we are trying to estimate lies on the interval from 0.60 to 0.80, we substitute $p = 0.60$.

EXAMPLE   Suppose we want to determine what proportion of the adult population have high blood pressure, and we want to be "99 percent sure" that the error of our estimate is less than 0.05. If we have no idea what the true proportion might be, we substitute $E = 0.05$ and $z_{0.005} = 2.58$ into the second of the two formulas for *n*, and we get

$$n = \frac{1}{4}\left[\frac{2.58}{0.05}\right]^2 = 665.64$$

Consequently, if we base our estimate on a random sample of size $n = 666$, we can assert with a probability of at least 0.99 that the sample proportion will be off by less than 0.05. On the other hand, if we knew that *p* is no greater than 0.20, substitution into the first of the two formulas for *n* would give

$$n = (0.20)(0.80)\left[\frac{2.58}{0.05}\right]^2 = 426.01$$

[†] See Exercise 48 on page 358.

so that a sample of size $n = 427$ would be large enough. This illustrates that knowledge about the possible values p might assume can appreciably reduce the size of the sample needed to attain a desired degree of precision.

12.2

The Estimation of Proportions (A Bayesian Method) ★

In the preceding section we looked upon the true proportion p (which we tried to estimate) as an unknown constant; in Bayesian estimation this parameter is looked upon as a random variable having a prior distribution which reflects one's belief about the values it can assume. As in the Bayesian estimation of means, we are thus faced with the problem of combining prior information with direct sample evidence.

EXAMPLE Consider a large company which routinely pays thousands of invoices submitted by its suppliers. Of course, it is of interest to know what proportion of these invoices might contain errors, and interviews of three of the company's executives reveal the following information: Mr. Martin feels that only 0.005 (or a half of one percent) of the invoices contain errors; Mr. Green, who is generally regarded to be as reliable in his estimates as Mr. Martin, feels that 0.01 (or one percent) of the invoices contain errors; and Mr. Jones, who is generally regarded to be twice as reliable in his estimates as either Mr. Martin or Mr. Green, feels that 0.02 (or two percent) of the invoices contain errors. Assuming that $p = 0.005$, $p = 0.01$, and $p = 0.02$ are the only possibilities, we thus have the following prior distribution for the proportion of invoices containing errors:

p	Prior probability
0.005	0.25
0.010	0.25
0.020	0.50

(The prior probabilities assigned to Mr. Martin's and Mr. Green's estimates are the same, and that assigned to Mr. Jones's estimate is twice as large.)

Now suppose that a random sample of 200 invoices is carefully checked, and that only one of them contains an error. The probabilities of this happening when $p = 0.005$, $p = 0.010$, $p = 0.020$

are, respectively,

$$\binom{200}{1}(0.005)^1(0.995)^{199} = 0.37$$

$$\binom{200}{1}(0.010)^1(0.990)^{199} = 0.27$$

and

$$\binom{200}{1}(0.020)^1(0.980)^{199} = 0.07$$

where we used the formula for the binomial distribution, and logarithms to perform the calculations. Combining these probabilities with the prior probabilities on page 303 using the formula for Bayes' theorem (see Section 5.7), we find that the **posterior probability** of $p = 0.005$ is

$$\frac{(0.25)(0.37)}{(0.25)(0.37) + (0.25)(0.27) + (0.50)(0.07)} = 0.47$$

while the corresponding posterior probabilities of $p = 0.010$ and $p = 0.020$ are

$$\frac{(0.25)(0.27)}{(0.25)(0.37) + (0.25)(0.27) + (0.50)(0.07)} = 0.35$$

and

$$\frac{(0.50)(0.07)}{(0.25)(0.37) + (0.25)(0.27) + (0.50)(0.07)} = 0.18$$

We have thus arrived at the following posterior distribution for the proportion of invoices containing errors:

p	Posterior probability
0.005	0.47
0.010	0.35
0.020	0.18

This distribution reflects the prior judgments as well as the direct sample evidence, and it should not come as a surprise that the highest posterior probability goes to $p = 0.005$—after all, the sample proportion actually equaled $\dfrac{x}{n} = \dfrac{1}{200} = 0.005$.

To continue, we could use the mean of the posterior distribution, namely, $(0.005)(0.47) + (0.010)(0.35) + (0.020)(0.18) = 0.009$ as a point estimate of the true proportion of invoices containing errors. In contrast to the mean of the prior distribution, which was $(0.005)(0.25) + (0.010)(0.25) + (0.020)(0.50) = 0.014$, the mean of the posterior distribution reflects the prior judgments as well as the direct sample evidence.

In the preceding example it was assumed that p had to be 0.005, 0.010, or 0.020 in order to simplify the calculations. The method of analysis would have been exactly the same, however, if we had considered ten different values of p or even a hundred. In fact, there exist Bayesian techniques, similar to that of Section 10.4, in which p can take on any value on the continuous interval from 0 to 1, and its prior distribution is a continuous curve.

EXERCISES

1 In a sample survey, 108 of 200 persons interviewed in a large city said that they opposed the construction of any more freeways in the city. Use Table V(a) to construct a 0.95 confidence interval for the corresponding true proportion.

2 Among 100 fish caught in a certain lake, 16 were inedible as a result of the chemical pollution of their environment. Use Table V(a) to construct a 0.95 confidence interval for the probability that a fish caught in this lake will be inedible for the given reason.

3 In a random sample of 60 claims filed against a company writing collision insurance on cars, 21 claims exceeded \$1,200. Use Table V(b) to construct a 0.99 confidence interval for the corresponding true proportion.

4 On page 297 we referred to a study which showed that a sample of 300 shoppers at a supermarket included 204 who regularly use cents-off coupons. Use Table V(b) to construct a 0.99 confidence interval for the corresponding true percentage.

5 On page 297 we referred to a study which showed that 54 of 120 cheerleaders, presumably a random sample, had suffered moderate to severe damage to their voices. Construct a 0.95 confidence interval for the corresponding true proportion for the population sampled using the large-sample formula on page 300.

6 In a random sample of 400 voters interviewed in a large city, only 228 felt that the President was doing a good job. Construct a 0.99 confidence interval for the true proportion of voters in this city who feel the same way using
   (a) Table V(b);
   (b) the large-sample formula on page 300.

7 In a random sample of 200 television viewers in a certain area, 76 had seen a certain controversial program. Construct a 0.95 confidence interval for the actual proportion of television viewers in that area who saw the program using
   (a) Table V(a);
   (b) the large-sample formula on page 300.

12.2 The Estimation of Proportions (A Bayesian Method)

8 In a random sample of 250 high school seniors in a large city, 165 said that they expect to continue their education at an in-state college or university. Construct a 0.99 confidence interval for the corresponding true proportion using
   (a) Table V(b);
   (b) the large-sample formula on page 300.

9 In a sample of 80 persons convicted in U.S. District Courts on narcotics charges, 36 received probation. If $\frac{36}{80} = 0.45$ is used as an estimate of the actual proportion for the population sampled, what can one assert with probability 0.95 about the maximum size of the error?

10 In a random sample of 140 supposed UFO sightings, 119 could easily be explained in terms of natural phenomena. If $\frac{119}{140} = 0.85$ is used as an estimate of the true proportion of UFO sightings that can easily be explained in terms of natural phenomena, what can one assert with probability 0.99 about the maximum size of the error?

11 In a random sample of 600 women over the age of 21 interviewed in a large city, 378 said that they held full- or part-time jobs. If $\frac{378}{600} = 0.63$ is used as an estimate of the actual proportion, what can we assert with probability 0.95 about the maximum size of the error?

12 If a sample constitutes at least 5 percent of a finite population, and the sample itself is large, we can use the finite population correction factor to reduce the width of confidence intervals for p. If we make this correction, the confidence limits for p become

Large-sample confidence limits for p (finite population)

$$\frac{x}{n} \pm z_{\alpha/2} \sqrt{\frac{\frac{x}{n}\left(1 - \frac{x}{n}\right)}{n}} \cdot \sqrt{\frac{N - n}{N - 1}}$$

where N is, as before, the size of the population.
   (a) A shoe manufacturer feels that unless her employees agree to a five percent wage reduction she cannot stay in business. If in a random sample of 60 of her 300 employees only 18 feel "kindly disposed" to the reduction, find a 0.95 confidence interval for the corresponding proportion for all of her employees.
   (b) Rework part (b) of Exercise 6, given that there are 6,000 registered voters in the given city.
   (c) Rework part (b) of Exercise 8, given that there are 800 high school seniors in the given city.

13 A private opinion poll is hired by a politician to estimate what proportion of the voters in his state favor the decriminalization of certain narcotics violations. How large a sample will the poll have to take to be able to assert with a probability of at least 0.95 that the sample proportion will be off by less than 0.02?

14 Suppose we want to estimate what percentage of the drivers are actually exceeding the 55-mph speed limit on a stretch of road between Indio and Blythe in California.
   (a) How large a sample will we need to be able to assert with a probability of at least 0.99 that the sample proportion and the true proportion will differ by less than 4 percent?

(b) If we have reason to believe that at least 60 percent of the drivers exceed the speed limit on that stretch of road, how will this affect the required sample size?

15 An automobile insurance company wants to estimate from a sample what proportion of its thousands of policyholders intend to buy a new car within the next twelve months.

    (a) How large a sample will they need to be able to assert with a probability of at least 0.95 that the sample proportion and the true proportion will differ by less than 0.03?

    (b) If they have reason to believe that the actual proportion is somewhere between 0.20 and 0.30, how will this affect the required size of the sample?

16 Show that if we add $z_{\alpha/2}\sqrt{np(1-p)}$ to the expressions on both sides of the inequality $np - z_{\alpha/2}\sqrt{np(1-p)} < x$ and then divide through by n, we get the inequality $p < \dfrac{x}{n} + \dfrac{1}{n} \cdot z_{\alpha/2}\sqrt{np(1-p)}$ and, hence,

$$p < \frac{x}{n} + z_{\alpha/2}\sqrt{\frac{p(1-p)}{n}}$$

Use a similar argument to change the other part of the double inequality on page 300.

17 Since the proportion of successes is simply the number of successes divided by n, the mean and the standard deviation of the sampling distribution of the proportion of successes may be obtained by dividing by n the mean and the standard deviation of the sampling distribution of the number of successes. Use this argument to verify the standard error formula given on page 300.

18 In planning the operation of a new school, one school board member claims that 4 out of 5 newly hired teachers will stay with the school for more than a year, while another school board member claims that it would be correct to say 7 out of 10. In the past, the two board members have been about equally reliable in their predictions, so that in the absence of direct information we would assign their judgments equal weight. What posterior probabilities would we assign to their claims if it were found that 11 of 12 newly hired teachers stayed with the school for more than a year? (*Hint:* Use Table VI.)

19 The landscaping plans for a new hotel call for a row of palm trees along the driveway. The landscape designer tells the owner that if he plants *Washingtonia filifera*, 20 percent of the trees will fail to survive the first heavy frost, the manager of the nursery which supplies the trees tells the owner that 10 percent of the trees will fail to survive the first heavy frost, and the owner's wife tells him that 30 percent of the trees will fail to survive the first heavy frost.

    (a) If the owner feels that in this matter the landscape designer is 10 times as reliable as his wife and the manager of the nursery is 9 times as reliable as his wife, what prior probabilities should he assign to these percentages?

    (b) If 13 of these palm trees are planted and 2 fail to survive the first heavy frost, what posterior probabilities should the manager assign to the three percentages? (*Hint:* Use Table VI.)

20 The method of Section 12.2 can also be used to find the posterior distribution of the parameter λ of the Poisson distribution. Suppose, for instance, that several brokers are discussing the sale of a large estate, and that one of them feels that a

newspaper ad should produce three serious inquiries, a second feels that it should produce five serious inquiries, and a third feels that it should produce six.

(a) If in the past the second broker has been twice as reliable as the first and the first has been three times as reliable as the third, what prior probabilities should we assign to their claims?

(b) How would these probabilities be affected if the ad actually produced only one serious inquiry and it can be assumed that the number of serious inquiries is a random variable having the Poisson distribution with either $\lambda = 3$, $\lambda = 5$, or $\lambda = 6$ according to the three claims.

21 The method of Section 12.2 can also be used to find the posterior distribution of the number of "successes" in a finite population. Suppose, for instance, that a coin dealer receives a shipment of five ancient coins from abroad, and that, on the basis of past experience, he feels that the probabilities that 0, 1, 2, 3, 4, or all 5 of them are counterfeits are 0.74, 0.11, 0.02, 0.01, 0.02, and 0.10. Since modern methods of counterfeiting have become greatly refined, the cost of authentication has risen sharply and the dealer decides to select one of the coins at random and send it away for authentication. If it turns out that the coin is a forgery, what posterior probabilities should he assign to the possibilities that 0, 1, 2, 3, or all 4 of the remaining coins are counterfeits?

12.3
Tests Concerning Proportions

In this section we shall be concerned with tests of hypotheses which enable us to decide, on the basis of sample data, whether the true value of a proportion (percentage, or probability) equals, is greater than, or is less than a given constant. These tests will make it possible, for example, to determine whether the true proportion of fifth graders who can name the governor of their state is 0.35, whether it is true that 10 percent of the answers which the IRS gives to taxpayers' telephone inquiries are in error, or whether the true probability is 0.25 that the downtime of a new computer will exceed two hours in any given week.

Questions of this kind are usually decided on the basis of either the observed number or the proportion of successes in n trials, and it will be assumed throughout this section that these trials are independent and that the probability of a success is the same for each trial. In other words, we shall assume that we can use the binomial distribution.

When n is small, tests concerning true proportions (the parameter p of binomial distributions) can be based directly on tables of binomial probabilities such as Table VI at the end of the book.

EXAMPLE Suppose we want to investigate the claim that at least 60 percent of the students attending a large university are opposed to a plan to increase student fees in order to build new parking facilities. Specifically, we shall want to use a random sample of size $n = 14$

(the opinions of 14 students selected at random) to test the null hypothesis $p = 0.60$ against the alternative hypothesis $p < 0.60$ at the level of significance $\alpha = 0.05$. In the criteria of Figure 10.8, where we dealt with continuous normal distributions, we could draw the dividing lines so that the probabilities associated with the "tails" were exactly α or $\alpha/2$. Since this cannot always be done when we deal with binomial distributions, we modify the criteria as follows: **We draw the dividing lines so that the probability of getting a value in the "tail" is as close as possible to the level of significance α (or to $\alpha/2$ in a two-tail test) without exceeding it.** Thus, for our example we observe from Table VI that for $p = 0.60$ and $n = 14$ the probability of getting at most 4 successes is

$$0.001 + 0.003 + 0.014 = 0.018$$

and the probability of getting at most 5 successes is

$$0.001 + 0.003 + 0.014 + 0.041 = 0.059$$

Since the first of these probabilities is less than $\alpha = 0.05$ and the second probability exceeds it, we reject the null hypothesis $p = 0.60$ (and accept the alternative hypothesis $p < 0.60$) when in a random sample of $n = 14$ students interviewed there are at most 4 who are opposed to the plan to increase student fees in order to build new parking facilities (see also Figure 12.3).

FIGURE 12.3
Binomial distribution with $p = 0.60$ and $n = 14$.

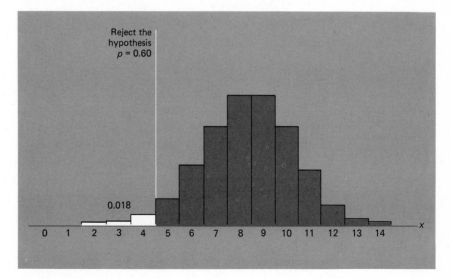

EXAMPLE To illustrate the use of a two-tail test, suppose we want to decide on the basis of a random sample of size $n = 12$ whether the true proportion of shoppers who can identify a highly advertised trade mark is 0.40, as claimed. Again using the level of significance $\alpha = 0.05$, we observe from Table VI that for $p = 0.40$ and $n = 12$ the probability of getting at most one success is

$$0.002 + 0.017 = 0.019$$

the probability of getting at most two successes is

$$0.002 + 0.017 + 0.064 = 0.083$$

the probability of getting 9 or more successes is

$$0.002 + 0.012 = 0.014$$

and the probability of getting 8 or more successes is

$$0.002 + 0.012 + 0.042 = 0.056$$

Since the first and third of these probabilities are less than 0.025 and the second and fourth probabilities exceed it, we reject the null

FIGURE 12.4

Binomial distribution with $p = 0.40$ and $n = 12$.

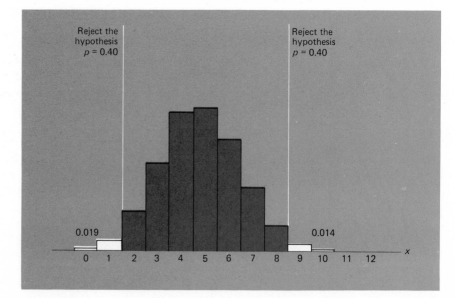

hypothesis $p = 0.40$ (and accept the alternative hypothesis $p \neq 0.40$) when in a random sample of $n = 12$ shoppers interviewed, less than 2 or more than 8 (namely, 0, 1, 9, 10, 11, or 12) can identify the trade mark (see also Figure 12.4).

When n is large, tests concerning proportions (percentages, or probabilities) are usually based on the normal-curve approximation to the binomial distribution of x, the observed number of successes, provided np and $n(1 - p)$ both exceed 5. Since the mean and the standard deviation of the binomial distribution are $\mu = np$ and $\sigma = \sqrt{np(1 - p)}$, we base tests of the null hypothesis $p = p_0$ on the result that, for large n,

$$z = \frac{x - np_0}{\sqrt{np_0(1 - p_0)}}$$

has approximately the standard normal distribution. Actually, we base the tests on the statistic

Statistic for test concerning proportion

$$z = \frac{(x \pm \frac{1}{2}) - np_0}{\sqrt{np_0(1 - p_0)}}$$

thus accounting for the continuity correction which we discussed in Section 8.4.[†] We use $x - \frac{1}{2}$ when x exceeds np_0 (z falls into the right-hand tail of the sampling distribution), and $x + \frac{1}{2}$ when x is less than np_0 (z falls into the left-hand tail of the sampling distribution). The test criteria, themselves, are again those of Figure 10.8 on page 273 with p substituted for μ. For the one-sided alternative $p < p_0$ we reject the null hypothesis when $z < -z_\alpha$; for the one-sided alternative $p > p_0$ we reject the null hypothesis when $z > z_\alpha$; and for the two-sided alternative $p \neq p_0$ we reject the null hypothesis when $z < -z_{\alpha/2}$ or $z > z_{\alpha/2}$. As before, α is the level of significance, and z_α and $z_{\alpha/2}$ are as defined on page 272.

EXAMPLE Suppose we want to investigate a nutritionist's claim that at least 75 percent of the preschool children in a certain country have protein-deficient diets, and that a sample survey reveals 206 of 300 preschool children in that country to have protein-deficient diets. To test the null hypothesis $p = 0.75$ against the alternative hypothesis $p < 0.75$ at the level of significance $\alpha = 0.01$, we substitute $x = 206$, $n = 300$, and $p_0 = 0.75$ into the above formula for z,

---

[†] When n is large, the continuity correction is usually omitted.

12.3 Tests Concerning Proportions

getting

$$z = \frac{206 + \frac{1}{2} - 300(0.75)}{\sqrt{300(0.75)(0.25)}} = -2.47$$

Since this is less than $-z_{0.01} = -2.33$, we reject the null hypothesis. The evidence does not support the claim and we conclude that less than 75 percent of the preschool children in this country have protein-deficient diets.

EXAMPLE A two-tail test is appropriate if we want to test a professor's claim that 80 percent of his students prefer writing an extra book report in lieu of taking a final examination. The information we have is that, in a random sample, 267 out of 320 of the professor's students preferred writing the extra book report. To test the null hypothesis $p = 0.80$ against the alternative hypothesis $p \neq 0.80$ at the level of significance $\alpha = 0.05$, we substitute $x = 267$, $n = 320$, and $p_0 = 0.80$ into the formula for z, getting

$$z = \frac{267 - \frac{1}{2} - 320(0.80)}{\sqrt{320(0.80)(0.20)}} = 1.47$$

Since this is not greater than $z_{0.025} = 1.96$, the null hypothesis cannot be rejected; that is, the evidence is not sufficient to reject the professor's claim.

EXERCISES

1 In a study of aviophobia, a psychologist claims that 30 percent of all women are afraid of flying. If 15 women are interviewed, how few or how many of them must be afraid of flying so that the null hypothesis $p = 0.30$ can be rejected against the alternative $p \neq 0.30$ at the level of significance $\alpha = 0.05$.

2 A physicist claims that at most 10 percent of all persons exposed to a certain amount of radiation will feel any effects. If 3 of 13 persons exposed to that much radiation felt some effects, can the null hypothesis $p = 0.10$ be rejected against the alternative hypothesis $p > 0.10$ at the level of significance $\alpha = 0.05$?

3 Suppose that 4 of 11 undergraduate engineering students, presumably a random sample of the undergraduate engineering students attending a certain university, say that they will go on to graduate school. Test the dean's claim that 70 percent of the undergraduate engineering students will go on to graduate school, using the alternative hypothesis $p \neq 0.70$ and the level of significance $\alpha = 0.05$.

4 Suppose we want to test the "honesty" of a coin on the basis of the number of heads we will get in 15 flips. Determine how few or how many heads we have to get so that we can reject the null hypothesis $p = 0.50$ against the alternative hypothesis $p \neq 0.50$
   (a) at the level of significance $\alpha = 0.05$;
   (b) at the level of significance $\alpha = 0.01$.

5 A television critic claims that at least 80 percent of all viewers find the noise level of a certain commercial objectionable. If 7 of 12 persons shown this commercial object to the noise level, what can we conclude about the claim at the level of significance $\alpha = 0.05$?

6 To check on an ambulance service's claim that at least 40 percent of its calls are life-threatening emergencies, a random sample was taken from its files, and it was found that only 49 of 150 calls were life-threatening emergencies. Test the null hypothesis $p = 0.40$ against a suitable alternative at the level of significance $\alpha = 0.01$.

7 The manufacturer of a spot remover claims that his product removes at least 90 percent of all spots. What can we conclude about this claim at the level of significance $\alpha = 0.05$, if the spot remover removed only 174 of 200 spots chosen at random from spots on clothes brought to a dry cleaning establishment?

8 A food processor wants to know whether the probability is really 0.65 that a customer will prefer a new kind of packaging to the old kind. What can he conclude at the level of significance $\alpha = 0.05$ if 74 of 100 randomly selected customers prefer the new kind of packaging and
 (a) the alternative hypothesis is $p \neq 0.65$?
 (b) the alternative hypothesis is $p > 0.65$?

9 In a random sample of 600 cars making a right turn at a certain intersection, 157 pulled into the wrong lane. Test the null hypothesis that the actual proportion of drivers who make this mistake (at the given intersection) is 0.30 against the alternative hypothesis that this figure is incorrect, using
 (a) the level of significance $\alpha = 0.05$;
 (b) the level of significance $\alpha = 0.01$.

10 In the construction of tables of random numbers there are various ways of testing for possible departures from randomness. For instance, there should be about as many even digits (0, 2, 4, 6, or 8) as there are odd digits (1, 3, 5, 7, or 9). Count the number of even digits among the 350 digits constituting the first ten rows of the table on page 471, and test at the level of significance $\alpha = 0.05$ whether, on the basis of this criterion, there is any reason to be concerned about the possibility that the random numbers are, in fact, not random.

11 In a random sample of 300 accidents in schools, it was found that 117 were due at least in part to improper supervision. Use the level of significance $\alpha = 0.05$ to decide whether this supports the claim that 35 percent of such accidents are due at least in part to improper supervision.

12.4
Differences Among Proportions

If we followed the pattern of Chapter 10, we would now consider methods which enable us to decide whether observed differences between two sample proportions are significant or whether they can be attributed to chance. For instance, if one method of producing rain by "seeding" clouds was successful in 19 of 60 attempts, while another method was successful in 31 of 80 attempts, we may have to decide whether the difference between the two proportions, $\frac{19}{60} = 0.32$ and $\frac{31}{80} = 0.39$, is significant or whether it is merely due to chance.

The method we shall study in this section is actually more general—it applies also to problems in which we want to decide whether differences among more than two sample proportions are significant, or whether they can be attributed to chance. For instance, if 24 of 200 brand A tires, 21 of 200 brand B tires, 16 of 200 brand C tires, and 33 of 200 brand D tires failed to last 20,000 miles, we may want to decide whether the differences among $\frac{24}{200} = 0.120$, $\frac{21}{200} = 0.105$, $\frac{16}{200} = 0.080$, and $\frac{33}{200} = 0.165$ are significant, or whether they may be due to chance.

EXAMPLE Suppose that a survey in which independent random samples of single, married, and widowed or divorced persons were asked whether "friends and social life" or "job or primary activity" contributes more to their general happiness, yielded the results shown in the following table:

	Single	Married	Widowed or divorced
Friends and social life	47	59	56
Job or primary activity	33	61	44
Total	80	120	100

The proportions of persons choosing "friends and social life" are $\frac{47}{80} = 0.59$, $\frac{59}{120} = 0.49$, and $\frac{56}{100} = 0.56$ (rounded to two decimals) for the three groups, and we want to decide at the 0.05 level of significance whether the differences among them can be attributed to chance.

To this end, we shall let p_1, p_2, and p_3 denote the true proportions of single, married, and widowed or divorced persons who would choose "friends and social life," and formulate the following hypotheses:

Null hypothesis: $p_1 = p_2 = p_3$
Alternative: p_1, p_2, and p_3 are not all equal

If the null hypothesis is true, the three samples come from populations having a common proportion p (of persons who would choose "friends and social life"), and we can combine the three samples and view them as one sample from one population. Also, we can estimate the common proportion p as

$$\frac{47 + 59 + 56}{80 + 120 + 100} = \frac{162}{300} = 0.54$$

With this estimate we would expect $80(0.54) = 43.2$ of the single persons, $120(0.54) = 64.8$ of the married persons, and $100(0.54) = 54.0$ of the widowed or divorced persons to choose "friends and social life." Subtracting these figures from the respective sizes of the three samples, we find that $80 - 43.2 = 36.8$ of the single persons, $120 - 64.8 = 55.2$ of the married persons, and $100 - 54.0 = 46.0$ of the widowed or divorced persons would be expected to choose "job or primary activity." These results are summarized in the following table, where the **expected frequencies** are shown in parentheses below the **observed frequencies**:

	Single	Married	Widowed or divorced
Friends and social life	47 (43.2)	59 (64.8)	56 (54.0)
Job or primary activity	33 (36.8)	61 (55.2)	44 (46.0)

To test the null hypothesis that the p's are all equal in a problem like this, we compare the frequencies which were actually observed with the frequencies we would expect if the null hypothesis were true. It stands to reason that the null hypothesis should be accepted if the discrepancies between the observed and expected frequencies are small. On the other hand, if the discrepancies between the two sets of frequencies are large, this suggests that the null hypothesis must be false.

Using the letter o for the observed frequencies and the letter e for the expected frequencies, we base their comparison on the following χ^2 (**chi-square**) **statistic**:

Statistic for test concerning differences among proportions

$$\chi^2 = \sum \frac{(o - e)^2}{e}$$

In words, χ^2 is the sum of the quantities obtained by dividing $(o - e)^2$ by e separately for each cell of the table.

12.4 Differences Among Proportions

EXAMPLE For our illustration we get

$$\chi^2 = \frac{(47 - 43.2)^2}{43.2} + \frac{(59 - 64.8)^2}{64.8} + \frac{(56 - 54.0)^2}{54.0}$$

$$+ \frac{(33 - 36.8)^2}{36.8} + \frac{(61 - 55.2)^2}{55.2} + \frac{(44 - 46.0)^2}{46.0}$$

$$= 2.016$$

and the question remains whether this value is large enough to enable us to reject the null hypothesis $p_1 = p_2 = p_3$.

If the null hypothesis that the population proportions are all equal is true, the sampling distribution of the χ^2 statistic is approximately the chi-square distribution, which we introduced in Section 11.1. Since the null hypothesis will be rejected only when the value obtained for χ^2 is too large to be accounted for by chance, we base our decision on the criterion of Figure 12.5, where χ_α^2 is such that the area under the chi-square distribution to its right equals α. The parameter of the chi-square distribution, the number of degrees of freedom (or simply the degrees of freedom), equals $k - 1$ when we compare k sample proportions. Intuitively, we can justify this formula with the argument that once we have calculated $k - 1$ of the expected frequencies in either row of the table, all of the other expected frequencies are also determined and they can be obtained by subtraction from the totals of the rows and columns (see Exercise 12 on page 320).

EXAMPLE Returning to our illustration, we find that $\chi^2 = 2.015$ does not exceed 5.991, the value of $\chi_{0.05}^2$ for $3 - 1 = 2$ degrees of freedom. Consequently, the null hypothesis cannot be rejected; if there are, in fact, differences among the actual proportions for the three groups (as may have seemed reasonable even before any data were collected), we have no evidence for it at the 0.05 level of significance.

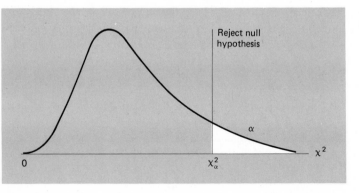

Reject null hypothesis

FIGURE 12.5

Test criterion.

In general, if we want to test the null hypothesis $p_1 = p_2 = \cdots = p_k$ on the basis of random samples from k populations, we combine the data, as in our example, and thus get the following estimate of the common population proportion p:

Estimate of common population proportion

$$\frac{x_1 + x_2 + \cdots + x_k}{n_1 + n_2 + \cdots + n_k}$$

where the n's are the sizes of the respective samples and the x's are the numbers of successes. We then multiply the n's by this estimate of p to get the expected frequencies for the first row of the table; after that we subtract these values from the respective sample sizes to get the expected frequencies for the second row of the table.[†] Next, we calculate χ^2 according to the formula on page 315, with $\frac{(o - e)^2}{e}$ determined separately for each of the $2k$ cells of the table, and reject the null hypothesis at the level of significance α if this value of χ^2 exceeds χ_α^2 for $k - 1$ degrees of freedom.

When we calculate the expected frequencies, we usually round them to the nearest integer or to one decimal. The entries in Table III are given to three decimals, in some instances more, but there is seldom any need to carry more than two decimals in calculating the value of the χ^2 statistic, itself. Also, the test we have been discussing is only an approximate test, since the sampling distribution of the χ^2 statistic is only approximately the chi-square distribution. Thus, it should not be used when one (or more) of the expected frequencies is less than 5. (If this is the case, we can sometimes combine two or more of the samples in such a way that none of the e's is less than 5.)

The method we have discussed here can be used only to test the null hypothesis $p_1 = p_2 = \cdots = p_k$ against the alternative hypothesis that the p's are not all equal. However, in the special case where $k = 2$ and we are interested in the significance of the difference between two sample proportions, we can also test against either of the alternative hypotheses $p_1 < p_2$ or $p_1 > p_2$ by using an equivalent test based on the normal curve rather than the chi-square distribution. This alternative procedure is discussed in Exercise 5 on page 318.

EXERCISES

1 In a random sample of visitors to a famous tourist attraction, 84 of 250 men and 156 of 250 women bought souvenirs. Use the level of significance $\alpha = 0.05$ to test the null hypothesis that there is no difference between the corresponding population proportions.

[†] The expected frequency for any one of the cells may also be obtained by multiplying the total of the row to which it belongs by the total of the column to which it belongs, and then dividing by the **grand total**, $n_1 + n_2 + \cdots + n_k$, for the entire table (see also page 323).

2 A study showed that 74 of 200 persons who saw a deodorant advertised during the telecast of a baseball game and 86 of 200 other persons who saw it advertised on a variety show remembered two hours later the name of the deodorant. Use the level of significance $\alpha = 0.05$ to test the null hypothesis that there is no difference between the corresponding true proportions.

3 In one random sample high school students were asked whether they would turn to their father or their mother for help with a homework assignment in mathematics, and in another random sample they were asked the same question with regard to a homework assignment in English. Using the results shown in the following table and the level of significance $\alpha = 0.01$, test the null hypothesis that there is no difference between the true proportions of students who would turn to their fathers rather than their mothers for help in these two subjects:

	Mathematics	English
Mother	59	85
Father	91	65

4 Out of 500 marriage license applications, chosen at random in 1971, 48 of the women were at least one year older than the men, and out of 500 marriage license applications, chosen at random in 1977, 85 of the women were at least one year older than the men. Use the level of significance $\alpha = 0.05$ to test the null hypothesis that there was no change from 1971 to 1977 in the actual proportion of women on marriage license applications who are at least one year older than the men.

5 When we compare two sample proportions, we are sometimes interested in testing the null hypothesis $p_1 = p_2$ against the one-sided alternative $p_1 < p_2$ or the one-sided alternative $p_1 > p_2$. In that case we base our decision on the statistic

Statistic for test concerning difference between two proportions

$$z = \frac{\dfrac{x_1}{n_1} - \dfrac{x_2}{n_2}}{\sqrt{p(1-p)\left(\dfrac{1}{n_1} + \dfrac{1}{n_2}\right)}} \quad \text{with} \quad p = \frac{x_1 + x_2}{n_1 + n_2}$$

whose sampling distribution is approximately the standard normal distribution so long as n_1 and n_2 are both large. The criteria for tests based on this statistic are again those of Figure 10.8 on page 273.

(a) In a random sample of 200 persons who skipped breakfast, 82 reported that they experienced midmorning fatigue, and in a random sample of 400 persons who ate breakfast, 116 reported that they experienced midmorning fatigue. Use the level of significance $\alpha = 0.01$ to test the null hypothesis that there is no difference between the corresponding population proportions against the alternative that midmorning fatigue is more prevalent among persons who skip breakfast.

(b) One mail solicitation for a charity brought 412 responses to 5,000 letters and another, more expensive, mail solicitation brought 311 responses to 3,000 letters. Use the level of significance $\alpha = 0.01$ to test the null hypothesis that the two solicitations are equally effective against the alternative that the more expensive one is more effective.

6 The z statistic of Exercise 5 can also be used when the alternative hypothesis is $p_1 \neq p_2$. In fact, it can be shown that such a test would be equivalent to the corresponding test based on the χ^2 statistic, and that the square of the value of z would equal the value of χ^2.

(a) Use this method to rework Exercise 1, and verify that the square of the value of z equals the value originally obtained for χ^2.

(b) Use this method to rework Exercise 2, and verify that the square of the value of z equals the value originally obtained for χ^2.

7 If we want to test the null hypothesis that the difference between two population proportions equals some constant δ (*delta*), not 0, we can base our decision on the statistic

Statistic for test concerning difference between two proportions

$$z = \frac{\dfrac{x_1}{n_1} - \dfrac{x_2}{n_2} - \delta}{\sqrt{\dfrac{\dfrac{x_1}{n_1}\left(1 - \dfrac{x_1}{n_1}\right)}{n_1} + \dfrac{\dfrac{x_2}{n_2}\left(1 - \dfrac{x_2}{n_2}\right)}{n_2}}}$$

whose sampling distribution is approximately the standard normal distribution so long as n_1 and n_2 are both large. The criteria for tests based on this statistic are again those of Figure 10.8 on page 273.

(a) In a true–false test, a test item is considered to be good if it discriminates between well-prepared and poorly prepared students. If 205 of 250 well-prepared students and 137 of 250 poorly prepared students answered a certain test item correctly, test at the level of significance $\alpha = 0.05$ whether for the given test item the percentage of correct answers will in general be at least 20 percent higher among well-prepared students than among poorly prepared students.

(b) Two groups of 80 patients each took part in an experiment in which one group received pills containing an anti-allergy drug, while the other group received a placebo (a pill containing no drug). If in the group given the drug 23 exhibited allergic symptoms while in the group given the placebo 51 exhibited such symptoms, test at the level of significance $\alpha = 0.01$ whether the percentage of patients exhibiting allergic symptoms is at least 15 percent less for those who actually receive the drug.

12.4 **Differences Among Proportions**

319

8 On page 314 we referred to a study in which 24 of 200 brand A tires, 21 of 200 brand B tires, 16 of 200 brand C tires, and 33 of 200 brand D tires failed to last 20,000 miles. Use the level of significance $\alpha = 0.05$ to test the null hypothesis that there is no difference in the quality of the four kinds of tires.

9 A market research study shows that among 100 men, 100 women, and 200 children interviewed, there were, respectively, 44, 53, and 133 who did not like the flavor of a new toothpaste. Use the level of significance 0.05 to test whether the differences among the corresponding sample proportions (0.44, 0.53, and 0.665) are significant.

10 The following table shows the results of a survey in which random samples of the parents of 12th graders in three school districts were asked whether they are for or against the removal of certain controversial books from a required reading list:

	District 1	District 2	District 3
For the removal of the books	8	13	12
Against the removal of the books	52	67	48

Use the level of significance $\alpha = 0.05$ to test whether the differences among the sample proportions (of parents favoring the removal of the controversial books from the reading list) are significant.

11 Suppose that the data shown in the following table were obtained in a study of methods of suicide attempt:

	White		Black	
	Female	Male	Female	Male
Ingestion or inhalation of drug	13	10	14	4
Violent self-injury	9	9	5	9

Use the level of significance $\alpha = 0.05$ to test whether the differences among the sample proportions (of suicide by violent self-injury) are significant.

12 Verify that if the expected numbers of successes for the k samples are determined as on page 315, the sum of the expected numbers of successes equals the sum of the observed numbers of successes.

12.5
Contingency Tables

The χ^2 statistic plays an important role in many other problems where information is obtained by counting or enumerating, rather than measuring. The method we shall describe here for analyzing such **count data** is an extension of the method of the preceding section, and it applies to two distinct kinds of problems. In the first kind of problem we deal with trials permitting more than two possible outcomes.

EXAMPLE Suppose that in the example of the preceding section the persons had been asked whether "friends and social life," "job or primary activity," or "health and physical condition" contributes most to their general happiness, with the results shown in the following table:

	Single	Married	Widowed or divorced
Friends and social life	41	49	42
Job or primary activity	27	50	33
Health and physical condition	12	21	25
Total	80	120	100

We refer to this kind of table as a 3×3 table (where 3×3 is read "3 by 3"), because it contains 3 rows and 3 columns; more generally, when there are c samples and each trial permits r alternatives, we refer to the resulting table as an $r \times c$ **table.** Here, as in the example of the preceding section, the column totals (representing the sizes of the different samples) are fixed. On the other hand, the row totals depend on the responses of the persons interviewed, and, hence, on chance.

In the second kind of problem where the method of this section applies, the column totals as well as the row totals are left to chance.

EXAMPLE Suppose we want to investigate whether there is a relationship between real-estate salespersons' standard of clothing and their success in their jobs (based on their total 1978 sales). Suppose, furthermore, that a random sample of 350 real-estate persons taken from among all those licensed in Southern California, yielded the results shown in the following table:

| | Success in job | | | |
	Below average	Average	Above average	Total
Poorly dressed	19	17	6	42
Well dressed	27	35	21	83
Very well dressed	30	137	58	225
Total	76	189	·85	350

This is also a 3×3 table, and it is mainly in connection with problems like this that $r \times c$ tables are referred to as **contingency tables.**

Before we demonstrate how $r \times c$ tables are analyzed, let us examine the null hypotheses we want to test. In the first problem, the one dealing with the different factors contributing to one's general happiness, we want to test the null hypothesis that the probabilities that a person will choose "friends and social life," "job or primary activity," or "health and physical condition" as the factor which contributes most to his or her general happiness are the same for each group. In other words, we want to test the null hypothesis that a person's choice of one of the three alternatives is independent of his or her being single, married, or widowed or divorced. In the second problem we are also concerned with a null hypothesis of independence; specifically, the null hypothesis that real estate salespersons' success in their job is independent of their standard of clothing.

EXAMPLE   To illustrate the analysis of an $r \times c$ table, let us refer to the second example and let us begin by calculating the **expected cell frequencies.** If the null hypothesis of independence is true, the probability of randomly selecting a real-estate salesperson who dresses poorly and whose 1978 sales were below average is given by the product of the probability that he or she dresses poorly and the probability that his or her 1978 sales were below average. Using the totals of the first row and the first column to estimate these two probabilities, we get $\dfrac{19 + 17 + 6}{350} = \dfrac{42}{350}$ for the probability that a real-estate salesperson (randomly selected from among those licensed in Southern California) dresses poorly, and $\dfrac{19 + 27 + 30}{350} = \dfrac{76}{350}$ for the probability that his or her 1978 sales were below average. Hence, we estimate the probability of choosing a real-estate salesperson who dresses poorly and whose 1978 sales were below average as $\dfrac{42}{350} \cdot \dfrac{76}{350}$, and in a sample of size 350 we would expect to find

$$350 \cdot \frac{42}{350} \cdot \frac{76}{350} = \frac{42 \cdot 76}{350} = 9.1$$

real-estate salespersons who fit this description.

In this last result, $\dfrac{42 \cdot 76}{350}$ is just the product of the total of the first row and the total of the first column divided by the grand total for the entire table. Indeed, the argument which led to this result can be used to show that in general **the expected frequency for any cell of a contingency table may be obtained by multiplying the total of the row to which it belongs by the total of the column to which it belongs and then dividing by the grand total for the entire table.**

EXAMPLE Using this rule to continue with our illustration, we get an expected frequency of $\dfrac{42 \cdot 189}{350} = 22.7$ for the second cell of the first row, and $\dfrac{83 \cdot 76}{350} = 18.0$ and $\dfrac{83 \cdot 189}{350} = 44.8$ for the first two cells of the second row.

It is not necessary to calculate all the expected cell frequencies in this way, as it can be shown that the sum of the expected frequencies for any row or column must equal the sum of the corresponding observed frequencies. Therefore, we can get some of the expected cell frequencies by subtraction from row and column totals.

EXAMPLE For our illustration we thus get by subtraction

$$42 - 9.1 - 22.7 = 10.2$$

for the third cell of the first row,

$$83 - 18.0 - 44.8 = 20.2$$

for the third cell of the second row, and

$$76 - 9.1 - 18.0 = 48.9$$
$$189 - 22.7 - 44.8 = 121.5$$

and

$$85 - 10.2 - 20.2 = 54.6$$

for the three cells of the third row. These results are summarized in the following table, where the expected frequencies are shown in parentheses below the corresponding observed cell frequencies:

	Success in job		
	Below average	*Average*	*Above average*
Poorly dressed	19 (9.1)	17 (22.7)	6 (10.2)
Well dressed	27 (18.0)	35 (44.8)	21 (20.2)
Very well dressed	30 (48.9)	137 (121.5)	58 (54.6)

From here on the work is like that of the preceding section; we calculate the χ^2 statistic according to the formula

Statistic for test of independence in contingency table

$$\chi^2 = \sum \frac{(o - e)^2}{e}$$

with $\dfrac{(o - e)^2}{e}$ calculated separately for each cell of the table. Then we reject the null hypothesis of independence at the level of significance α if the value obtained for χ^2 exceeds χ^2_α for $(r - 1)(c - 1)$ degrees of freedom, where r is the number of rows and c the number of columns. To justify this formula for the degrees of freedom, observe that once we give (or calculate) the expected frequencies for the $(r - 1)(c - 1)$ cells of the first $r - 1$ rows and $c - 1$ columns, all the remaining expected cell frequencies are automatically determined. This follows from the result stated earlier that the sum of the expected frequencies for any row or column must equal the sum of the corresponding observed frequencies. For our example the number of degrees of freedom is $(3 - 1)(3 - 1) = 4$, and it should be observed that after we had calculated four of the expected cell frequencies, we got all the others by subtraction from the totals of appropriate rows and columns.

EXAMPLE Returning to our illustration and specifying that $\alpha = 0.01$, we find from the table on page 324 that

$$\chi^2 = \frac{(19 - 9.1)^2}{9.1} + \frac{(17 - 22.7)^2}{22.7} + \frac{(6 - 10.2)^2}{10.2}$$

$$+ \frac{(27 - 18.0)^2}{18.0} + \frac{(35 - 44.8)^2}{44.8} + \frac{(21 - 20.2)^2}{20.2}$$

$$+ \frac{(30 - 48.9)^2}{48.9} + \frac{(137 - 121.5)^2}{121.5} + \frac{(58 - 54.6)^2}{54.6}$$

$$= 30.10$$

Since this exceeds 13.277, the value of $\chi^2_{0.01}$ for 4 degrees of freedom, the null hypothesis must be rejected. Thus, we have shown at the 0.01 level of significance that there is a dependence (or relationship) between real estate salespersons' standard of clothing and their success in their jobs.

The method we have used to analyze this problem applies also to problems in which the column totals are fixed sample sizes (as in the example on page 321) and do not depend on chance. The expected frequencies are determined in the same way, but the formula according to which we multiply the row total by the column total and then divide by the grand total has to be justified in a different way (see Exercise 9 on page 330).

As in the method described in Section 12.4, the χ^2 statistic we are using here has only approximately a chi-square distribution. Thus, as we explained on page 317, it should not be used when any of the expected frequencies are less than 5. When there are expected cell frequencies less than 5, it is often possible to combine some of the cells, subtract 1 degree of freedom for each cell eliminated, and then perform the test just as it has been described.

12.6
Goodness of Fit

In this section we shall treat a further application of the χ^2 criterion, in which we compare an observed frequency distribution with a distribution we might expect according to theory or assumptions. We refer to such a comparison as a test of **goodness of fit**.

EXAMPLE Table IX is a table of random numbers which is supposed to have been constructed in such a way that each digit is a value of a random variable which takes on the values 0, 1, 2, 3, 4, 5, 6, 7, 8, and 9 with equal probabilities of 0.10. To see whether it is reasonable to maintain that this is, indeed, the case, we might count how many times each digit appears in the table or part of the table; specifically, let us take the 250 digits in the first five columns of the table on page 470. This yields the values shown in the "observed frequency" column of the following table:

Digit	Probability	Observed frequency o	Expected frequency e
0	0.10	22	25
1	0.10	17	25
2	0.10	23	25
3	0.10	26	25
4	0.10	27	25
5	0.10	31	25
6	0.10	26	25
7	0.10	23	25
8	0.10	29	25
9	0.10	26	25
		250	250

The expected frequencies in the right-hand column were obtained by multiplying each of the probabilities of 0.10 by 250, the total number of digits counted.

To test whether the discrepancies between the observed and expected frequencies can be attributed to chance, we use the same chi-square statistic as in the two preceding sections:

Statistic for test of goodness of fit

$$\chi^2 = \sum \frac{(o - e)^2}{e}$$

calculating $\frac{(o - e)^2}{e}$ separately for each class of the distribution. Then, if the value we get for χ^2 exceeds χ^2_α, we reject the null hypothesis (on which the expected frequencies are based) at the level of significance α. The number of degrees of freedom is $k - m$, where k is the number of terms $\frac{(o - e)^2}{e}$ added in the formula for χ^2, and m is the number of quantities we must obtain from the observed data to calculate the expected frequencies.

EXAMPLE For the illustration dealing with the random digits, let us test at the level of significance $\alpha = 0.05$ whether the table of random numbers is truly random. We get

$$\chi^2 = \frac{(22 - 25)^2}{25} + \frac{(17 - 25)^2}{25} + \frac{(23 - 25)^2}{25} + \frac{(26 - 25)^2}{25}$$

$$+ \frac{(27 - 25)^2}{25} + \frac{(31 - 25)^2}{25} + \frac{(26 - 25)^2}{25} + \frac{(23 - 25)^2}{25}$$

$$+ \frac{(29 - 25)^2}{25} + \frac{(26 - 25)^2}{25}$$

$$= 5.60$$

and since this value does not exceed 16.919, the value of $\chi^2_{0.05}$ for $10 - 1 = 9$ degrees of freedom, we find that the null hypothesis cannot be rejected—the table of random numbers has "passed the test." The number of degrees of freedom is 9, since there are ten terms in the formula for χ^2, and the only quantity needed from the observed data to calculate the expected frequencies is the total frequency of 250.

The method we have illustrated here is used quite generally to test how well distributions, expected on the basis of theory or assumptions, fit, or describe, observed data. Thus, in the exercises which follow we shall test whether observed distributions have (at least approximately) the shape of binomial, Poisson, and normal distributions. As in the tests of the two preceding sections, the sampling distribution of the χ^2 statistic is only approximately a chi-square distribution when it is used for tests of goodness of fit. So, if any one of the expected frequencies is less than 5, we shall again have to combine some of the data; in this case, we would combine adjacent classes of the distribution.

EXERCISES 1 Analyze the 3 × 3 table on page 321 and decide at the level of significance $\alpha = 0.05$ whether the probabilities of choosing the three alternatives are the same for the three groups.

2 A market research organization wants to determine, on the basis of the following information, whether there is a relationship between the size of a tube of toothpaste which a shopper buys and the number of persons in the shopper's household:

| | Number of persons in household | | | |
	1–2	3–4	5–6	7 or more
Giant	23	116	78	43
Large	54	25	16	11
Small	31	68	39	8

Size of tube bought

At the level of significance $\alpha = 0.01$, is there a relationship?

3 Suppose that for 120 mental patients who did not receive psychotherapy and 120 mental patients who received psychotherapy a panel of psychiatrists determined after six months whether their condition had deteriorated, remained unchanged, or improved. Based on the results shown in the following table, test at the level of significance $\alpha = 0.05$ whether the therapy is effective:

	No therapy	Therapy
Deteriorated	6	11
Unchanged	65	31
Improved	49	78

4 With reference to the data of Exercise 3, suppose that somebody observes that even though the therapy seems to be effective, the proportion of patients whose condition deteriorated nearly doubled. What can we say in reply, if we combine the "unchanged" and "improved" categories, and then analyze the resulting 2×2 table?

5 Tests of the fidelity and the selectivity of 190 radios produced the results shown in the following table:

| | | Fidelity | | |
		Low	Average	High
	Low	7	12	31
Selectivity	**Average**	35	59	18
	High	15	13	0

Use the level of significance $\alpha = 0.01$ to test the null hypothesis that there is no relationship (dependence) between fidelity and selectivity.

6 A sample survey, designed to show where persons living in different parts of the country buy nonprescribed medicines, yielded the following results:

	North-east	North central	South	West
Drugstores	218	200	183	179
Grocery stores	39	52	87	62
Others	43	48	30	59

Use the level of significance $\alpha = 0.05$ to test the null hypothesis that, so far as nonprescribed medicines are concerned, the buying habits of persons living in the given parts of the country are the same.

7 In a study of students' parents' feelings about a required course in sex education, a random sample of 360 parents are classified according to whether they have one, two, or three or more children in the school system, and also whether they feel that the course is poor, adequate, or good. Supposing that the results are as shown in the following table, test at the $\alpha = 0.05$ level of significance whether there is a relationship between parents' reaction to the course and the number of children they have in school:

	Number of children		
	1	2	3 or more
Poor	48	40	12
Adequate	55	53	29
Good	57	46	20

8 If the analysis of a contingency table shows that there is a relationship between the two variables under consideration, the strength of the relationship can be measured by the **contingency coefficient**

Contingency coefficient

$$C = \sqrt{\frac{\chi^2}{\chi^2 + n}}$$

where χ^2 is the value of the chi-square statistics obtained for the table and n is the grand total of all the frequencies. This coefficient takes on values between 0 (corresponding to independence) and a maximum value less than 1 depending on the size of the table; for instance, it can be shown that for a 3×3 table the maximum value of C is $\sqrt{2/3} = 0.82$.

    (a) Calculate C for the example in the text which deals with the real estate salespersons' standard of clothing and their success in their jobs.

    (b) Calculate C for the contingency table of Exercise 3.

9 Use an argument similar to that on page 323 to show that the rule for calculating the expected cell frequencies (dividing the product of the row total and the column total by the grand total) applies also when the column totals are fixed sample sizes and do not depend on chance.

10 Four coins are tossed 160 times and 0, 1, 2, 3, and 4 heads showed 19, 54, 58, 23, and 6 times. Using the probabilities given on page 155 and the level of significance $\alpha = 0.05$, test whether it is reasonable to suppose that the coins are balanced and randomly tossed.

11 Ten years' data show that in a given Western city there were no bank robberies in 51 months, one bank robbery in 39 months, two bank robberies in 21 months, and three or more bank robberies in 9 months. At the level of significance $\alpha = 0.01$, does this substantiate the claim that the monthly number of bank robberies in this city is a random variable having the Poisson distribution with $\lambda = 1$, namely, that the probabilities of 0, 1, 2, or 3 or more bank robberies are 0.37, 0.37, 0.18, and 0.08?

12 To see whether a die is balanced, it is rolled 360 times and the following results are obtained: 1 showed 55 times, 2 showed 48 times, 3 showed 67 times, 4 showed 51 times, 5 showed 73 times, and 6 showed 66 times. At the level of significance $\alpha = 0.05$, do these results support the hypothesis that the die is balanced?

13 A quality control engineer takes daily samples of four tractors coming off an assembly line and on 200 consecutive working days he obtains the data summarized in the following table:

Number of tractors requiring adjustments	Number of days
0	102
1	78
2	19
3	1
4	0

To test the claim that 10 percent of all the tractors coming off this assembly line require adjustments, look up the corresponding probabilities in Table VI, calculate the expected frequencies, and perform the chi-square test at the level of significance $\alpha = 0.01$.

14 The following is the distribution of the sulfur oxides emission data obtained on page 14:

Tons of sulfur oxides	Frequency
5.0–8.9	3
9.0–12.9	10
13.0–16.9	14
17.0–20.9	25
21.0–24.9	17
25.0–28.9	9
29.0–32.9	2
	80

As we showed on page 53, the mean of this distribution is $\bar{x} = 18.85$ and its standard deviation is $s = 5.55$.

(a) Find the probabilities that a random variable having a normal distribution with $\mu = 18.85$ and $\sigma = 5.55$ takes on a value less than 8.95, between 8.95 and 12.95, between 12.95 and 16.95, between 16.95 and 20.95, between 20.95 and 24.95, between 24.95 and 28.95, and greater than 28.95.

(b) Multiply the probabilities obtained in part (a) by the total frequency, $n = 80$, thus getting the expected normal-curve frequencies corresponding to the seven classes of the given distribution.

(c) Using the level of significance $\alpha = 0.05$, test the null hypothesis that the given data may be looked upon as a random sample from a normal population by comparing the observed and expected frequencies with an appropriate χ^2 statistic. Explain why the number of degrees of freedom for this χ^2 test is $k - 3$, where k is the number of terms in the χ^2 statistic.

BIBLIOGRAPHY The theory which underlies the various tests of this chapter is discussed in most textbooks on mathematical statistics; for instance, in the author's book listed on page 238.

13

Most of the tests discussed in the last three chapters required specific assumptions about the population, or populations, sampled. Among other things, we often assumed that these populations can be approximated closely by normal distributions, that their standard deviations are known, or are known to be equal, and that the samples are independent. Since there are many situations in which these assumptions cannot be met, statisticians have developed alternative techniques based on less stringent assumptions, which have become known as **nonparametric methods.**

Aside from the fact that nonparametric methods can be used under more general conditions, they are often easier to explain and easier to understand than the standard methods which they replace. Moreover, in many nonparametric methods the computational burden is so light that they come under the heading of "quick and easy" or "short-cut" techniques. For these reasons, nonparametric methods have become quite popular, and extensive literature is devoted to their theory and application.

Nonparametric Methods

13.1

The One-Sample Sign Test

The small-sample tests concerning means and differences between means, which we studied in Chapter 10, are all based on the assumption that the populations sampled have roughly the shape of normal distributions. When this assumption is untenable in particular situations, these standard tests can be replaced by any one of several nonparametric alternatives, among them the **sign test**, which we shall study in this section and the next.

The **one-sample sign test** applies when we sample a continuous symmetrical population, so that the probability of getting a sample value less than the mean and the probability of getting a sample value greater than the mean are both $\frac{1}{2}$. To test the null hypothesis $\mu = \mu_0$ against an appropriate alternative on the basis of a random sample of size n, we replace each sample value greater than μ_0 with a plus sign and each sample value less than μ_0 with a minus sign, and then we test the null hypothesis that these plus and minus signs are values of a random variable having the binomial distribution with $p = \frac{1}{2}$.[†] (If a sample value equals μ_0, which is not unlikely since we usually deal with rounded data, we simply discard it.)

To perform a one-sample sign test when the sample is small, we refer directly to tables of binomial probabilities such as Table VI at the end of the book; when the sample is large, we use the normal approximation to the binomial distribution.

EXAMPLE To illustrate the small-sample case, let us consider the following 15 measurements of the octane rating of a certain kind of gasoline:

$$97.0 \quad 94.7 \quad 96.8 \quad 95.5 \quad 96.3 \quad 99.8 \quad 96.9 \quad 94.8$$
$$92.7 \quad 98.6 \quad 95.6 \quad 97.1 \quad 97.7 \quad 98.0 \quad 94.4$$

To test the null hypothesis $\mu = 98.0$ against the alternative hypothesis $\mu < 98.0$ at the level of significance $\alpha = 0.01$, we first replace each value greater than 98.0 with a plus sign, each value less than 98.0 with a minus sign, and discard the one value which actually equals 98.0. Thus, we get

$$- \quad - \quad - \quad - \quad - \quad + \quad - \quad - \quad - \quad + \quad - \quad - \quad - \quad -$$

and it remains to be seen whether 2 plus signs observed in 14 trials support the null hypothesis $p = \frac{1}{2}$ or the alternative hypothesis

---

[†] If we cannot assume that the population is symmetrical, we can still use the one-sample sign test, but we have to test the null hypothesis $\tilde{\mu} = \tilde{\mu}_0$, where $\tilde{\mu}$ is the population median.

$p < \frac{1}{2}$. From Table VI we find that for $n = 14$ and $p = \frac{1}{2}$ the probability of 2 or fewer successes is $0.001 + 0.006 = 0.007$, and since this is less than $\alpha = 0.01$, the null hypothesis must be rejected. We conclude that the mean octane rating of the given kind of gasoline is less than 98.0.

Since np and $n(1 - p)$ were both greater than 5 for $n = 14$ and $p = \frac{1}{2}$, we could have used the normal approximation in this example, but we shall illustrate this technique next with a different example.

EXAMPLE   Suppose that the air pollution data on page 14 were obtained to decide at the 0.05 level of significance whether or not the plant's average daily emission of sulfur oxides is greater than 17.5 tons. That is, we want to test the null hypothesis $\mu = 17.5$ against the alternative hypothesis $\mu > 17.5$. There are 50 plus signs and 29 minus signs in this case, and we must see whether, on the basis of this, we can reject the null hypothesis that the probability of getting a plus sign is $\frac{1}{2}$ and conclude that this probability is greater than $\frac{1}{2}$. Substituting $n = 79$ and $p_0 = \frac{1}{2}$ into the formula for the large-sample statistic on page 311, we get

$$z = \frac{50 - \dfrac{1}{2} - 79 \cdot \dfrac{1}{2}}{\sqrt{79 \cdot \dfrac{1}{2} \cdot \dfrac{1}{2}}} = 2.25$$

Since this exceeds $z_{0.05} = 1.64$, it follows that the null hypothesis must be rejected, and we conclude that the plant's true average daily emission of sulfur oxides exceeds 17.5 tons.

13.2

The Two-Sample Sign Test

The sign test has important applications in problems where we deal with paired data. In these problems, each pair of sample values can be replaced with a plus sign if the first value is greater than the second, a minus sign if the first value is smaller than the second, or be discarded if the two values are equal. In connection with this, there are two kinds of situations, depending on whether the data are actually given as pairs or whether they consist of two independent samples whose values are randomly paired.

EXAMPLE   To illustrate the first case, suppose that in a study designed to determine the effectiveness of a new traffic control system, the following data were collected on the number of accidents that occurred at 12 dangerous intersections, presumably a random sample, during three months before and three months after the installation of the new system: 3 and 1, 5 and 2, 2 and 0, 3 and 2, 3 and 2, 3 and 1, 0 and 2, 4 and 3, 1 and 3, 5 and 3, 4 and 2, and 1 and 0. This yields

$$+ \quad + \quad + \quad + \quad + \quad + \quad - \quad + \quad - \quad + \quad + \quad +$$

and the question is whether "10 successes in 12 trials" supports the null hypothesis $p = \frac{1}{2}$ or the alternative hypothesis $p > \frac{1}{2}$ (the alternative hypothesis that the probability of a decrease in the number of accidents exceeds $\frac{1}{2}$), say, at the level of significance $\alpha = 0.05$. Referring to Table VI, we find that for $n = 12$ and $p = \frac{1}{2}$ the probability of "10 or more successes" is $0.016 + 0.003 = 0.019$, and it follows at the 0.05 level of significance that the null hypothesis must be rejected. We conclude that the new traffic control system is effective in reducing the number of accidents at dangerous intersections.

EXAMPLE To illustrate the case where we have two independent random samples, let us consider the following grades which 25 male freshmen and 25 female freshmen (random samples of the students entering a major university) received on a foreign language placement text:

Women: 60, 80, 81, 88, 58, 64, 78, 57, 80, 84, 72, 84, 65
46, 75, 85, 79, 86, 85, 50, 78, 66, 89, 61, 68

Men: 62, 44, 62, 82, 54, 39, 51, 76, 51, 66, 63, 56, 77
33, 59, 88, 76, 50, 76, 51, 25, 63, 22, 51, 64

We want to test, at the level of significance $\alpha = 0.05$, whether the mean grades are equal for the two populations sampled, but since there is much more variability among the grades of the men students, the standard t test of Section 10.12 cannot be used. To use the two-sample sign test as a nonparametric alternative, we randomly pair each woman's grade with a man's grade, and doing so with the use of random numbers, we get the results shown in the following table:

Women	Men		Women	Men	
60	33	+	46	51	−
80	63	+	75	76	−
81	56	+	85	64	+
88	39	+	79	51	+
58	62	−	86	59	+
64	76	−	85	50	+
78	76	+	50	62	−
57	54	+	78	88	−
80	51	+	66	63	+
84	44	+	89	82	+
72	51	+	61	25	+
84	22	+	68	77	−
65	66	−			

There are 17 plus signs and 8 minus signs, and we want to know whether "17 successes in 25 trials" supports the null hypothesis $p = \frac{1}{2}$ (the null hypothesis that the true average grades are the same for male and female freshmen entering the university) or the alternative hypothesis $p \neq \frac{1}{2}$. Since np and $n(1 - p)$ are both greater than 5 for $n = 25$ and $p = \frac{1}{2}$, we can use the normal-curve approximation to the binomial distribution, and substitution into the formula for the test statistic on page 311 yields

$$z = \frac{17 - \frac{1}{2} - 25 \cdot \frac{1}{2}}{\sqrt{25 \cdot \frac{1}{2} \cdot \frac{1}{2}}} = 1.60$$

Since this does not exceed $z_{0.025} = 1.96$, the null hypothesis cannot be rejected. This result may come as a surprise since there is a substantial difference between the mean grades of the two groups— the women averaged 72.8 and the men averaged 57.6—but it illustrates the fact that nonparametric methods are potentially wasteful of information. For instance, the difference between 60 and 33 (which was replaced by a plus sign) counts only as much as the difference between 65 and 66 (which was replaced by a minus sign). Since the standard t test cannot be used, the only alternative is to try a nonparametric test which is generally less wasteful of information (see Exercise 2 on page 343). To be fair, though, we must decide which test to use before any calculations are made.

In the preceding example the two samples were of the same size, so that the values could all be paired. If the sample sizes are not equal, some of the values of the larger sample, left over after the random pairing, have to be discarded.

EXERCISES

1 On 15 occasions, Mr. Jones had to wait 4, 8, 7, 7, 2, 6, 8, 5, 9, 6, 1, 5, 6, 5, and 9 minutes for the bus he takes to work. Use the sign test at the level of significance $\alpha = 0.05$ to test the bus company's claim that on the average he should not have to wait more than 5 minutes for a bus.

2 Use the normal approximation to the binomial distribution to rework the example on page 333, which dealt with the measurement of the octane rating of a gasoline.

3 In a laboratory experiment, 18 determinations of the coefficient of friction between leather and metal yielded the following results: 0.59, 0.48, 0.53, 0.57, 0.55, 0.61, 0.56, 0.60, 0.55, 0.58, 0.52, 0.55, 0.56, 0.51, 0.57, 0.56, 0.59, and 0.58. Use the sign test at the level of significance $\alpha = 0.05$ to test the null hypothesis $\mu = 0.55$ against the alternative hypothesis $\mu \neq 0.55$.

4 Looking upon the weights of Exercise 9 on page 23 as a random sample (of the weights of all the rats supplied to the research facilities), use the sign test at the level of significance $\alpha = 0.01$ to test the null hypothesis $\mu = 120$ gr against the alternative hypothesis $\mu \neq 120$ gr.

5 Playing four rounds of golf at the Paradise Valley Country Club, ten professionals totaled 280, 282, 287, 279, 278, 283, 280, 279, 280, and 283. Use the sign test at the level of significance $\alpha = 0.05$ to test the null hypothesis that professional golfers average $\mu = 284$ for four rounds on this course against the alternative hypothesis $\mu < 284$.

6 With reference to Exercise 10 on page 23, use the sign test at the level of significance $\alpha = 0.01$ to test the null hypothesis that on the average the company's cabs spend \$50.00 a week on gasoline against the alternative hypothesis that this figure is too high.

7 Use the two-sample sign test to rework Exercise 26 on page 284.

8 Use the two-sample sign test to rework Exercise 27 on page 284.

9 The following are the numbers of artifacts dug up by two archaeologists at an ancient cliff dwelling on 30 days: 1 and 0, 0 and 0, 2 and 1, 3 and 0, 1 and 2, 0 and 0, 2 and 0, 2 and 1, 3 and 1, 0 and 2, 1 and 0, 1 and 1, 4 and 2, 1 and 1, 2 and 1, 1 and 0, 3 and 2, 5 and 2, 2 and 6, 1 and 0, 3 and 2, 2 and 3, 4 and 0, 1 and 2, 3 and 1, 2 and 0, 0 and 1, 2 and 0, 4 and 1, and 2 and 0. Use the sign test at the level of significance $\alpha = 0.01$ to test the null hypothesis that the two archaeologists are equally good at finding artifacts against the alternative hypothesis that the first one is better.

10 The following are the numbers of speeding tickets issued by two policemen on 30 days: 7 and 10, 11 and 13, 10 and 11, 14 and 14, 11 and 15, 12 and 9, 6 and 10, 9 and 13, 8 and 11, 10 and 11, 11 and 15, 13 and 11, 7 and 10, 6 and 12, 10 and 14, 8 and 8, 11 and 12, 9 and 14, 9 and 7, 10 and 12, 6 and 7, 12 and 14, 9 and 11, 12 and 10, 11 and 13, 12 and 15, 7 and 9, 10 and 9, 11 and 13, and 8 and 10. Use the sign test at the level of significance $\alpha = 0.01$ to test the null hypothesis that on the average the two policemen issue equally many speeding tickets against the alternative hypothesis that on the average the second policeman issues more speeding tickets than the first.

11 Random samples of 20 freshmen and 20 sophomores attending a certain university spent the following amounts of money (in dollars) on books:

Freshmen:	36, 74, 33, 62, 75, 55, 73, 93, 65, 50, 83, 62, 77, 74, 67, 44, 70, 99, 87, 61
Sophomores:	89, 77, 99, 72, 79, 91, 92, 97, 69, 75 68, 89, 93, 81, 92, 57, 86, 84, 88, 73

Randomly pair each amount spent by a sophomore with an amount spent by a freshman, and then perform a sign test of the null hypothesis that on the average freshman and sophomores spend equal amounts of money on books. Use the alternative hypothesis that on the average sophomores spend more money on books and the level of significance $\alpha = 0.05$.

12 For comparison, two kinds of feed are fed to samples of pigs, and the following are their gains in weight (in pounds) after a fixed period of time:

Feed C:	13.8, 13.4, 12.0, 12.9, 15.2, 14.2, 13.1, 14.6, 13.3, 12.5, 10.4, 14.1, 15.0, 12.4, 13.5, 14.8, 15.5, 13.2, 12.3, 12.0, 15.6, 14.1
Feed D:	11.4, 13.8, 11.7, 14.3, 12.6, 11.1, 13.6, 12.2, 13.8, 10.5, 12.8, 14.0, 10.2, 12.8, 12.7, 15.4, 11.3, 12.6, 11.8, 13.1, 10.8, 11.9, 14.0, 12.4, 12.7, 11.8, 13.1

Randomly pair each value obtained for feed C with a value obtained for feed D, and perform a sign test of the null hypothesis that the true average gain in weight is the same for both feeds. Use the two-sided alternative that the true average gain in weight is not the same for both feeds and the level of significance $\alpha = 0.01$.

13.3

Rank-Sum Tests: The Mann–Whitney Test

In the preceding section we showed how the two-sample sign test can be used as a nonparametric alternative to the standard small-sample t test for the difference between two means. Another nonparametric test which can be used for this purpose is the **Mann–Whitney test** (or the U **test**). This test is usually less wasteful of information than the sign test and it applies under very general conditions; in fact, it requires only that the populations sampled are continuous, and in practice even the violation of this assumption is seldom of any consequence.

EXAMPLE To illustrate how the Mann–Whitney test is performed, let us suppose that we want to test, at the level of significance $\alpha = 0.01$, whether the average grain size of sand obtained from two different locations on the moon is the same. The measurements that are available for this purpose are the following diameters (in millimeters):

Location A: 0.30, 0.43, 1.34, 0.18, 0.41, 0.47,
0.15, 0.38, 0.82, 0.49, 0.45, 0.10,
0.16, 0.33, 0.61

Location B: 1.09, 0.87, 0.99, 0.90, 0.37, 0.94,
1.11, 0.56, 1.05, 0.46, 0.51, 0.86,
0.24, 0.52

The means of these two samples are 0.44 and 0.75, and their difference is large, but it remains to be seen whether it is significant.

To perform the Mann–Whitney test, we first rank the data jointly, as though they comprised a single sample, in either an increasing or decreasing order of magnitude. The low-to-high ranking of our data is shown below, and underneath each item the letter A or B serves to indicate the location from which it came:

0.10	0.15	0.16	0.18	0.24	0.30	0.33	0.37	0.38	0.41
A	A	A	A	B	A	A	B	A	A
0.43	0.45	0.46	0.47	0.49	0.51	0.52	0.56	0.61	0.82
A	A	B	A	A	B	B	B	A	A
0.86	0.87	0.90	0.94	0.99	1.05	1.09	1.11	1.34	
B	B	B	B	B	B	B	B	A	

Assigning the data in this order the ranks 1, 2, 3, ..., and 29, we find that the values of the first sample (location A) occupy ranks 1, 2, 3, 4, 6, 7, 9, 10, 11, 12, 14, 15, 19, 20, and 29, while those of the second sample (location B) occupy ranks 5, 8, 13, 16, 17, 18, 21, 22,

23, 24, 25, 26, 27, and 28. There are no ties here among values belonging to different samples, but if there were, we would assign each of the tied observations the mean of the ranks which they jointly occupy. (For instance, if the third and fourth values were identical, we would assign each the rank $\frac{3 + 4}{2} = 3.5$; if the ninth, tenth, and eleventh values were identical, we would assign each the rank $\frac{9 + 10 + 11}{3} = 10$.)

The null hypothesis we want to test in a problem such as this is that the two samples come from identical populations. If this hypothesis is true, it seems reasonable to suppose that the means of the ranks assigned to the values of the two samples should be more or less the same. Under the alternative hypothesis the means of the two populations are not equal, and if this is the case and the difference is pronounced, most of the smaller ranks will go to the values of one sample, while most of the higher ranks will go to those of the other sample.

In the Mann–Whitney test we base the test of this null hypothesis on the statistic

U statistic

$$U = n_1 n_2 + \frac{n_1(n_1 + 1)}{2} - R_1$$

where n_1 and n_2 are the sample sizes and R_1 is the sum of the ranks assigned to the values of the first sample. In practice we find whichever rank sum can be most easily obtained, since it is immaterial which sample is called the "first" sample.

Under the null hypothesis that the two samples come from identical populations (or that the $n_1 + n_2$ observations come from one population, which is the same thing), it can be shown that the mean and the standard deviation of the sampling distribution of U are[†]

Mean and standard deviation of sampling distribution of U

$$\mu_U = \frac{n_1 n_2}{2}$$

and

$$\sigma_U = \sqrt{\frac{n_1 n_2 (n_1 + n_2 + 1)}{12}}$$

[†] If there are ties in rank, these formulas provide only approximations, but if the number of ties is small, there is usually no need to make any corrections.

13.3   **Rank-Sum Tests: The Mann–Whitney Test**

Furthermore, if n_1 and n_2 are sufficiently large (in this case, both greater than 8), the sampling distribution of U can be approximated closely with a normal curve. Thus, we reject the null hypothesis that the two samples come from identical populations when

Statistic for Mann–Whitney test

$$z = \frac{(U \pm \frac{1}{2}) - \mu_U}{\sigma_U}$$

is less than $-z_{\alpha/2}$ or exceeds $z_{\alpha/2}$, where α is the level of significance and $z_{\alpha/2}$ is defined as on page 248. In the continuity correction, $\pm \frac{1}{2}$, we use the minus sign when U exceeds μ_U, and the plus sign when U is less than μ_U; it is not used, however, when U is not an integer due to ties in rank. If either n_1 or n_2 is so small that this normal-curve approximation to the sampling distribution of U cannot be used, exact tests may be based on special tables (see Bibliography on page 349).

EXAMPLE   For the problem we have been discussing, let us call the sample from location A the first sample, so that $n_1 = 15$, $n_2 = 14$, $R_1 = 1 + 2 + 3 + 4 + 6 + 7 + 9 + 10 + 11 + 12 + 14 + 15 + 19 + 20 + 29 = 162$, and

$$U = 15 \cdot 14 + \frac{15 \cdot 16}{2} - 162 = 168$$

Also we find that

$$\mu_U = \frac{15 \cdot 14}{2} = 105$$

$$\sigma_U = \sqrt{\frac{15 \cdot 14 \cdot (15 + 14 + 1)}{12}} = 22.9$$

and hence that

$$z = \frac{168 - \frac{1}{2} - 105}{22.9} = 2.73$$

Since this value exceeds $z_{0.005} = 2.58$, we reject the null hypothesis at the level of significance $\alpha = 0.01$ and conclude that there is a difference between the true average grain size of the sand at the two locations on the moon.

An interesting feature of the Mann–Whitney test is that, when there is no difference between the means of two populations, it can be used to test the null hypothesis that two samples come from identical populations against the alternative that the two populations have unequal dispersions. The test is performed in the same way as before, except for a modification in the way in which the data are ranked. Again, the data are arranged in an increasing (or decreasing) order of magnitude, but now they are ranked from both ends toward the middle. We assign rank 1 to the smallest value, ranks 2 and 3 to the largest and second largest, ranks 4 and 5 to the second and third smallest, ranks 6 and 7 to the third and fourth largest, and so on. Subsequently, U is calculated and the test is performed exactly as before. With the new kind of ranking, however, a small rank sum for one sample suggests that the population from which it came has a greater variation than the other population since the sample values occupy the more extreme positions. We shall not illustrate this technique with reference to our sand-grain-size example, because this test loses its sensitivity for detecting differences in variability when the populations have unequal means; see, however, Exercises 6 and 7 on page 343.

13.4

Rank-Sum Tests: The Kruskal–Wallis Test

The **Kruskal–Wallis test** (or H **test**) is conducted in a way similar to the U test, and it serves to test the null hypothesis that k independent random samples come from identical populations against the alternative that the means of these populations are not all equal. Unlike the standard test which it replaces, the **one-way analysis of variance** which we shall discuss in Chapter 16, it does not require the assumptions that the samples come from (approximately) normal populations having the same standard deviation σ.

As in the Mann–Whitney test, the data are ranked jointly from low to high, or high to low, as though they constituted a single sample; then, letting R_i be the sum of the ranks assigned to the n_i observations in the ith sample, the Kruskal–Wallis test is based on the statistic

Statistic for Kruskal–Wallis test

$$H = \frac{12}{n(n + 1)} \sum_{i=1}^{k} \frac{R_i^2}{n_i} - 3(n + 1)$$

where $n = n_1 + n_2 + \cdots + n_k$. If the null hypothesis is true and each sample has at least five observations, the sampling distribution of H can be approximated closely with a chi-square distribution with $k - 1$ degrees of freedom. Consequently, we can reject the null hypothesis $\mu_1 = \mu_2 = \cdots = \mu_k$ against the alternative that these μ's are not all equal at the level of significance α, if H exceeds χ_α^2 for $k - 1$ degrees of freedom. If any sample has less than five observations so that this approximation cannot be used, exact tests may be based on special tables (see the Bibliography on page 349).

EXAMPLE Suppose that three groups of students are taught German by three different methods (classroom instruction and language laboratory, only classroom instruction, and only self-study in language laboratory), and at the end of the course they are all given the same final examination. The following are their scores on this examination and, on the basis of these, we want to determine at the level of significance $\alpha = 0.05$ whether the three methods of instruction are equally effective:

First method:	94, 87, 91, 74, 87, 97
Second method:	85, 82, 79, 84, 61, 72, 80
Third method:	89, 67, 72, 76, 69

After arranging the 18 scores according to size and assigning rank 1 to the largest value, rank 2 to the second largest value, ..., and rank 18 to the smallest value, we find that $R_1 = 1 + 2 + 3 + 5 + 6 + 13 = 30$, $R_2 = 7 + 8 + 9 + 10 + 11 + 14.5 + 18 = 77.5$, and $R_3 = 4 + 12 + 14.5 + 16 + 17 = 63.5$. (There is one tie, and the two tied scores which occupy ranks 14 and 15 are both ranked $\frac{14 + 15}{2} = 14.5$.) Substituting the values of R_1, R_2, and R_3 together with $n_1 = 6$, $n_2 = 7$, $n_3 = 5$, and $n = 18$ into the formula for H, we get

$$H = \frac{12}{18 \cdot 19}\left(\frac{30^2}{6} + \frac{77.5^2}{7} + \frac{63.5^2}{5}\right) - 3 \cdot 19$$

$$= 6.67$$

Since this exceeds 5.991, the value of $\chi^2_{0.05}$ for $k - 1 = 3 - 1 = 2$ degrees of freedom, we reject the null hypothesis and conclude that the means of the three populations sampled are not all equal. Assuming that the trainees were randomly assigned to the three methods of instruction, it appears that the methods are not all equally effective.

EXERCISES

1 In a study connected with the revision of mortality tables, the following ages at time of death were obtained for random samples of the members of two ethnic groups:

Group 1:	78, 54, 67, 69, 57, 75, 92, 71, 77, 79, 73, 83
	60, 85, 55
Group 2:	65, 84, 76, 82, 72, 81, 53, 89, 56, 70, 66, 74

Use the Mann–Whitney test at the level of significance $\alpha = 0.05$ to test the null hypothesis that the two samples come from identical populations against the alternative that the populations have unequal means.

2 Apply the Mann–Whitney test at the level of significance $\alpha = 0.01$ to the sample data on page 335 to test the null hypothesis that there is no difference between the average scores (on the foreign language placement test) of the two populations sampled.

3 Comparing two kinds of emergency flares, a consumer testing service obtained the following burning times (rounded to the nearest tenth of a minute):

Brand X: 17.5, 21.2, 20.3, 14.4, 15.2, 19.3, 18.8, 21.2,
19.1, 18.1, 14.6, 17.2

Brand Y: 13.4, 9.9, 13.5, 11.3, 22.5, 14.3, 13.6, 15.2,
11.8, 13.7, 8.0, 13.6

Use the Mann–Whitney test at the level of significance $\alpha = 0.01$ to see whether it is reasonable to say that there is no difference between the true average burning times of the two kinds of flares.

4 Use the Mann–Whitney test instead of the sign test to rework Exercise 11 on page 337.

5 The following are the lifetimes (in hours) of random samples of two kinds of light bulbs in continuous use:

Brand 1: 407, 426, 453, 378, 434, 396, 441, 373, 393, 386,
415, 418

Brand 2: 403, 424, 383, 445, 439, 417, 412, 462, 439, 432,
413, 433

Use the Mann–Whitney test at the level of significance $\alpha = 0.01$ to determine whether it is reasonable to claim that there is no difference between the true average lifetimes of the two kinds of light bulbs.

6 With reference to Exercise 1, use the Mann–Whitney test (modified as suggested on page 341) to test the null hypothesis that the two samples come from identical populations against the alternative that the populations have unequal dispersions. Use the level of significance $\alpha = 0.05$.

7 The following are the weekly food expenditures (in dollars) of ten families with two children chosen at random from the suburbs of a large city:

Suburb A: 66.15, 59.19, 67.00, 62.89, 65.78, 62.95, 81.45,
70.38, 65.50, 73.60

Suburb B: 56.75, 63.76, 85.59, 75.91, 61.12, 79.72, 89.19,
50.35, 83.16, 55.63

(a) Verify that if we use the Mann–Whitney test at the level of significance $\alpha = 0.05$ to test the null hypothesis that the two samples come from identical populations against the alternative that the populations have unequal means, the null hypothesis cannot be rejected.

(b) Use the Mann–Whitney test (modified as suggested on page 341) to test, at the level of significance $\alpha = 0.05$, whether the populations are identical or have unequal dispersions.

13.4 Rank-Sum Tests: The Kruskal–Wallis Test

343

8 The following are the miles per gallon which a test driver got for ten tankfuls each of three kinds of gasoline:

Gasoline 1:   20, 31, 24, 33, 23, 24, 28, 16, 19, 26
Gasoline 2:   29, 18, 29, 19, 20, 21, 34, 33, 30, 23
Gasoline 3:   19, 31, 16, 26, 31, 33, 28, 28, 25, 30

Use the Kruskal–Wallis test at the level of significance $\alpha = 0.05$ to test the null hypothesis that there is no difference in the average mileage yield of the three kinds of gasoline.

9 To compare four bowling balls, a professional bowler bowls five games with each ball and gets the following results:

Bowling ball D: 221, 232, 207, 198, 212
Bowling ball E: 202, 225, 252, 218, 226
Bowling ball F: 210, 205, 189, 196, 216
Bowling ball G: 229, 192, 247, 220, 208

Use the Kruskal–Wallis test at the level of significance $\alpha = 0.05$ to test the null hypothesis that the bowler performs equally well with the four bowling balls.

10 Three groups of guinea pigs were injected, respectively, with 0.5 mg, 1.0 mg, and 1.5 mg of a tranquilizer, and the following are the numbers of seconds it took them to fall asleep:

0.5-mg dose: 8.2, 10.0, 10.2, 13.7, 14.0, 7.8, 12.7, 10.9
1.0-mg dose: 9.7, 13.1, 11.0, 7.5, 13.3, 12.5, 8.8, 12.9, 7.9, 10.5
1.5-mg dose: 12.0, 7.2, 8.0, 9.4, 11.3, 9.0, 11.5, 8.5

Use the Kruskal–Wallis test at the level of significance $\alpha = 0.05$ to test the null hypothesis that the differences in dosage have no effect.

13.5

Tests of Randomness: Runs

All the methods of inference discussed so far are based on the assumption that the samples are random; yet, there are many applications in which it is difficult to decide whether this assumption is justifiable. This is true, particularly, when we have little or no control over the selection of the data. For instance, if we want to predict a department store's sales volume for a given month, we have no choice but to use past sales data and, perhaps, collateral information about economic conditions in general. None of this information constitutes a random sample in the strict sense. Also, we have no choice but to rely on whatever records are available if we want to make long-range predictions of the weather, estimate the mortality rate of a disease, or study the frequency of traffic accidents at a dangerous intersection.

There are several methods of judging the randomness of a sample on the basis of the order in which the observations are taken, and we can thus test, after the data have been collected, whether patterns that look suspiciously nonrandom may be attributed to chance. The technique we shall describe here and in Section 13.6 is based on the **theory of runs.**

A **run** is a succession of identical letters (or other kinds of symbols) which is followed and preceded by different letters or no letters at all.

EXAMPLE   To illustrate, consider the following arrangement of healthy, H, and diseased, D, elm trees that were planted many years ago along a country road:

$$\underline{H\ H\ H\ H\ H}\ \ \underline{D\ D\ D}\ \ \underline{H\ H\ H\ H\ H\ H\ H\ H\ H}$$
$$\underline{D\ D}\ \ \underline{H\ H}\ \ \underline{D\ D\ D\ D\ D}\ \ \underline{H\ H\ H\ H}$$

Using underlines to combine the letters which constitute the runs, we find that first there is a run of five H's, then a run of three D's, then a run of nine H's, then a run of two D's, then a run of two H's, then a run of five D's, and finally a run of four H's.

The **total number of runs** appearing in an arrangement of this kind is often a good indication of a possible lack of randomness. If there are too few runs we might suspect a definite grouping or clustering, or perhaps a trend; if there are too many runs, we might suspect some sort of repeated alternating pattern. In the above example there seems to be a definite clustering; that is, the diseased trees seem to come in groups. It remains to be seen, however, whether this is significant or whether it can be attributed to chance.

The test we shall use to put this decision on a precise basis utilizes the fact that for arrangements of n_1 letters of one kind and n_2 letters of another kind the sampling distribution of u, the total number of runs, has the mean

Mean of sampling distribution of u

$$\mu_u = \frac{2n_1 n_2}{n_1 + n_2} + 1$$

and the standard deviation

Standard deviation of sampling distribution of u

$$\sigma_u = \sqrt{\frac{2n_1 n_2 (2n_1 n_2 - n_1 - n_2)}{(n_1 + n_2)^2 (n_1 + n_2 - 1)}}$$

Furthermore, if n_1 and n_2 are sufficiently large (in this case, neither n_1 nor n_2 less than 10), the sampling distribution of u can be approximated closely with a normal curve. Thus, we reject the null hypothesis of randomness at

13.5   Tests of Randomness: Runs

the level of significance α when

Statistic for test of randomness

$$z = \frac{(u \pm \frac{1}{2}) - \mu_u}{\sigma_u}$$

is less than $-z_{\alpha/2}$ or exceeds $z_{\alpha/2}$. (In the continuity correction we use the minus sign when u exceeds μ_u and the plus sign when u is less than μ_u.) If the alternative hypothesis asserts that there is a trend which would account for a small number of runs, or a repeated cyclical pattern which would account for a large number of runs, we would reject the null hypothesis for values of z which are, respectively, less than $-z_\alpha$ or greater than z_α. Also, if either n_1 or n_2 is so small that the normal-curve approximation to the sampling distribution of u cannot be used, exact tests may be based on special tables (see the Bibliography on page 349).

EXAMPLE Returning to the problem concerning the healthy and diseased elm trees, we find that $n_1 = 20$, $n_2 = 10$, and $u = 7$, so that

$$\mu_u = \frac{2 \cdot 20 \cdot 10}{20 + 10} + 1 = 14.33$$

$$\sigma_u = \sqrt{\frac{2 \cdot 20 \cdot 10(2 \cdot 20 \cdot 10 - 20 - 10)}{(20 + 10)^2(20 + 10 - 1)}} = 2.38$$

and

$$z = \frac{7 + \frac{1}{2} - 14.33}{2.38} = -2.87$$

Since this value is less than $-z_{0.005} = -2.58$, we reject the null hypothesis of randomness at the level of significance $\alpha = 0.01$. There is a strong indication that the diseased trees come in non-random clusters.

13.6

Tests of Randomness: Runs Above and Below the Median

The method of the preceding section is not limited to tests of the randomness of series of attributes (such as the H's and D's of our example). Any sample consisting of numerical measurements or observations can be treated similarly by using the letters a and b to denote, respectively, values falling above and below the median of the sample. Numbers equal to the median are omitted. The resulting series of *a*'s and *b*'s (representing the data in their original order) can then be tested for randomness on the basis of the total number of runs of *a*'s and *b*'s, the total number of **runs above and below the median.**

EXAMPLE To illustrate this technique, suppose that on 40 successive runs between two cities, a bus carried 21, 19, 24, 33, 23, 22, 26, 29, 31, 13, 28, 24, 21, 10, 28, 18, 19, 27, 28, 30, 23, 27, 32, 23, 20, 26, 28, 21, 24, 17, 23, 14, 27, 22, 35, 17, 24, 21, 28, and 18 passengers. Using the level of significance $\alpha = 0.05$, we want to test the null hypothesis that these data may be treated as if they were a random sample. The median of this sample is 23.5, so that we get the following arrangement of *a*'s and *b*'s:

$$b\,b\,a\,a\,b\,b\,a\,a\,a\,b\,a\,a\,b\,b\,a\,b\,b\,a\,a\,a\,b\,a\,a\,b\,b\,a\,a\,b\,a\,b\,b\,b\,a\,b\,a\,b\,a\,b\,a\,b$$

Since $n_1 = 20$, $n_2 = 20$, and $u = 25$, we get

$$\mu_u = \frac{2 \cdot 20 \cdot 20}{20 + 20} + 1 = 21$$

$$\sigma_u = \sqrt{\frac{2 \cdot 20 \cdot 20(2 \cdot 20 \cdot 20 - 20 - 20)}{(20 + 20)^2(20 + 20 - 1)}} = 3.12$$

and

$$z = \frac{25 - \frac{1}{2} - 21}{3.12} = 1.12$$

Since this value falls between $-z_{0.025} = -1.96$ and $z_{0.025} = 1.96$, the null hypothesis cannot be rejected; there is no real evidence to suggest that the data may not be treated as if they constituted a random sample.

The method of runs above and below the median is especially useful in testing for trends or cyclical patterns in economic data. If there is an upward trend there will be first mostly *b*'s and later mostly *a*'s, if there is a downward trend there will be first mostly *a*'s and later mostly *b*'s, and if there is a cyclical pattern there will be a systematic alternating of *a*'s and *b*'s and, probably, too many runs.

13.7
Some Further Considerations

One thing to remember about nonparametric methods is that statistical methods which require no (or virtually no) assumptions about the populations sampled are usually

less efficient than the corresponding standard techniques. To illustrate what we mean by "less efficient," let us refer to the calculations in Chapter 9, where we showed that, by using Chebyshev's theorem, we can assert with a probability of at least 0.75 that the mean of a random sample of size $n = 64$ drawn from an infinite population with $\sigma = 20$ will differ from the population mean by less than 5. As we also showed, however, this probability becomes 0.9544 (instead of "at least 0.75") when we assume that the population sampled has roughly the shape of a normal distribution. To put it another way, nonparametric methods tend to be wasteful of information. It is generally true that **the less one assumes, the less one can infer from a set of data,** but it must also be recognized that **the more one assumes, the more one limits the applicability of one's methods.**

EXERCISES

1 The following arrangement indicates whether 50 consecutive persons interviewed by a pollster are for, F, or against, A, an increase in the state gasoline tax to build more roads:

$$A\,A\,A\,F\,A\,F\,A\,F\,A\,A\,A\,A\,F\,F\,A\,A\,F\,A\,A\,A\,A\,A\,F\,A\,A\,F\,F$$
(cont.) $A\,A\,F\,A\,A\,A\,A\,F\,A\,F\,F\,A\,A\,A\,A\,A\,F\,A\,A\,F\,A\,A\,A\,A\,F$

Use the level of significance $\alpha = 0.05$ to test whether this arrangement of A's and F's may be regarded as random.

2 Test at the level of significance $\alpha = 0.05$ whether the following arrangement of defective, d, and nondefective, n, pieces produced in the given order by a certain machine may be regarded as random:

$$n\,n\,d\,d\,d\,n\,n\,n\,d\,d\,n\,n\,n\,n\,n\,n\,n\,d\,d\,d\,n\,n\,n\,d\,n\,n\,n\,n\,d\,d\,d$$

3 Choose any two complete columns of random digits from Table IX (100 digits in all), represent each 0, 2, 4, 6, and 8 by the letter E, each 1, 3, 5, 7, and 9 by the letter O, and test for randomness at the level of significance $\alpha = 0.05$.

4 To test whether a radio signal contains a message or constitutes random noise, an interval of time is subdivided into a number of very short intervals and for each of these it is determined whether the signal strength exceeds, E, or does not exceed, N, a certain level of background noise. Test at the level of significance $\alpha = 0.05$ whether the following arrangement, thus obtained, may be regarded as random, and hence that the signal contains no message and may be regarded as random noise:

$$E\,N\,N\,N\,N\,E\,N\,E\,N\,N\,N\,E\,E\,N\,N\,N\,E\,E\,N\,E\,N\,E\,N\,N\,N\,E\,E\,N\,N\,N$$
(cont.) $N\,N\,E\,E\,N\,E\,N\,N\,E\,N\,N\,N\,E\,E\,E\,N\,N\,N\,E\,N\,E\,N\,N\,N\,N\,E\,N$

5 Making use of the fact that their median is 19.05 and reading successive rows across, test the randomness of the sulfur oxides emission data on page 14 at the level of significance $\alpha = 0.05$.

6 The following are the numbers of students absent from a school on 24 consecutive days: 29, 25, 31, 28, 30, 28, 33, 31, 35, 29, 31, 33, 35, 28, 36, 30, 33, 26, 30, 28, 32, 31, 38, and 27. Test for randomness at the level of significance $\alpha = 0.01$.

7 Making use of the fact that the median is 14 and reading successive rows across, test the randomness of the wildlife counts of Exercise 7 on page 65 at the level of significance $\alpha = 0.05$.

8 The total numbers of retail stores opening for business and also quitting business within the calendar years 1950–1978 in a large city were 109, 107, 125, 142, 147, 122, 116, 153, 144, 162, 143, 126, 145, 129, 134, 137, 143, 150, 148, 152, 125, 106, 112, 139, 132, 122, 138, 148, and 155. Test at the level of significance $\alpha = 0.05$ whether there is a significant trend in this series of data.

BIBLIOGRAPHY

Further information about the nonparametric tests discussed in this chapter and many others may be found in

CONOVER, W. J., *Practical Nonparametric Statistics*. New York: John Wiley & Sons, Inc., 1971.

GIBBONS, J. D., *Nonparametric Statistical Inference*. New York: McGraw-Hill Book Co., 1971.

LEHMANN, E. L., *Nonparametrics: Statistical Methods Based on Ranks*. San Francisco: Holden-Day, Inc., 1975.

MOSTELLER, F., and ROURKE, R. E. K., *Sturdy Statistics, Nonparametrics and Order Statistics*. Reading, Mass.: Addison-Wesley Publishing Co., Inc., 1973.

NOETHER, G. E., *Elements of Nonparametric Statistics, 2nd ed*. New York: John Wiley & Sons, Inc., 1976.

SIEGEL, S., *Nonparametric Statistics for the Behavioral Sciences*. New York: McGraw-Hill Book Co., 1956.

The tables that are needed to perform various nonparametric tests for small samples are given in some of the aforementioned books, including the one by S. Siegel, and, among others, in

OWEN, D. B., *Handbook of Statistical Tables*. Reading, Mass.: Addison-Wesley Publishing Co., Inc., 1962.

Further Exercises for Chapters

10, 11, 12, and 13

1 It is desired to test the null hypothesis $\mu = 0$ against the alternative hypothesis $\mu > 0$ on the basis of the mean of a random sample of size $n = 9$ from a normal population with $\sigma = 5$. If the probability of a Type I error is to be $\alpha = 0.05$,
   (a) verify that the null hypothesis must be rejected when $\bar{x} > 2.73$;
   (b) calculate β for $\mu = 2.50, 5.00,$ and 7.50, and draw a rough sketch of the OC-curve.

2 Mentally simulate a hundred flips of a coin by writing down a sequence of one hundred H's (heads) and T's (tails). Test for randomness at the level of significance $\alpha = 0.05$.

3 In a study of the effectiveness of certain exercises in weight reduction, a group of 16 persons engaged in these exercises for one month and showed the following results:

Weight before	Weight after	Weight before	Weight after
211	198	172	166
180	173	155	154
171	172	185	181
214	209	167	164
182	179	203	201
194	192	181	175
160	161	245	233
182	182	146	142

Use the method of Exercise 26 on page 284 and the level of significance $\alpha = 0.05$ to test the null hypothesis that the exercises are not effective in reducing weight.

4 Use the sign test to rework Exercise 3.

5 At an airport, random samples of persons who had just got off a plane and persons who were about to board a plane were asked whether they are afraid of flying. If 54 of 200 persons who had just got off a plane said that they were afraid of flying, and 35 of 250 who were about to board a plane said that they were afraid of flying, is the difference between the corresponding proportions significant at the level of significance $\alpha = 0.01$?

6 In a random sample of 40 persons eating dinner by themselves in a French restaurant, 28 had wine with their dinner. Find a 0.95 confidence interval for the actual proportion of persons eating by themselves in that restaurant who have wine with their dinner.

7 The refractive indices of 18 pieces of glass (randomly selected from a large shipment purchased by an optical firm) have a standard deviation of 0.011. Construct a 0.95 confidence interval for σ, the true standard deviation of the population sampled.

8 When asked to identify Jean Jacques Rousseau, 28 of 120 persons said that he is a deep-sea diver. Use the level of significance $\alpha = 0.05$ to test the claim that 30 percent of all persons make this mistake.

9 Test runs with six models of an experimental engine showed that they operated, respectively, for 27, 33, 21, 34, 29, and 26 minutes with a gallon of a certain kind of fuel. Estimate the standard deviation of the population sampled using
   (a) the sample standard deviation;
   (b) the sample range and the table of Exercise 8 on page 289.

10 Suppose that a law firm has one secretary whom it suspects of making more mistakes than the average of all its secretaries.
   (a) If the law firm decides that it will let the secretary go, provided this suspicion is confirmed on the basis of observations made on the secretary's performance, what hypothesis and alternative should the law firm set up?
   (b) If the law firm decides to let the secretary go unless he can prove himself significantly better than the average of all its secretaries, what hypothesis and alternative should the law firm set up?

11 In a study of the relationship between family size and intelligence, 40 "only children" had an average IQ of 101.5 with a standard deviation of 6.7 and 50 "first-borns" in two-child families had an average IQ of 105.9 with a standard deviation of 5.8. Test whether the difference between these two means is significant at $\alpha = 0.05$.

12 In a random sample of 200 retired persons, 137 stated that they prefer living in an apartment to living in a one-family home. At the 0.05 level of significance, does this refute the claim that at most 60 percent of all retired persons prefer living in an apartment to living in a one-family home?

13 A random sample of 50 cans of pineapple slices has a mean weight of 29.75 ounces and a standard deviation of 0.22 ounce. If this mean is used as an estimate of the true mean weight of all the cans of pineapple slices from which this sample came, with what probability can we assert that this estimate is "off" by at most 0.05 ounce?

14 A random sample of 62 shirts worn by soldiers in a tropical climate had an average (useful) life of 63.4 washings with a standard deviation of 4.5. Under moderate

weather conditions, such shirts are known to have an average (useful) life of 81.6 washings. Can we conclude at the level of significance $\alpha = 0.01$ that their use in a tropical climate reduces the average (useful) life of such shirts?

15   The following table shows how samples of the residents of three federally financed housing projects replied to the question whether they would continue to live there if they had the choice:

	Project 1	Project 2	Project 3
Yes	63	84	69
No	37	16	31

Test at the level of significance $\alpha = 0.05$ whether the differences among the three sample proportions (of "Yes" answers) may be attributed to chance.

16   Given that 40 one-gallon cans of a certain kind of paint covered on the average 463.3 square feet with a standard deviation of 26.6 square feet, construct a 0.95 confidence interval for σ.

17   An efficiency expert wants to determine the average time it takes a person to buy a week's groceries at a supermarket. If preliminary studies have shown that it is reasonable to let $\sigma = 2.8$ minutes, how large a sample will be needed to be able to assert with probability 0.95 that the mean of the sample will be "off" by at most 0.2 minute?

18   A mental health study yielded the results shown in the following table:

	Men	Women
Never been in therapy	125	144
Therapy for six months or less	54	42
Therapy for more than six months	21	14

Analyze this 3×2 table by performing an appropriate chi-square test at the level of significance $\alpha = 0.05$.

19   A sales manager's feelings about the average monthly demand for one of her company's products may be described by means of a normal distribution with $\mu_0 = 4,300$ units and $\sigma_0 = 260$ units.
    (a) What probability does she, thus, assign to the true average (monthly demand) being somewhere on the interval from 4,000 to 4,500 units?

(b) If data for ten months show an average demand for 4,202 units with a standard deviation of 380 units, what is the mean of the posterior distribution?

(c) How would the probability of part (a) be modified in the light of the information given in part (b)?

20 It is desired to test the null hypothesis $\mu = 50$ against the alternative hypothesis $\mu > 50$ on the basis of the mean of a random sample from a normal population with $\sigma = 12$. If the probability of a Type I error is to be 0.05 and the probability of a Type II error is to be 0.10 when $\mu = 55$, use the formula of Exercise 8 on page 280 to find the required size of the sample.

21 To compare two kinds of bumper guards, six of each kind were mounted on Ford Pintos. Then each car was run into a concrete wall at 5 miles per hour, and the following are the costs of the repairs (in dollars):

Bumper guard 1:   89, 72, 105, 93, 98, 77
Bumper guard 2:   93, 101, 80, 71, 75, 84

Use the level of significance $\alpha = 0.05$ to test whether the difference between the mean repair costs is significant.

22 The **two-sample median test** is another nonparametric test of the null hypothesis that two independent random samples come from identical populations against the alternative that the two populations have unequal medians or means. To perform this test we first find the median of the combined data, determine how many of the values in each sample fall above and below the median, and then analyze the resulting 2 × 2 table by the method of Section 12.4.

(a) Use the median test to rework the example an page 335, which dealt with the scores obtained by 25 male freshmen and 25 female freshmen on a foreign language placement test.

(b) Use the median test to rework Exercise 12 on page 337.

(c) Use the median test to rework the example on page 338, which dealt with the grain size of sand obtained from two different locations on the moon.

23 The **median test** of Exercise 22 can easily be generalized so that it applies to independent random samples from k populations. As before, we find the median of the combined data, determine how many of the values in each sample fall above and below the median, and then analyze the resulting 2 × k table by the method of Section 12.4.

(a) The following are the scores which a golf professional made on the same eighteen-hole course with three different sets of clubs:

Set 1:   77, 76, 71, 73, 68, 74, 70, 71, 67, 75,
         72, 70, 74, 72, 69, 73
Set 2:   73, 75, 70, 78, 71, 72, 70, 70, 69, 68,
         70, 72, 73, 66, 72, 69, 73
Set 3:   77, 75, 73, 70, 69, 76, 74, 73, 73, 71
         72, 78, 75, 74, 71

Use the median test at the level of significance $\alpha = 0.05$ to test the null hypothesis that the golf professional plays equally well with the three sets of clubs.

(b) The following are data on the percentage kill of four kinds of insecticides used against mosquitos:

> Insecticide A: 44.8, 38.0, 40.0, 39.3, 42.2, 42.5,
> 41.5, 43.3, 46.0, 38.5, 44.1, 40.3
>
> Insecticide B: 41.9, 46.9, 44.6, 43.9, 42.0, 44.0,
> 41.0, 43.1, 39.0, 45.2, 44.6, 42.0
>
> Insecticide C: 45.7, 39.8, 42.8, 41.2, 45.0, 40.2,
> 40.2, 41.7, 37.4, 38.8, 41.7, 38.7
>
> Insecticide D: 42.6, 41.4, 39.7, 45.4, 40.1, 40.8,
> 38.2, 44.4, 40.1, 43.7, 39.6, 42.6

Use the median test at the level of significance $\alpha = 0.05$ to test the null hypothesis that the four kinds of insecticides are equally effective.

24 Asked whether milk helps in providing healthy skin, which it does not, 266 of 350 persons interviewed in a sample survey answered incorrectly that it does. Construct a 0.98 confidence interval for the true proportion of persons in the population sampled who are thus misinformed about the value of milk.

25 Use the Kruskal–Wallis test to rework part (b) of Exercise 23.

26 In ten test runs, a truck operated for 14, 13, 17, 12, 15, 12, 15, 14, 15, and 13 miles with one gallon of a certain gasoline. Is this evidence at the 0.05 level of significance that the truck is not operating at an average of 15.5 miles per gallon with this gasoline?

27 During the investigation of an alleged unfair trade practice, the Federal Trade Commission has an investigator take a random sample of 49 "8-ounce" candy bars from a large shipment. The mean of the weights of these candy bars is 7.94 ounces and the standard deviation is 0.12 ounce. At the level of significance $\alpha = 0.01$, does the commission have grounds upon which to proceed against the manufacturer of the candy bars on the unfair practice of short-weight selling?

28 A distributor of soft-drink vending machines feels that the prior distribution of the average number of drinks one of his machines will dispense per week has the mean 845 and the standard deviation 12.4. So far as any one of the machines is concerned, the number of drinks it dispenses will vary from week to week, and this variation is measured by a standard deviation of 43.6. If the distributor puts one of these vending machines into a new supermarket and it averages 827 drinks per week during the first 50 weeks, find a Bayesian estimate, the posterior mean, of the number of drinks this machine can be expected to dispense per week.

29 Two independent random samples of size 12 from normal populations have the respective standard deviations $s_1 = 8.2$ and $s_2 = 14.3$. Test at the level of significance $\alpha = 0.02$ whether the two populations have equal standard deviations, so that we can use the method of Section 10.12 to compare the two sample means.

30 If we solve the first of the two double inequalities on page 300 for p, we obtain the confidence limits

$$\frac{x + \frac{1}{2} z_{\alpha/2}^2 \pm z_{\alpha/2} \sqrt{\dfrac{x(n - x)}{n} + \dfrac{1}{4} z_{\alpha/2}^2}}{n + z_{\alpha/2}^2}$$

(a) Use this formula to rework Exercise 1 on page 305;

(b) Use this formula to rework Exercise 4 on page 305;

(c) Use this formula to rework Exercise 5 on page 305.

31 To see whether a newly discovered manuscript can be attributed to a famous nineteenth-century historian, his literary style was analyzed statistically, and it was found, among other things, that in many samples of 1,000 words of text material he used the word "from" 0 to 4 times 43 percent of the time, 5 to 7 times 32 percent of the time, and 8 or more times 25 percent of the time. Now, if in 20 samples of 1,000 words of text material from the newly discovered manuscript, the word "from" is used 0 to 4 times, 5 to 7 times, and 8 or more times in 15, 3, and 2 of the samples, is this evidence at the level of significance $\alpha = 0.05$ that the manuscript is not the work of this historian?

32 The following are measurements of the amount of chloroform (micrograms per liter) in the drinking water of a city on 25 consecutive days:

47	58	51	44	49	59	60	63	67	71	66	67	63
(cont.) 69	62	65	72	69	61	59	57	49	53	58	56	

Test for randomness at the level of significance $\alpha = 0.05$.

33 The proportion of uncollectible charge accounts of a certain department store is $p = 0.008$, and to see whether this figure can be reduced, the store's credit manager tries a more threatening kind of collection letter on a random sample of the store's delinquent accounts.

(a) Against what alternative should he test the null hypothesis $p = 0.008$ if he is a very careful man and does not want to introduce the more threatening collection letters unless they really work?

(b) Against what alternative should he test the null hypothesis $p = 0.008$ if he wants to use the new letter unless it actually makes things worse?

34 In an experiment, an interviewer of job applicants is asked to write down his first impression (favorable or unfavorable) after two minutes and his final impression at the end of the interview. Use the following sample data and the level of significance $\alpha = 0.01$ to test the interviewer's claim that his first impression and his final impression are the same at least 90 percent of the time:

		First impression	
		Favorable	Unfavorable
Final impression	Favorable	186	33
	Unfavorable	54	127

35 Asked about the centerfolds of a certain magazine, 39 of 300 men and 44 of 200 women felt that the nudes were too explicit. Is the difference between the two sample proportions significant at $\alpha = 0.05$?

36 Among 80 baseball players, 47 improved their batting average after eating a

certain breakfast food regularly for several weeks, the other 33 did not. Is this evidence at the level of significance $\alpha = 0.05$ that the probability is 0.75 that the breakfast food will improve the performance of a baseball player?

37 The weights of 30 full-grown Gambel's quails (trapped in the Sonora desert) have a standard deviation of 5.8 grams. Find a 0.95 confidence interval for the corresponding population standard deviation, using
    (a) the small-sample technique based on the chi-square distribution;
    (b) the large-sample technique based on the normal distribution.

38 A general achievement test is standardized so that eighth graders should average 79.4 with a standard deviation of 4.8. If the superintendent of a school district hopes to show that eighth graders in her district average better than that, and she has the test given to a random sample of 35 eighth graders in her district, by how much must their average exceed 79.4 to make the difference significant at $\alpha = 0.01$?

39 To compare some of the features of the confidence interval formula on page 250 and the test described on page 272, suppose that a random sample of size n is taken from a normal population with the mean μ and the standard deviation σ. What relationship is there between the set of values covered by the $1 - \alpha$ confidence interval for μ and the set of values of μ_0 for which the null hypothesis $\mu = \mu_0$ cannot be rejected at the level of significance α against the two-sided alternative $\mu \neq \mu_0$?

40 A private opinion poll is hired by a political leader to estimate what percentage of the voters in his state believes a rumor about him spread by his opponent. How large a sample will the poll have to take to be able to assert with a probability of at least 0.90 that the sample percentage will be within 5 percent of the correct value?

41 In a restaurant, the chef receives 27, 22, 18, 24, 21, and 20 orders for coq au vin on six different nights. Construct a 0.98 confidence interval for the number of orders of coq au vin which this chef can expect per night.

42 A sanitation inspector for a national chain of gasoline stations obtained the following results in a sample survey in which he inspected 600 of the chain's gasoline stations in three parts of the country:

Standard of sanitation

	Below average	*Average*	*Above average*
East	31	93	58
Midwest	38	112	50
West	46	104	68

Use the level of significance $\alpha = 0.05$ to test the null hypothesis that there is no difference in the standard of sanitation of the chain's gasoline stations in the three parts of the country.

43 Five containers of a commercial solvent randomly selected from a large production lot weigh 24.9, 24.2, 25.2, 24.5, and 24.7 pounds.

(a) What can we assert with a probability of 0.95 about the possible size of our error in estimating the population mean as 24.7 pounds?

(b) Construct a 0.99 confidence interval for the mean weight of all the containers of the solvent from which this sample came.

44 Among 210 persons with alcohol problems admitted to the psychiatric emergency room of a hospital, 26 were admitted on a Monday, 23 on a Tuesday, 19 on a Wednesday, 25 on a Thursday, 33 on a Friday, 44 on a Saturday, and 40 on a Sunday. Use the level of significance $\alpha = 0.05$ to test the null hypothesis that the psychiatric emergency room can expect equally many persons with alcohol problems on each day of the week.

45 The technique of Exercise 25 on page 283 can also be used to construct confidence intervals for the difference, $\delta = \mu_1 - \mu_2$, between the means of two populations. In the large-sample case, we modify the z-statistic on page 277 as indicated in that exercise, substitute it for the middle term of $-z_{\alpha/2} < z < z_{\alpha/2}$, and manipulate this double inequality algebraically so that the middle term is δ; in the small-sample case, we proceed similarly, with the t-statistic on page 278.

(a) With reference to the example on page 277, construct a 0.95 confidence interval for the difference between the true average heights of adult females born in the two countries.

(b) With reference to the example on page 279, construct a 0.99 confidence interval for the difference between the true average heat-producing capacities of coal from the two mines.

(c) With reference to Exercise 19 on page 282, construct a 0.99 confidence interval for the difference between the true average times it takes to repair failures of the two kinds of photocopying equipment.

(d) With reference to Exercise 22 on page 282, construct a 0.95 confidence interval for the difference between the true average heights of the two kinds of citrus trees.

46 Suppose we want to test the null hypothesis $p = 0.30$ on the basis of the number of successes, x, in 14 trials. For what values of x would we reject the null hypothesis if the level of significance and the alternative hypothesis are, respectively,

(a) $\alpha = 0.05$ and $p \neq 0.30$?

(b) $\alpha = 0.05$ and $p > 0.30$?

(c) $\alpha = 0.05$ and $p < 0.30$?

(d) $\alpha = 0.10$ and $p \neq 0.30$?

In each case state the *actual* level of significance.

47 The statistic of Exercise 7 on page 319 can also be used to construct confidence intervals for the difference, δ, between two population proportions. We substitute the expression for z into $-z_{\alpha/2} < z < z_{\alpha/2}$, and then manipulate this double inequality algebraically so that the middle term is δ.

(a) With reference to Exercise 1 on page 317, construct a 0.95 confidence interval for the difference between the true proportions of men and women who buy souvenirs at the tourist attraction.

(b) With reference to Exercise 2 on page 318, construct a 0.99 confidence interval for the difference between the true proportions of viewers of the two telecasts who remember the name of the deodorant.

Further Exercises for Chapters 10, 11, 12, and 13

(c) With reference to Exercise 4 on page 318, construct a 0.95 confidence interval for the difference between the true proportions of marriage license applicants in the two years in which the women were at least one year older than the men.

48 Verify that $p(1 - p) = \frac{1}{4} - (p - \frac{1}{2})^2$ and use this result to justify the second formula for n on page 302.

14

The main objective of many statistical investigations is to establish relationships which make it possible to predict one or more variables in terms of others. For instance, studies are made to predict the future sales of a product in terms of its price, a person's weight loss in terms of the number of weeks he or she will stay on an 800-calories-per-day diet, family expenditures on medical care in terms of family income, the per capita consumption of certain food items in terms of their nutritional value and the amount of money spent advertising them on television, and so forth.

Of course, it would be ideal if we could predict one quantity exactly in terms of another, but this is seldom possible. In most instances we must be satisfied with predicting averages or expected values. For instance, we cannot predict exactly how much money a specific college graduate will earn ten years after graduation, but given suitable data we can predict the average earnings of all college graduates ten years after graduation. Similarly, we can predict the average yield of a variety of wheat in terms of the total rainfall in July, and we can predict the expected grade-point average of a student starting law school in terms of his or her IQ. In this chapter we shall concentrate on this problem of predicting the average value of one variable in terms of the known value of another variable (or the known values of other variables), called the problem of **regression.** This term dates back to Sir Francis Galton (1822–1911), who used it first in connection with a study of the relationship between the heights of fathers and sons.

Regression

14.1
Curve Fitting

Whenever possible, we strive to express, or approximate, relationships between known quantities and quantities that are to be predicted in terms of mathematical equations. This approach has been very successful in the natural sciences: In physics it is known, for example, that at a constant temperature the relationship between the volume, y, of a gas and its pressure, x, is given by the formula $y = \dfrac{k}{x}$, where k is a numerical constant. Similarly, in biology it has been discovered that the size of a culture of bacteria, y, can be expressed in terms of the time, x, it has been exposed to certain favorable conditions by means of the formula $y = a \cdot b^x$, where a and b are numerical constants. More recently, equations like these have also been used to describe relationships in the social sciences and other fields. For instance, the first of the above formulas is often used in economics to describe the relationship between price and demand, and the second may be used to describe the growth of one's vocabulary or the accumulation of wealth.

Whenever we want to use observed data to arrive at a mathematical equation and use it to predict the value of one variable in terms of a given value of another—a procedure known as **curve fitting**—there are basically three kinds of problems. First, we must decide what kind of "predicting" equation we want to use; then we must find the particular equation which is "best" in some sense; and finally we must investigate certain questions regarding the goodness of the particular equation, or of the predictions made from it.

With respect to the first of these problems of curve fitting, there are many different kinds of curves (and their equations) that can be used for predictive purposes. The choice of one of them is sometimes decided for us by theoretical considerations, but usually it is decided by direct inspection of the data. We plot the data on ordinary (arithmetic) graph paper, sometimes on special graph paper with special scales (see Section 14.3), and we decide by visual inspection upon the kind of curve (a straight line, a parabola, . . .) which best describes the over-all pattern of the data. There exist various methods for putting this decision on a more objective basis, but they will not be discussed in this book. So far as the second and third problems of curve fitting are concerned, we shall study the second in some detail in Section 14.2, and the third in Section 14.5.

14.2
Linear Regression

Of the many equations that can be used to predict values of one variable, y, from given values of another variable, x, simplest and most widely used is the **linear equation in two unknowns,** which is of the form

$$y = a + bx$$

where a is the y-intercept (the value of y for $x = 0$) and b is the slope of the line (namely, the change in y which accompanies a change of one unit in x). Ordinarily, the values of a and b are estimated from given data, and once they have been determined, we can substitute a value of x into the equation and calculate the corresponding predicted value of y. **Linear equations are useful and important not only because many relationships are actually of this form, but also because they often provide close approximations to relationships which would otherwise be difficult to describe in mathematical terms.**

The term "linear equation" arises from the fact that, when plotted on ordinary (arithmetic) graph paper, all pairs of values of x and y which satisfy an equation of the form $y = a + bx$ fall on a straight line.

EXAMPLE Suppose we want to predict the bushels-per-acre yield of wheat in a certain Midwestern county, y, in terms of the county's annual rainfall (in inches, measured from September through August), x. Based on past experience, we have the predicting equation

$$y = 0.23 + 4.42x$$

whose graph is shown in Figure 14.1, and any pair of values of x and y which are such that $y = 0.23 + 4.42x$ forms a point (x, y) that falls on the line. Substituting $x = 6$, for instance, we find that when there is an annual rainfall of 6 inches we can expect a yield of $0.23 + 4.42 \cdot 6 = 26.75$ bushels per acre; and when there is an annual rainfall of 12 inches we can expect a yield of $0.23 + 4.42 \cdot 12 = 53.27$ bushels per acre.

Once we have decided to fit a straight line, we are faced with the second kind of problem of curve fitting, namely, that of finding the equation

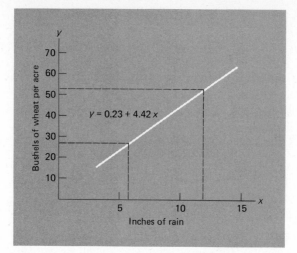

FIGURE 14.1

Graph of linear equation.

of the particular line which in some sense provides the best possible fit to the observed data.

EXAMPLE Let us consider the following sample data obtained in a study of the relationship between the number of years that applicants for certain foreign service jobs studied German in high school or college and the grades which they received in a proficiency test in that language:

Number of years x	Grade in test y
3	57
4	78
4	72
2	58
5	89
3	63
4	73
5	84
3	75
2	48

If we plot the points which correspond to these ten pairs of values as in Figure 14.2, we observe that, even though the points do not all fall on a straight line, the overall pattern of the relationship is reasonably well described by the white line. There is no noticeable departure from linearity in the scatter of the points, so we feel jus-

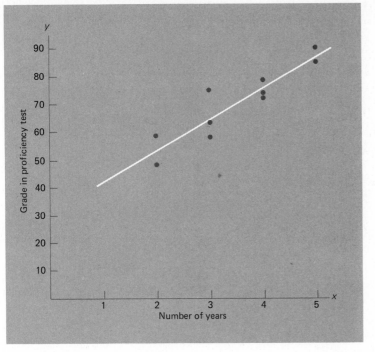

FIGURE 14.2

Data on number of years studied and grades in test.

tified in deciding that a straight line is a suitable description of the underlying relationship.

We now face the problem of finding the equation of the line which in some sense provides the best fit to the data and which, it is hoped, will later yield the best possible predictions of y from x. Logically speaking, there is no limit to the number of straight lines which can be drawn on a piece of graph paper. Some of these lines would fit the data so poorly that we could not consider them seriously, but many others would seem to provide more or less "good" fits, and the problem is to find the one line which fits the data "best" in some well-defined sense.[†] If all the points actually fell on a straight line there would be no problem, but this is an extreme case which we rarely encounter in practice. In general, we have to be satisfied with a line having certain desirable properties, short of perfection.

The criterion which, nowadays, is used almost exclusively for defining a "best" fit dates back to the early part of the nineteenth century and the

---

[†] Anyone who ever had to analyze paired data which were plotted as points on a piece of graph paper, probably felt the urge to take a ruler and draw a line which, to the eye, presents a fairly good fit. There is no law which says that this cannot be done, but it is certainly not very "scientific." Also, freehand curve fitting like this is too subjective to evaluate the goodness of subsequent predictions.

14.2 Linear Regression

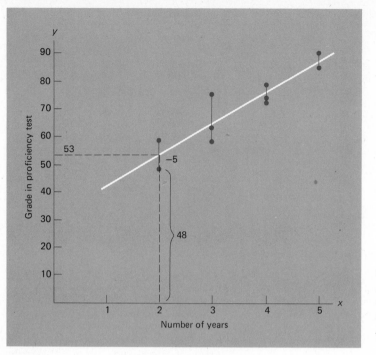

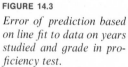

FIGURE 14.3

Error of prediction based on line fit to data on years studied and grade in proficiency test.

French mathematician Adrien Legendre; it is known as the **method of least squares.** As it will be used here, **this method requires that the line which we fit to our data be such that the sum of the squares of the vertical deviations of the points from the line is a minimum.**

EXAMPLE   For the problem concerning the proficiency grades in German, the method of least squares requires that the sum of the squares of the distances represented by the solid-line segments of Figure 14.3 be as small as possible. To explain why this is done, let us consider one of the applicants for the foreign service jobs, say, the one who studied German for two years and received a 48 in the test. If we mark $x = 2$ on the horizontal scale and read the corresponding value of y off the line of Figure 14.3, we find that the corresponding grade is about 53; therefore, the error in the prediction based on the line, represented by the vertical distance from the point to the line, is $48 - 53 = -5$. Altogether, there are ten such errors in our example, and the least-squares criterion requires that we minimize the sum of their squares.

We minimize the sum of the squares of the deviations and not the sum of the deviations, themselves, for the same reason why we squared the deviations

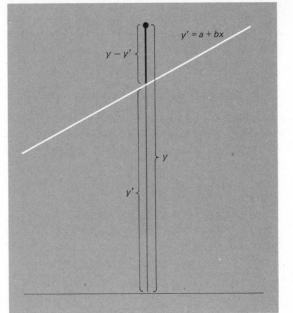

FIGURE 14.4
Least-squares line.

from the mean in the definition of the standard deviation. We are interested in the magnitude of the vertical deviations from the line, and not in their signs.

To show how a **least-squares line** is actually fitted to a set of data, let us consider n pairs of numbers (x_1, y_1), (x_2, y_2), ..., and (x_n, y_n), which might represent such things as the thrust and the speed of n rockets, the height and weight of n persons, the reading rate and reading comprehension of n students, or the number of persons unemployed in two cities in n years.

If we write the equation of the line as $y' = a + bx$, where the symbol y' is used to distinguish between the observed values of y and the corresponding values y' on the line, the least-squares criterion requires that we minimize the sum of the squares of the differences between the y's and the y''s (see Figure 14.4). This means that we must find the numerical values of the constants a and b appearing in the equation $y' = a + bx$ for which

$$\sum (y - y')^2 = \sum [y - (a + bx)]^2$$

is as small as possible. We shall not go through the derivation of the two equations, called the **normal equations,** which provide the solution to this problem.[†] Instead, we merely state that minimizing $\sum (y - y')^2$ yields the

---

[†] It takes calculus to derive these equations, or a rather tedious algebraic process called "completing the square."

following system of two linear equations in the unknowns a and b:

Normal equations

$$\sum y = na + b(\sum x)$$
$$\sum xy = a(\sum x) + b(\sum x^2)$$

In these equations, whose solution gives the least-squares values of a and b, n is the number of pairs of observations, $\sum x$ and $\sum y$ are the sums of the observed x's and y's, $\sum x^2$ is the sum of the squares of the x's, and $\sum xy$ is the sum of the products obtained by multiplying each x by the corresponding y.

EXAMPLE   Returning to our numerical illustration and copying the original pairs of data in the first two columns, we obtain the sums needed for substitution into the normal equations by performing the calculations shown in the following table:

Number of years x	Test grade y	x^2	xy
3	57	9	171
4	78	16	312
4	72	16	288
2	58	4	116
5	89	25	445
3	63	9	189
4	73	16	292
5	84	25	420
3	75	9	225
2	48	4	96
35	697	133	2,554

(There are many desk calculators, or hand-held calculators, on which the various sums can be accumulated directly, so that there is no need to fill in all the details.) Substituting $\sum x = 35$, $\sum y = 697$, $\sum x^2 = 133$, $\sum xy = 2,554$, and $n = 10$ into the two normal equations, we get

$$697 = 10a + 35b$$

$$2,554 = 35a + 133b$$

and we must now solve these two simultaneous linear equations for a and b. There are several ways in which this can be done; for instance, by the method of elimination or by the use of determinants. In either case, we get $a = 31.5$ and $b = 10.9$.

As an alternative to these procedures we can use the following formulas, which result when we symbolically solve the two normal equations for a and b (see Exercise 8 on page 370):

Solutions of
normal equations

$$a = \frac{(\sum y)(\sum x^2) - (\sum x)(\sum xy)}{n(\sum x^2) - (\sum x)^2}$$

$$b = \frac{n(\sum xy) - (\sum x)(\sum y)}{n(\sum x^2) - (\sum x)^2}$$

EXAMPLE Using these formulas in our illustration, we get as before

$$a = \frac{(697)(133) - (35)(2{,}554)}{10(133) - (35)^2} = \frac{3{,}311}{105} = 31.5$$

and

$$b = \frac{10(2{,}554) - (35)(697)}{105} = \frac{1{,}145}{105} = 10.9$$

To simplify the calculations, we sometimes use the second of the above formulas to calculate b, and then substitute the result into the first of the two normal equations on page 366 to get the value of a. For this purpose, we solve the first of the two normal equations for a, getting $a = \dfrac{\sum y - b(\sum x)}{n}$. A further simplification, which applies when the x's are equally spaced, is explained in Exercise 6 on page 370.

EXAMPLE We can now write the equation of the least-squares line for our numerical illustration as

$$y' = 31.5 + 10.9x$$

This is the "predicting" equation, or the equation we use to predict the proficiency score of an applicant for the foreign service job who has studied German in high school or college for x years. For

$x = 2$, for instance, we get $y' = 31.5 + 10.9(2) = 53.3$, and this is the best prediction we can make in the least-squares sense.

When we make a prediction like this, we cannot really expect that we will always hit the answer right on the nose; in fact, we cannot possibly be right when the answer has to be a whole number, as in our illustration, and our prediction is 53.3. With reference to this example, it would be very unreasonable to expect that every applicant who has studied German for a given number of years will get the same grade in the proficiency test; indeed, the data on page 362 show that this is not the case. Thus, to make meaningful predictions based on least-squares lines, we must look upon the values of y' obtained by substituting given values of x as averages, or expected values. Interpreted in this way, we refer to least-squares lines as **regression lines,** or better as **estimated regression lines,** since the values of a and b are estimated on the basis of sample data. Questions relating to the goodness of these estimates will be discussed in Section 14.5.

In the discussion of this section we have considered only the problem of fitting a straight line to paired data. More generally, the method of least squares can also be used to fit other kinds of curves and to derive predicting equations in more than two unknowns. The problem of fitting some curves other than straight lines by the method of least squares will be discussed briefly in Section 14.3, and a simple example of a predicting equation in more than two unknowns will be given in Section 14.4.

EXERCISES

1 The following data show the improvement (gain in reading speed) of six students in a speed-reading program, and the number of weeks they have been in the program

Number of weeks	Speed gain (words per minute)
3	65
4	90
1	22
3	81
5	132
7	180

(a) Fit a least-squares line from which we can predict the gain in reading speed in terms of the number of weeks a person has been in the program. (See also Exercise 9 on page 371.)

(b) Use the equation of the least-squares line to predict the gain in reading speed of a person who has been in the program for two weeks.

2   The following data pertain to the chlorine residual in a swimming pool at various times after it has been treated with chemicals:

Number of hours	Chlorine residual (parts per million)
2	1.8
4	1.5
6	1.4
8	1.1
10	1.1
12	0.9

(a) Fit a least-squares line from which we can predict the chlorine residual in terms of the number of hours since the pool has been treated with chemicals.

(b) Use the equation of the least-squares line to estimate the chlorine residual in the pool five hours after it has been treated with chemicals.

3   The following sample data show the demand for a product (in thousands of units) and its price (in cents) charged in six different market areas:

Price	18	10	14	11	16	13
Demand	9	125	57	90	22	79

(a) Fit a least-squares line from which we can predict the demand for the product in terms of its price.

(b) Estimate the demand for the product in a market area where it is priced at 15 cents.

4   Verify that the equation of the example on page 361 can be obtained by fitting a least-squares line to the following data:

Rainfall (inches)	Yield of wheat (bushels per acre)
12.9	62.5
7.2	28.7
11.3	52.2
18.6	80.6
8.8	41.6
10.3	44.5
15.9	71.3
13.1	54.4

5 Raw material used in the production of a synthetic fiber is stored in a place which has no humidity control. Measurements of the relative humidity in the storage place and the moisture content of a sample of the raw material (both in percentages) on 12 days yielded the following results:

Humidity	Moisture content
46	12
53	14
37	11
42	13
34	10
29	8
60	17
44	12
41	10
48	15
33	9
40	13

(a) Fit a least-squares line from which we can predict the moisture content in terms of the humidity.

(b) Use the least-squares line to estimate the moisture content when the relative humidity is 38 percent.

6 When the x's are equally spaced (that is, when the differences between successive values of x are all equal), the calculation of a and b can be simplified by coding the x's by assigning them the values ..., -3, -2, -1, 0, 1, 2, 3, ..., when n is odd, or the values ..., -5, -3, -1, 1, 3, 5, ..., when n is even. With this coding, the sum of the x's is zero, and the formulas for a and b on page 367 become

$$a = \frac{\sum y}{n} \quad \text{and} \quad b = \frac{\sum xy}{\sum x^2}$$

Of course, the equation of the resulting least-squares line expresses y in terms of the coded x's, and we have to account for this when we use the equation to make predictions.

(a) The following are the gross incomes from sales of the IBM Corporation for the years 1971–1975: 2.2, 2.9, 3.4, 4.3, and 4.5 (billion dollars). Fit a least-squares line and, assuming that the trend continues, predict this corporation's gross income from sales in 1980.

(b) For the years 1969–1974, income from manufacturing in Arizona (in millions of dollars) totaled 1,275, 1,437, 1,385, 1,916, 2,160, and 2,270. Fit a least-squares line and, assuming that the trend continues, predict the 1982 income from manufacturing in Arizona.

7 Use the method of Exercise 6 to rework Exercise 2.

8 Use the method of elimination to solve the two normal equations on page 366, and thus verify the symbolic solutions for a and b on page 367. (*Hint*: Multiply

the expressions on both sides of the first normal equation by $\sum x$, multiply those of the second normal equation by n, eliminate a by subtraction, and solve for b. Then, substitute the expression obtained for b into the first normal equation, and solve for a.)

9　Sometimes it is known that the line we fit to a set of data must pass through the origin (namely, that $a = 0$), and in that case the method of least squares yields

$$b = \frac{\sum xy}{\sum x^2}$$

(a)　Fit this kind of line to the data of Exercise 1, and plot it together with the line obtained in that exercise.

(b)　Fit this kind of line to the data of Exercise 4, and plot it together with the line obtained in that exercise.

14.3

Nonlinear Regression ★

When data depart more or less widely from linearity, we must consider fitting some curve other than a straight line. In this section, we shall first give two cases where the relationship is not linear but the method of Section 14.2 can nevertheless be employed; then we shall give an example of **polynomial curve fitting** by fitting a **parabola** whose equation is

$$y' = a + bx + cx^2$$

It is common practice to plot paired data on various kinds of graph paper, to see whether there are scales for which the points fall close to a straight line. Of course, when this is the case for ordinary (arithmetic) graph paper, we proceed as in Section 14.2. If it is the case when we use **semi-log paper** (with equal subdivisions for x and a logarithmic scale for y, as shown in Figure 14.6), this indicates that an **exponential curve** will provide a good fit. The equation of such a curve is

$$y' = a \cdot b^x$$

or in logarithmic form[†]

$$\log y' = \log a + x(\log b)$$

which is a linear equation in x and $\log y'$. (Writing A, B, and Y for $\log a$, $\log b$, and $\log y'$, the equation becomes $Y = A + Bx$, which is the usual equation of a straight line.) For fitting an exponential curve, the two normal

---

[†] In this notation, log stands for logarithm to the base 10.

equations on page 366 become

equations on page 366 become

Normal equations
(exponential curve)

$$\sum \log y = n(\log a) + (\log b)(\sum x)$$
$$\sum x(\log y) = (\log a)(\sum x) + (\log b)(\sum x^2)$$

and for any set of paired data they can be solved for log *a* and log *b*, and hence for *a* and *b*.

EXAMPLE To illustrate this technique, let us fit an exponential curve to the following data on a company's net profits during the first six years that it has been in business:

Year	Net profits (thousands of dollars)
1	112
2	149
3	238
4	354
5	580
6	867

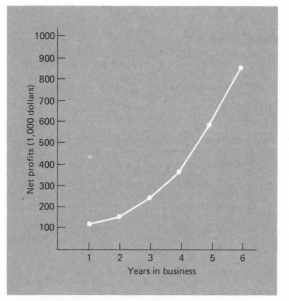

FIGURE 14.5

Net profits plotted on ordinary graph paper.

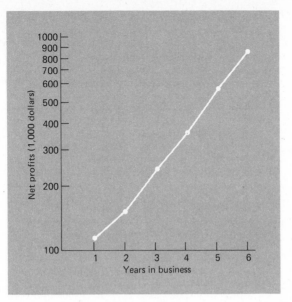

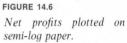

FIGURE 14.6

Net profits plotted on semi-log paper.

Plotting these data on ordinary graph paper in Figure 14.5, we find that the relationship is definitely not linear. However, the over-all pattern is remarkably well "straightened out" when we use a logarithmic scale for the net profits, y, as in Figure 14.6. Thus, it seems reasonable to fit an exponential curve to the given data.

Looking up the required logarithms in Table XI, we develop the sums needed for substitution into the two normal equations in the following table[†]:

x	y	$\log y$	$x \cdot \log y$	x^2
1	112	2.0492	2.0492	1
2	149	2.1732	4.3464	4
3	238	2.3766	7.1298	9
4	354	2.5490	10.1960	16
5	580	2.7634	13.8170	25
6	867	2.9380	17.6280	36
21		14.8494	55.1664	91

Substituting the appropriate column totals and $n = 6$ into the two

---

[†] Since the x's are equally spaced, these calculations could be simplified considerably by using the coding suggested in Exercise 6 on page 370 (see Exercise 3 on page 382).

14.3   Nonlinear Regression

373

normal equations, we get

$$14.8494 = 6(\log a) + 21(\log b)$$

$$55.1664 = 21(\log a) + 91(\log b)$$

To solve these equations, we substitute into formulas analogous to those for a and b on page 367, and we find that

$$\log a = \frac{14.8494(91) - 21(55.1664)}{6(91) - (21)^2} = 1.8362$$

$$\log b = \frac{6(55.1664) - 21(14.8494)}{6(91) - (21)^2} = 0.1825$$

and hence that $a = 68.6$ and $b = 1.52$. Thus, the equation of the exponential curve which best describes the relationship between the company's net profits and the number of years it has been in business is given by

$$y' = 68.6(1.52)^x$$

where y' is in thousands of dollars.

To estimate (predict) the company's net profit, say, for the eighth year it will have been in business, it is more convenient to use the logarithmic form of the equation; namely,

$$\log y' = 1.8362 + x(0.1825)$$

Substituting $x = 8$, we get

$$\log y' = 1.8362 + 8(0.1825)$$

$$= 3.2962$$

and hence $y' = 1,980$ (or \$1,980,000).

If points representing paired data fall close to a straight line when plotted on **log-log paper** (with logarithmic scales for both x and y), this indicates that an equation of the form

$$y = a \cdot x^b$$

will provide a good fit. In the logarithmic form, the equation of such a **power function**

$$\log y' = \log a + b(\log x)$$

which is a linear equation in $\log x$ and $\log y'$. (Writing A, X, and Y for $\log a$, $\log x$, and $\log y'$, the equation becomes $Y = A + bX$, which is the usual equation of a straight line.) For fitting a power function, the two normal equations on page 366 become

<table>
<tr><td rowspan="2">*Normal equations*
(*power function*)</td><td>$\sum \log y = n(\log a) + b(\sum \log x)$</td></tr>
<tr><td>$\sum (\log x)(\log y) = (\log a)(\sum \log x) + b(\sum \log^2 x)$</td></tr>
</table>

and for any set of paired data they can be solved for $\log a$ and b, and hence for a and b. In these equations, $\sum (\log x)(\log y)$ is the sum of the products obtained by multiplying the logarithm of each observed value of x by the logarithm of the corresponding value of y, and $\sum \log^2 x$ is the sum of the squares of the logarithms of the x's. Since the work required to fit a power function is very similar to that required to fit an exponential curve, we shall not give an example, but in Exercises 4 and 5 on page 382 the reader will find data to which the method can be applied.

When the values of y first increase and then decrease, or first decrease and then increase, a **parabola** having the equation

$$y' = a + bx + cx^2$$

will often provide a good fit. In fitting a parabola by the method of least squares, we must determine a, b, and c so that $\sum (y - y')^2 = \sum [y - (a + bx + cx^2)]^2$ is a minimum. Again stating merely the results, we must solve the following set of normal equations for a, b, and c:

<table>
<tr><td rowspan="3">*Normal equations*
(*parabola*)</td><td>$\sum y = na + b(\sum x) + c(\sum x^2)$</td></tr>
<tr><td>$\sum xy = a(\sum x) + b(\sum x^2) + c(\sum x^3)$</td></tr>
<tr><td>$\sum x^2 y = a(\sum x^2) + b(\sum x^3) + c(\sum x^4)$</td></tr>
</table>

In these equations, $\sum x^2 y$ is the sum of the products obtained by multiplying the square of each value of x by the corresponding observed value of y, and $\sum x^3$ and $\sum x^4$ are the sums of the third and fourth powers of the x's.

14.3 Nonlinear Regression

375

To illustrate this technique, let us fit a parabola to the following data on the drying time of a varnish and the amount of a certain chemical additive:

Amount of additive (grams) x	Drying time (hours) y
1	7.2
2	6.7
3	4.7
4	3.7
5	4.7
6	4.2
7	5.2
8	5.7

As can be seen from these data, and also from Figure 14.7, the values of y first decrease and then increase, and this suggests that a parabola may well give a good fit.

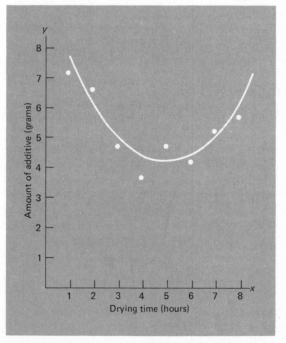

FIGURE 14.7

Parabola fitted to varnish-additive-drying-time data.

To calculate a, b, and c from the normal equations requires that we find n, $\sum x$, $\sum x^2$, $\sum x^3$, $\sum x^4$, $\sum y$, $\sum xy$, and $\sum x^2 y$. The work needed to find these quantities may conveniently be arranged as follows[†]:

x	y	x^2	x^3	x^4	xy	$x^2 y$
1	7.2	1	1	1	7.2	7.2
2	6.7	4	8	16	13.4	26.8
3	4.7	9	27	81	14.1	42.3
4	3.7	16	64	256	14.8	59.2
5	4.7	25	125	625	23.5	117.5
6	4.2	36	216	1,296	25.2	151.2
7	5.2	49	343	2,401	36.4	254.8
8	5.7	64	512	4,096	45.6	364.8
36	42.1	204	1,296	8,772	180.2	1,023.8

Substituting the appropriate column totals and $n = 8$ into the three normal equations, we get

$$42.1 = 8a + 36b + 204c$$

$$180.2 = 36a + 204b + 1{,}296c$$

$$1{,}023.8 = 204a + 1{,}296b + 8{,}772c$$

and the solution of this system of equations (rounded to one decimal) is $a = 9.2$, $b = -2.0$, and $c = 0.2$, as can be verified by the method of elimination or by using determinants. Thus, the equation of the parabola is

$$y' = 9.2 - 2.0x + 0.2x^2$$

Having obtained this equation, let us use it to determine the expected drying time of the varnish when 6.5 grams of the chemical are added, and also the amount of the additive for which we can expect the drying time to be a minimum. Substituting $x = 6.5$ into the equation of the parabola, we get

$$y' = 9.2 - 2.0(6.5) + 0.2(6.5)^2 = 4.65 \text{ hours}$$

[†] Since the x's are equally spaced, these calculations could be simplified considerably by using the coding suggested in Exercise 6 on page 370 (see Exercise 7 on page 382).

for the drying time we can expect when 6.5 grams of the chemical are added to the varnish. Making use of the fact that a parabola of the form $y = a + bx + cx^2$ has a minimum (or maximum) at $x = -\dfrac{b}{2c}$, which is $x = -\dfrac{-2.0}{2(0.2)} = 5.0$ grams for our example, we find that the corresponding drying time is $y' = 9.2 - 2.0(5.0) + 0.2(5.0)^2 = 4.2$ hours. Actually, one of the observed values was less than this, but this is no contradiction. The parabola which we fitted to the data must be looked upon as a regression curve, and we have shown merely that the average, or expected, drying time is least when 5.0 grams of the chemical are added to the varnish.

We have introduced the parabola, also called a **second-degree polynomial equation,** as a curve which bends once; that is, its values first increase and then decrease, or first decrease and then increase. As curves which bend more than once, polynomial equations of higher degree in x than two, such as $y' = a + bx + cx^2 + dx^3$ and $y' = a + bx + cx^2 + dx^3 + ex^4$, can also be fitted by the method of least squares. In practice, we also use parts of such curves, especially parts of parabolas, when there is only a slight curvature in the pattern we want to describe (see Exercise 6 on page 382).

14.4
Multiple Regression ⋆

Although there are many problems in which one variable can be predicted quite accurately in terms of another, it stands to reason that predictions should improve if one considers additional relevant information. For instance, we should be able to make a better prediction of a student's grade in a final examination if we consider not only her IQ but also the number of hours she studied for the examination. Similarly, we should be able to make a better prediction of the attendance at a theater if we consider not only the quality of the performers, but also the size of the community, its wealth, and perhaps the number of competing attractions that are available to the public on the same night.

Many mathematical formulas can serve to express relationships between more than two variables, but most commonly used in statistics (partly for reasons of convenience) are linear equations of the form

$$y' = b_0 + b_1 x_1 + b_2 x_2 + \cdots + b_k x_k$$

Here y' is the variable which is to be predicted, $x_1, x_2, \ldots,$ and x_k are the k known variables on which predictions are to be based, and $b_0, b_1, b_2, \ldots,$

and b_k are numerical constants which must be determined from the observed data.

EXAMPLE   The following equation was derived in a study of the demand for different meats:

$$y' = 3.489 - 0.090x_1 + 0.064x_2 + 0.019x_3$$

where y denotes the total consumption of federally inspected beef and veal in millions of pounds, x_1 denotes a composite retail price of beef in cents per pound, x_2 denotes a composite retail price of pork in cents per pound, and x_3 denotes income as measured by a certain payroll index. With this equation, we can predict the total consumption of federally inspected beef and veal on the basis of specified values of x_1, x_2, and x_3.

The main problem in deriving a linear equation in more than two variables which best describes a given set of data is that of finding numerical values for b_0, b_1, b_2, ..., and b_k. This is usually done by the method of least squares; that is, we minimize the sum of squares $\sum (y - y')^2$, where as before the y's are the observed values and the y''s are the values calculated by means of the linear equation. In principle, the problem of determining the values of b_0, b_1, b_2, ..., and b_k is the same as it is in the two-variable case, but manual solutions may be very tedious because the method of least squares requires that we solve as many normal equations as there are unknown constants b_0, b_1, b_2, ..., and b_k. For instance, when there are two independent variables x_1 and x_2, and we want to fit the equation $y' = b_0 + b_1x_1 + b_2x_2$, we must solve the three normal equations

Normal equations (two independent variables)

$$\sum y = n \cdot b_0 + b_1\left(\sum x_1\right) + b_2\left(\sum x_2\right)$$

$$\sum x_1 y = b_0\left(\sum x_1\right) + b_1\left(\sum x_1^2\right) + b_2\left(\sum x_1 x_2\right)$$

$$\sum x_2 y = b_0\left(\sum x_2\right) + b_1\left(\sum x_1 x_2\right) + b_2\left(\sum x_2^2\right)$$

Here $\sum x_1 y$ is the sum of the products obtained by multiplying each given value of x_1 by the corresponding value of y, $\sum x_1 x_2$ is the sum of the products obtained by multiplying each given value of x_1 by the corresponding value of x_2, and so on.

EXAMPLE   To illustrate this technique, let us consider the following data showing the number of bedrooms and the number of baths of eight one-family homes, and the prices at which they were sold recently

14.4   **Multiple Regression**

379

in a suburban area:

Number of bedrooms x_1	Number of baths x_2	Price (dollars) y
3	2	33,800
2	1	29,300
4	3	38,800
2	1	29,200
3	2	34,700
2	2	29,900
5	3	43,400
4	2	37,900

Based on these data, we want to find an equation of the form $y' = b_0 + b_1 x_1 + b_2 x_2$, which will enable us to predict the average price of a one-family home in this area in terms of the number of bedrooms and baths. To get the sums needed for substitution into the three normal equations, we perform the calculations shown in the following table:

x_1	x_2	y	$x_1 y$	$x_2 y$	x_1^2	$x_1 x_2$	x_2^2
3	2	33,800	101,400	67,600	9	6	4
2	1	29,300	58,600	29,300	4	2	1
4	3	38,000	155,200	116,400	16	12	9
2	1	29,200	58,400	29,200	4	2	1
3	2	34,700	104,100	69,400	9	6	4
2	2	29,900	59,800	59,800	4	4	4
5	3	43,400	217,000	130,200	25	15	9
4	2	37,900	151,600	75,800	16	8	4
25	16	277,000	906,100	577,700	87	55	36

Then, substituting the appropriate column totals and $n = 8$ into the normal equations, we get

$$277,000 = 8b_0 + 25b_1 + 16b_2$$
$$906,100 = 25b_0 + 87b_1 + 55b_2$$
$$577,700 = 16b_0 + 55b_1 + 36b_2$$

and the solution of this system of linear equations is $b_0 = 20,197$, $b_1 = 4,149$, and $b_2 = 731$, as can be verified by the method of elimination or by using determinants. Thus, the least-squares equation is

$$y' = 20,197 + 4,149x_1 + 731x_2$$

and this tells us that, in this study, each extra bedroom adds on the average \$4,149 to the price of a house, and each extra bath adds on the average \$731. To estimate (predict) the average price of three-bedroom houses with two baths, for example, we substitute $x_1 = 3$ and $x_2 = 2$, and get

$$y' = 20,197 + 4,149 \cdot 3 + 731 \cdot 2 = \$34,106$$

or approximately \$34,100.

EXERCISES

1 Fit an exponential curve to the following data on the percentage of the radial tires made by a certain manufacturer that are still usable after having been driven the given numbers of miles:

Miles driven (thousands) x	Percentage usable y
1	97.2
2	91.8
5	82.5
10	64.4
20	41.0
30	29.9
40	17.6
50	11.3

Also estimate the percentage of the tires we can expect to be usable after they have been driven for 25,000 miles.

2 The following data pertain to the growth of a colony of bacteria in a culture medium:

Days since inoculation x	Bacteria count (thousands) y
2	112
4	148
6	241
8	363
10	585

Fit an exponential curve and use it to estimate the bacteria count at the end of the fifth day.

3   When the x's are equally spaced, the calculations that are needed to fit an exponential curve can be simplified by using the coding of Exercise 6 on page 370; that is, by assigning the x's the values ..., -3, -2, -1, 0, 1, 2, 3, ... when n is odd, or the values ..., -5, -3, -1, 1, 3, 5, ... when n is even.

    (a) Use this kind of coding to rework the example on page 372. What coded value of x do we have to substitute into the least-squares equation to predict the company's net profit for the eighth year that it will have been in business? Make this substitution and compare the result with that obtained on page 374.

    (b) In the years 1970–1974, Argentina's Index of Consumer Prices (with 1967 = 100) was 142, 191, 303, 486, and 604. Suitably coding the years, fit an exponential curve to these data and use it to estimate the 1980 value of this index.

4   The following data pertain to the demand for a product (in thousands of units) and its price (in cents) charged in five different market areas:

Price	20	16	10	11	14
Demand	22	41	120	89	56

Fit a power function and use it to estimate the demand when the price of the product is 12 cents.

5   The following data pertain to the volume of a gas (in cubic inches) and its pressure (in pounds per square inch), when the gas is compressed at a constant temperature:

Volume x	Pressure y
50	16.0
30	40.1
20	78.0
10	190.5
5	532.2

Fit a power function and use it to estimate the pressure of this gas when it is compressed to a volume of 15 cubic inches.

6   Fit a parabola to the data of Exercise 4 and use it to estimate the demand when the price of the product is 12 cents. Explain why the parabola cannot be used, for example, to estimate the demand when the price of the product is 40 cents.

7   When the x's are equally spaced, the calculations that are needed to fit a parabola can be simplified by using the coding of Exercise 6 on page 370; that is, by assigning the x's the values ..., -3, -2, -1, 0, 1, 2, 3, ... when n is odd, or ..., -5, -3, -1, 1, 3, 5, ... when n is even.

    (a) Since this kind of coding makes $\sum x$ and $\sum x^3$ equal to 0, show that b can be found directly with the formula $b = \dfrac{\sum xy}{\sum x^2}$, and a and c can be found by solving the system of equations

$$\sum y = na + c\left(\sum x^2\right)$$
$$\sum x^2 y = a\left(\sum x^2\right) + c\left(\sum x^4\right)$$

(b) Use this kind of coding to rework the example on page 376. What coded value of x do we have to substitute into the least-squares equation to estimate the average drying time of the varnish when 6.5 grams of the chemical are added. Make this substitution and compare the result with that obtained on page 377.

(c) In the years 1968–1972 the installed capacity of nuclear electric energy in the United States was 2,817, 3,980, 6,493, 8,687, and 15,301 thousands of kilowatts. Fit a parabola to these data and plot it together with the original data on ordinary (arithmetic) graph paper.

8 The following are data on the ages and incomes of a random sample of five executives working for a large multinational company, and the number of years each went to college:

Age x_1	Years college x_2	Income (dollars) y
38	4	31,700
46	0	27,300
39	5	35,500
43	2	30,800
32	4	25,900

Fit an equation of the form $y' = b_0 + b_1 x_1 + b_2 x_2$ to these data, and use it to estimate the average income of 39-year-old executives with four years of college working for this company.

9 The following data were collected to determine the relationship between two processing variables and the hardness of a certain kind of steel:

Hardness (Rockwell 30-T) y	Copper content (percent) x_1	Annealing temperature (degrees F) x_2
78.9	0.02	1,000
55.2	0.02	1,200
80.9	0.10	1,000
57.4	0.10	1,200
85.3	0.18	1,000
60.7	0.18	1,200

Fit an equation of the form $y' = b_0 + b_1 x_1 + b_2 x_2$ to these data.

10 Rework Exercise 9 after coding the three x_1 values -1, 0, and 1, and the two x_2 values -1 and 1.

14.4 Multiple Regression

383

14.5

Regression Analysis ★

In Section 14.2 we used a least-squares line to predict that someone who has studied German in high school or college for two years will score 53.3 in the proficiency test, but even if we interpret the line correctly as a regression line (that is, treat the predictions made from it as averages or expected values), several questions remain to be answered.

> *How good are the values we found for the constants a and b in the equation $y' = a + bx$? After all, $a = 31.5$ and $b = 10.9$ are only estimates based on sample data, and if we had based our work on a sample of ten other applicants for the foreign service jobs, the method of least squares would probably have led to different values of a and b.*

> *How good an estimate is 53.3 of the true average score of persons who have had two years of German in high school or college?*

> *How can we obtain limits (two numbers) for which we can assert with a given probability that they will contain the score of a person who has studied German in high school or college for two years?*

When, in the first of these questions, we said that $a = 31.5$ and $b = 10.9$ are "only estimates based on sample data," we implied the existence of corresponding true values, usually designated α and β, and therefore of a true regression line $y' = \alpha + \beta x$. Another way to ask the first question is: How good are the estimates a and b which we obtained for the **regression coefficients** α and β.[†] To distinguish between a and b, and α and β, we refer to the former as the **estimated regression coefficients.**

To clarify the idea of a true regression line $y' = \alpha + \beta x$, let us consider Figure 14.8, in which we have drawn the distributions of y for several values of x. With reference to our example, these curves should be looked upon as the distributions of the proficiency scores of persons who have had, respectively, one, two, or three years of German in high school or college. To complete the picture, we can visualize similar distributions for all other values of x within the range of values under consideration.

In **linear regression analysis** we assume that the x's are constants, not values of random variables, and that for each value of x the variable to be predicted has a certain distribution (as in Figure 14.8) whose mean is $\alpha + \beta x$. In **normal regression analysis** we assume, furthermore, that these distributions are all normal distributions having the same standard deviation σ.

[†] The coefficients α and β should not be confused with the probabilities α and β of rejecting a true hypothesis or accepting a false one.

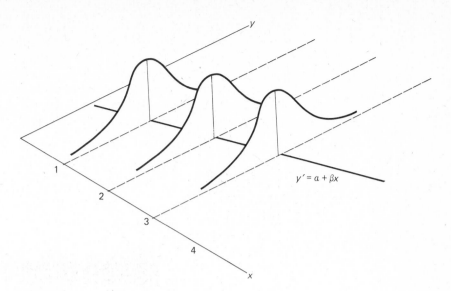

FIGURE 14.8

Distributions of y for given values of x.

In other words, the distributions pictured in Figure 14.8, as well as those we add mentally, are normal curves with the means $\alpha + \beta x$ and the standard deviation σ.

If we make all these assumptions, inferences about the regression coefficients α and β can be based on the statistics

Statistics for tests about regression coefficients

$$t = \frac{a - \alpha}{s_e \sqrt{\dfrac{1}{n} + \dfrac{n \cdot \bar{x}^2}{n(\sum x^2) - (\sum x)^2}}}$$

$$t = \frac{b - \beta}{s_e} \sqrt{\frac{n(\sum x^2) - (\sum x)^2}{n}}$$

whose sampling distributions are t distributions with $n - 2$ degrees of freedom. Here α and β are the regression coefficients we want to estimate or test, and a and b are their estimates calculated from a given set of data by the method of least squares. Also, n, $\bar{x}$, $\sum x$, and $\sum x^2$ come from the original data, and s_e is an estimate of σ, the common standard deviation of the normal distributions pictured in Figure 14.8, given by

$$s_e = \sqrt{\frac{\sum (y - y')^2}{n - 2}}$$

14.5   Regression Analysis

385

Here again, y is an observed value of y and y' is the corresponding value on the least-squares line. It is customary to refer to s_e as the **standard error of estimate,** and it should be observed that its square, s_e^2, is the sum of the squares of the vertical deviations (distances) from the points to the line in Figure 14.3 divided by $n - 2$. (We lose two degrees of freedom, so to speak, and divide by $n - 2$, because α and β are replaced by estimates, a and b, in the formula for s_e.) In practice, it is easier to calculate s_e by means of the formula

Standard error of estimate

$$s_e = \sqrt{\frac{\sum y^2 - a(\sum y) - b(\sum xy)}{n - 2}}$$

EXAMPLE   To illustrate how the two t statistics are used to make inferences about the regression coefficients α and β, suppose that someone claims that $\beta = 12.5$ in the example dealing with the applicants for the foreign service jobs and their proficiency in German, and that the study was undertaken partly to test this claim. Note that β is the slope of the regression line; more specifically, it is the true change in y associated with an increase of one unit in x. Hence, the hypothesis $\beta = 12.5$ asserts that each additional year of German studied in high school or college adds another 12.5 to the average proficiency score.

Among the various quantities needed for substitution into the formula for the second of the two t statistics on page 385, we already have $n = 10$, $\sum x = 35$, $\sum x^2 = 133$, $\sum y = 697$, $\sum xy = 2,554$, and $a = 31.5$ and $b = 10.9$ from the work of Section 14.2. By hypothesis, $\beta = 12.5$, and to calculate s_e, the only other quantity needed, we first determine $\sum y^2 = 50,085$ from the original data. We now have

$$s_e = \sqrt{\frac{50,085 - (31.5)(697) - (10.9)(2,554)}{10 - 2}}$$

$$= 6.0$$

and, hence,

$$t = \frac{10.9 - 12.5}{6.0} \sqrt{\frac{10(133) - (35)^2}{10}}$$

$$= -0.86$$

To test the null hypothesis $\beta = 12.5$ against the two-sided alternative $\beta \neq 12.5$ at the level of significance $\alpha = 0.05$, we compare the value obtained for t with $-t_{0.025}$ and $t_{0.025}$ for $n - 2 = 10 - 2 = 8$ degrees of freedom. Table II shows that $t_{0.025} = 2.306$ for 8 degrees of freedom, and since $t = -0.86$ falls between -2.306 and 2.306, the null hypothesis cannot be rejected. In other words, we conclude that the difference between $b = 10.9$ and $\beta = 12.5$ may well be due to chance.

Tests concerning the regression coefficient α are performed in the same way, except that we use the first, instead of the second, of the two t statistics on page 385. In most practical applications, the regression coefficient α is not of as much interest as the regression coefficient β; α is just the y-intercept (that is, the value of y which corresponds to $x = 0$), and in many cases it has no real meaning. In our illustration, α is the mean proficiency score of persons who have not studied any German in high school or college.

To construct confidence intervals for the regression coefficients α and β, we substitute into the middle term of $-t_{\alpha/2} < t < t_{\alpha/2}$ the appropriate t statistic from page 385. Then, simple algebra leads to the formulas

Confidence limits for regression coefficients

$$a \pm t_{\alpha/2} \cdot s_e \sqrt{\frac{1}{n} + \frac{n \cdot \bar{x}^2}{n(\sum x^2) - (\sum x)^2}}$$

and

$$b \pm \frac{t_{\alpha/2} \cdot s_e}{\sqrt{\dfrac{n(\sum x^2) - (\sum x)^2}{n}}}$$

where the degree of confidence is $1 - \alpha$ and $t_{\alpha/2}$ is the entry in Table II for $n - 2$ degrees of freedom.

EXAMPLE To illustrate the calculation of a confidence interval for a regression coefficient, let us find a 0.95 confidence interval for β for the example dealing with the applicants for the foreign service jobs and their proficiency in German. Having all the quantities needed for substitution from our previous work, we substitute them into the confidence-limits formula for β and get

$$10.9 \pm \frac{(2.306)(6.0)}{\sqrt{\dfrac{10(133) - (35)^2}{10}}}$$

Thus, the 0.95 confidence interval is

$$6.6 < \beta < 15.2$$

This confidence interval for the average increase in the proficiency score for each additional year of German studied in high school or college is rather wide, and this is due to two things—the large variation in the proficiency scores of applicants who have studied German for the same number of years and the very small sample size.

To answer the other two questions asked on page 384, those concerning the "goodness" of estimates made from a least-squares equation $y' = a + bx$, we use methods very similar to the ones just discussed. They are based on two more t statistics, which are given in Exercises 8 and 9.

EXERCISES

1 With reference to the illustration in the text, use the level of significance 0.05 to test the null hypothesis $\alpha = 20.0$ against the one-sided alternative $\alpha > 20.0$.

2 With reference to the illustration in the text, construct a 0.95 confidence interval for the regression coefficient α.

3 In each part, explain in words what hypothesis is being tested, and perform the indicated tests:
   (a) With reference to Exercise 1 on page 368, test the null hypothesis $\beta = 30$ against the alternative hypothesis $\beta \neq 30$ at the 0.05 level of significance.
   (b) With reference to Exercise 2 on page 369, test the null hypothesis $\beta = -0.12$ against the alternative hypothesis $\beta > -0.12$ at the 0.05 level of significance.
   (c) With reference to Exercise 4 on page 369, test the null hypothesis $\beta = 6.0$ against the alternative hypothesis $\beta < 6.0$ at the level of significance 0.01.

4 With reference to Exercise 2 on page 369, test the null hypothesis $\alpha = 2.1$ against the alternative hypothesis $\alpha < 2.1$ at the 0.05 level of significance. Also explain in words what hypothesis is being tested.

5 The following data show the advertising expenses (expressed as a percentage of total expenses) and the net operating profits (expressed as a percentage of total sales) in a random sample of six drug stores:

Advertising expenses	Net operating profits
1.5	3.6
1.0	2.8
2.8	5.4
0.4	1.9
1.3	2.9
2.0	4.3

(a) Fit a least-squares line and use it to predict the net operating profit of such a store (expressed as a percentage of total sales) when its advertising expenses are 1.2 percent of its total expenses.

(b) Test the null hypothesis $\alpha = 0.8$ against the alternative hypothesis $\alpha > 0.8$ at the 0.05 level of significance. Also explain in words what hypothesis is being tested.

6 For each of the following, construct a confidence interval for β at the indicated degree of confidence:

(a) Exercise 1 on page 368, degree of confidence 0.99;

(b) Exercise 3 on page 369, degree of confidence 0.95;

(c) Exercise 5 above, degree of confidence 0.98.

7 For each of the following, construct a confidence interval for α at the indicated degree of confidence:

(a) Exercise 2 on page 369, degree of confidence 0.95;

(b) Exercise 5 above, degree of confidence 0.99

8 To answer the second of the three questions we asked in the beginning of this section, we can use the following $1 - \alpha$ confidence interval for the mean $\alpha + \beta x_0$ of the distribution of y when $x = x_0$:

Confidence interval for mean of y when $x = x_0$

$$(a + bx_0) \pm t_{\alpha/2} \cdot s_e \sqrt{\frac{1}{n} + \frac{n(x_0 - \bar{x})^2}{n(\sum x^2) - (\sum x)^2}}$$

As before, $t_{\alpha/2}$ may be read from Table II, and the number of degrees of freedom is $n - 2$. To illustrate, let us refer again to the example dealing with the applicants for the foreign service jobs and their proficiency in German, and let us find a 0.95 confidence interval for the true average score of persons who have studied German in high school or college for two years. Since $a + bx_0 = 31.5 + (10.9)(2) = 53.3$, $s_e = 6.0$, and $\bar{x} = \frac{35}{10} = 3.5$, we get

$$53.3 \pm (2.306)(6.0) \sqrt{\frac{1}{10} + \frac{10(2 - 3.5)^2}{10(133) - (35)^2}}$$

$$53.3 \pm 13.84\sqrt{0.314}$$

and, hence, the 0.95 confidence interval $45.5 - 61.1$.

(a) With reference to Exercise 1 on page 368, find a 0.95 confidence interval for the average gain in reading speed of persons who have been in the program for three weeks.

(b) With reference to Exercise 4 on page 369, find a 0.99 confidence interval for the true average yield of wheat in the county when there are 11 inches of rain during the year.

9 The third of the questions asked in the beginning of this section differs from the other two in that it does not involve an estimate of a population parameter, but a prediction of a single future observation. Two values for which we can assert with a given probability that they will contain such an observation are called **limits of prediction**. Appropriate limits for predicting with probability $1 - \alpha$ a

value of y when $x = x_0$ are given by

Limits of prediction

$$(a + bx_0) \pm t_{\alpha/2} \cdot s_e \sqrt{1 + \frac{1}{n} + \frac{n(x_0 - \bar{x})^2}{n(\sum x^2) - (\sum x)^2}}$$

Again, $t_{\alpha/2}$ may be read from Table II, and the number of degrees of freedom is $n - 2$. To illustrate, let us refer again to the language proficiency example, and let us find 0.95 limits of prediction for the proficiency score of someone who has studied German in high school or college for two years. Noting that the only difference between the above limits and the confidence limits of Exercise 8 is that we add 1 to the quantity under the radical, we can immediately write the limits of prediction as

$$53.3 \pm 13.84\sqrt{1.314}$$

Thus, the 0.95 limits of prediction are 37.4 and 69.2.

(a) Referring to Exercise 1 on page 368, find 0.95 limits of prediction for the gain in reading speed of a person who has been in the program for three weeks.

(b) With reference to Exercise 4 on page 369, find 0.99 limits of prediction for the county's yield of wheat when there are 11 inches of rain during the year.

BIBLIOGRAPHY Methods of deciding which kind of curve to fit to a given set of paired data may be found in books on numerical analysis and in more advanced texts in statistics. Further information about the material of this chapter may be found in

DRAPER, N. R., and SMITH, H., *Applied Regression Analysis*. New York: John Wiley & Sons, Inc., 1966.

EZEKIEL, M., and FOX, K. A., *Methods of Correlation and Regression Analysis, 3rd ed.* New York: John Wiley & Sons, Inc., 1959.

MOSTELLER, F., and TUKEY, J. W., *Data Analysis and Regression*. Reading, Mass.: Addison-Wesley Publishing Co., Inc., 1977.

15

Correlation

Having learned how to fit a least-squares line to paired data, we turn now to the problem of determining how well such a line actually fits the data. Of course, we can get some idea by inspecting, say, a diagram like that of Figure 14.2, but to show how we can be more objective, let us refer back to the original data of the example dealing with the foreign service job applicants' proficiency in German, namely,

Years studied German x	Grade in test y
3	57
4	78
4	72
2	58
5	89
3	63
4	73
5	84
3	75
2	48

As can be seen from this table, there are considerable differences among the y's, with the smallest being 48 and the largest being 89. However, we also see that the grade of 48 was obtained by a person who had studied German for two years, while the grade of 89 was obtained by a person who had studied German for five years, and this suggests that the differences among the grades may well be due, at least in part, to the fact that the job applicants did not all study German for the same number of years. These observations raise the following question, which we shall answer in this chapter: **Of the total variation among the y's, how much can be attributed to chance and how much can be attributed to the relationship between the two variables x and y; that is, to the fact that the observed y's correspond to different values of x.**

15.1

The Coefficient of Correlation

With regard to the question raised in the introduction, we are essentially faced here with an **analysis-of-variance** problem, and a study of Figure 15.1 will help to understand what this means. With reference to this figure, we see that the deviation of any observed value y from its mean, $y - \bar{y}$, can be written as the sum of two parts: $y' - \bar{y}$, the deviation of the value on the line (corresponding to the observed value of x) from the mean of the y's, and $y - y'$, the deviation of the observed value of y from the corresponding value on the line. Symbolically, we write

$$(y - \bar{y}) = (y' - \bar{y}) + (y - y')$$

for any observed value y, and if we square the expressions on both sides of this identity and sum over all n values of y, we find that algebraic simplifications lead to

$$\sum (y - \bar{y})^2 = \sum (y' - \bar{y})^2 + \sum (y - y')^2$$

As our measure of the total variation of the y's we use the quantity on the left above, $\sum (y - \bar{y})^2$, called the **total sum of squares**; it is just $n - 1$ times the variance of the y's, and, as the equation shows, it has been partitioned

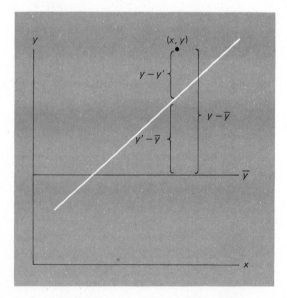

FIGURE 15.1

Illustration that $y - \bar{y} = (y' - \bar{y}) + (y - y')$.

x and y is such that small values of y tend to go with small values of x and large values of y tend to go with large values of x. Also, r is negative when the least-squares line has a downward slope; that is, when large values of y tend to go with small values of x and small values of y tend to go with large values of x. Geometrically, the ideas of a **positive correlation** and a **negative correlation** are illustrated in the first two diagrams of Figure 15.2; the third diagram illustrates the case where the least-squares line is horizontal, $r = 0$ and there is **no correlation**.

Since part of the variation of the y's cannot exceed their total variation, $\sum (y - y')^2$ cannot exceed $\sum (y - \bar{y})^2$, and it follows from the formula defining r that a correlation coefficient must lie on the interval from -1 to $+1$. If all the points actually fall on a straight line, the residual sum of squares, $\sum (y - y')^2$, is zero, and the resulting value of r, -1 or $+1$, is indicative of the perfect fit. If, however, the scatter of the points is such that the least-squares line is a horizontal line coincident with $\bar{y}$ (that is, a line with slope 0 which intersects the y-axis at $a = \bar{y}$), then $\sum (y - y')^2$ equals $\sum (y - \bar{y})^2$ and $r = 0$. In this case none of the variation of the y's can be attributed to their relationship with x, and the fit is so poor that knowledge of x is of no help in predicting y—the predicted value of y is $\bar{y}$ regardless of x.

The formula which defines r shows clearly the nature, or essence, of the coefficient of correlation, but in actual practice it is much easier to calculate r from the following computing formula[†]:

Computing formula for coefficient of correlation

$$r = \frac{n(\sum xy) - (\sum x)(\sum y)}{\sqrt{n(\sum x^2) - (\sum x)^2}\sqrt{n(\sum y^2) - (\sum y)^2}}$$

[†] This computing formula follows directly from an alternative, but equivalent, definition of the coefficient of correlation based on the **sample covariance**

$$s_{xy} = \frac{\sum (x - \bar{x})(y - \bar{y})}{n - 1}$$

In this formula, we add the products obtained by multiplying the deviation of each x from $\bar{x}$ by the deviation of the corresponding y from $\bar{y}$, and divide by $n - 1$. In this way we literally measure the way in which the values of x and y vary together. If the relationship between the x's and the y's is such that large values of x tend to go with large values of y, and small values of x with small values of y, the deviations $x - \bar{x}$ and $y - \bar{y}$ tend to be both positive or both negative, so that most of the products $(x - \bar{x})(y - \bar{y})$ and, hence, the covariance, are positive. On the other hand, if the relationship between the x's and the y's is such that large values of x tend to go with small values of y and vice versa, the deviations $x - \bar{x}$ and $y - \bar{y}$ tend to be of opposite sign, so most of the products and, hence, the covariance, are negative. Using the covariance and the standard deviations, s_x and s_y, of the x's and the y's, we can define the coefficient of correlation as

$$r = \frac{s_{xy}}{s_x \cdot s_y}$$

The sum of products $\sum (x - \bar{x})(y - \bar{y})$ divided by n is called a **product moment**, and this explains the term "product-moment coefficient of correlation."

15.1 The Coefficient of Correlation

This formula may look imposing but, with the exception of $\sum y^2$, the quantities needed for substitution are the same ones which are required to calculate the coefficients a and b of a least-squares line.

EXAMPLE Squaring and summing the y's (proficiency scores) in the table on page 362, we get $\sum y^2 = 50{,}085$, and substituting this value together with $\sum x = 35, \sum x^2 = 133, \sum y = 697, \sum xy = 2{,}554$, and $n = 10$ into the above formula, we get

$$ r = \frac{10(2{,}554) - (35)(697)}{\sqrt{10(133) - (35)^2}\sqrt{10(50{,}085) - (697)^2}} $$

$$ = 0.91 $$

This agrees, as it should, with the result obtained on page 394. Observe that the sign of r is automatically determined when we use this computing formula.

15.2
The Interpretation of r

When r equals $+1$, -1, or 0, there is no problem about the interpretation of the coefficient of correlation. As we have already indicated, it is $+1$ or -1 when all the points actually fall on a straight line, and it is zero when the fit of the least-squares line is so poor that knowledge of x does not help in the prediction of y. In general, $100r^2$ gives the percentage of the total variation of the y's which is explained by, or is due to, their relationship with x. This itself is an important measure of the relationship between two variables; beyond this, it permits valid comparisons of several coefficients of correlation.

EXAMPLE If $r = 0.80$ in one study and $r = 0.40$ in another study, it would be misleading to say that the correlation of 0.80 is "twice as good" or "twice as strong" as the correlation of 0.40. When $r = 0.80$, then 64 percent of the variation of the y's is accounted for by the relationship with x, and when $r = 0.40$, only 16 percent of the variation of the y's is accounted for by the relationship with x. Thus, in the sense of "percentage of variation accounted for" we can say that a correlation of 0.80 is four times as strong as a correlation of 0.40. In the same way, we say that a correlation of 0.60 is nine times as strong as a correlation of 0.20, and so on.

There are several pitfalls in the interpretation of the coefficient of correlation. First, it is often overlooked that r measures only the strength of linear relationships; second, a strong correlation does not necessarily imply a cause–effect relationship.

If r is calculated indiscriminantly, for instance, for the three sets of data of Figure 15.3, we get $r = 0.75$ in each case, but it is a meaningful measure of the strength of the relationship only in the first case. In the

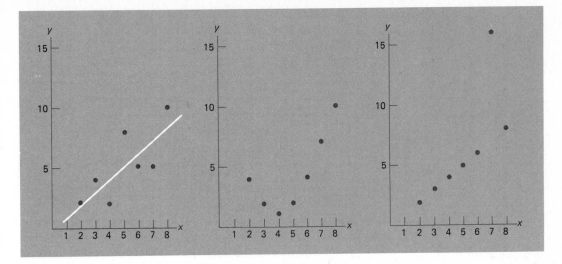

FIGURE 15.3
Three sets of paired data for which r = 0.75.

second case there is a very strong curvilinear relationship between the two variables, and in the third case six of the seven points actually fall on a straight line, but the seventh point is so far off that it suggests the possibility of a gross error of measurement or an error in recording the data. Thus, before we calculate *r*, we should always plot the data to see whether there is reason to believe that the relationship is, in fact, linear.

The fallacy of interpreting a high value of *r* (that is, a value close to +1 or −1) as an indication of a cause–effect relationship, is best explained with a few examples. Frequently used as an illustration, is the high positive correlation between the annual sales of chewing gum and the incidence of crime in the United States. Obviously, one cannot conclude that crime might be reduced by prohibiting the sale of chewing gum; both variables depend upon the size of the population, and it is this mutual relationship with a third variable (population size) which produces the positive correlation. Another example is the strong positive correlation which was observed between the number of storks seen nesting in English villages and the number of children born in the same villages. We leave it to the reader's ingenuity to explain why there might be a strong correlation in this case in the absence of any cause–effect relationship.

15.3

Correlation Analysis

When *r* is calculated on the basis of sample data, we may get a strong positive or negative correlation purely by chance, even though there is actually no relationship whatever between the two variables under consideration.

EXAMPLE Suppose we take a pair of dice, one red and one green, roll them five times, and get the following results:

Red die x	Green die y
4	5
2	2
4	6
2	1
6	4

Calculating r for these data, we get the surprisingly high value $r = 0.66$, and this raises the question whether anything is wrong with presuming that there is really no relationship between x and y—after all, one die does not know what the other die is doing. To answer this question, we shall have to see whether this high value of r may be attributed to chance.

When a correlation coefficient is calculated from sample data, as in the above example, the value we obtain for r is only an estimate of a corresponding parameter, the **population correlation coefficient,** which we denote ρ (the Greek lowercase letter *rho*). What r measures for a sample, ρ measures for a population. To test the null hypothesis of no correlation, namely, the hypothesis $\rho = 0$, we shall have to make several assumptions about the distribution of the random variables whose values we observe. In **normal correlation analysis** we make the same assumptions as in normal regression analysis (see page 384), except that now the x's are also values of a random variable having a normal distribution. If all these assumptions are met and the null hypothesis $\rho = 0$ is true, it can be shown that

Statistic for test of null hypothesis $\rho = 0$

$$t = \frac{r\sqrt{n - 2}}{\sqrt{1 - r^2}}$$

has the t distribution with $n - 2$ degrees of freedom.[†]

EXAMPLE To test whether the value of r which we obtained for the two dice is significant at $\alpha = 0.05$, we substitute $r = 0.66$ and $n = 5$ into the

---

[†] See Exercise 5 on page 445 for a method, based on this statistic, to calculate the smallest value of r that is significant for given values of α and n.

formula for t, and get

$$t = \frac{(0.66)\sqrt{5 - 2}}{\sqrt{1 - (0.66)^2}} = 1.52$$

Since this value does not exceed 3.182, the value of $t_{0.025}$ for $5 - 2 = 3$ degrees of freedom, the null hypothesis of no correlation cannot be rejected. In other words, the correlation coefficient of 0.66 which we obtained for the two dice is not significant.

Since the method we have discussed applies only when the null hypothesis is $\rho = 0$, let us present an alternative technique, based on an approximation, which can be used more generally to test the null hypothesis $\rho = \rho_0$ or to construct confidence intervals for ρ. It is based on the **Fisher Z transformation**, a change of scale from r to Z, which is given by

$$Z = \frac{1}{2} \cdot \ln \frac{1 + r}{1 - r}$$

where ln denotes "natural logarithm," that is, logarithm to the base e, where $e = 2.71828. \ldots$. This transformation is named after R. A. Fisher, a prominent statistician, who showed that under the assumptions of normal correlation analysis and for any value of ρ, the distribution of Z is approximately normal with

$$\mu_Z = \frac{1}{2} \cdot \ln \frac{1 + \rho}{1 - \rho} \quad \text{and} \quad \sigma_Z = \frac{1}{\sqrt{n - 3}}$$

Hence,

Statistic for inferences about ρ

$$\boxed{z = \frac{Z - \mu_Z}{\sigma_Z} = (Z - \mu_Z)\sqrt{n - 3}}$$

has approximately the standard normal distribution. The application of this theory is greatly facilitated by the use of Table VII at the end of the book, which gives the values of Z corresponding to $r = 0.00, 0.01, 0.02, 0.03, \ldots$, and 0.99. Observe that only positive values are given in this table; if r is negative, we simply look up $-r$ and take the negative of the corresponding Z. Note also that the formula for μ_Z is like that for Z with r replaced by ρ; therefore, Table VII can be used to look up values of μ_Z.

EXAMPLE Let us refer again to the proficiency scores of the applicants for the foreign service jobs, where we had $n = 10$ and $r = 0.91$, and let us test the null hypothesis $\rho = 0.70$ against the alternative hypothesis $\rho > 0.70$ at the level of significance $\alpha = 0.05$. Since the values of Z corresponding to $r = 0.91$ and $\rho = 0.70$ are, respectively, 1.528 and 0.867 according to Table VII, substitution into the above formula for z yields

$$z = (1.528 - 0.867)\sqrt{10 - 3} = 1.75$$

Since this exceeds $z_{0.05} = 1.64$ (for the standard normal distribution), the null hypothesis must be rejected; in other words, we conclude that $\rho > 0.70$ for the proficiency scores of applicants to the foreign service jobs and the number of years that they have studied German in high school or college.

To construct a confidence interval for ρ, we first construct a confidence interval for μ_Z, and then convert to r and ρ by means of Table VII. The confidence interval for μ_Z may be obtained by substituting

$$z = (Z - \mu_Z)\sqrt{n - 3}$$

for the middle term of the double inequality $-z_{\alpha/2} < z < z_{\alpha/2}$, and then manipulating the terms algebraically so that the middle term is μ_Z. This leads to the following $1 - \alpha$ confidence interval for μ_Z:

Confidence interval for μ_Z

$$Z - \frac{z_{\alpha/2}}{\sqrt{n - 3}} < \mu_Z < Z + \frac{z_{\alpha/2}}{\sqrt{n - 3}}$$

EXAMPLE Suppose that $r = 0.62$ for a set of paired data (say, the final examination grades of 30 students taking both history and English). Reading the value of Z which corresponds to $r = 0.62$ from Table VII, we get 0.725, and substituting this value, $n = 30$, and $z_{\alpha/2} = 1.96$ into the above formula, we get the 0.95 confidence interval

$$0.725 - \frac{1.96}{\sqrt{27}} < \mu_Z < 0.725 + \frac{1.96}{\sqrt{27}}$$

or

$$0.348 < \mu_Z < 1.102$$

Finally, looking up the values of r which come closest to $Z = 0.348$ and $Z = 1.102$ in Table VII, we get the 0.95 confidence interval

$$0.33 < \rho < 0.80$$

for the true strength of the linear relationship between final examination grades in the two subjects.

EXAMPLE To give an example which involves negative values of r and Z, suppose that $n = 40$ and $r = 0.20$ for a given set of paired data. Since Table VII shows that $Z = 0.203$ corresponds to $r = 0.20$, we first get the 0.95 confidence interval

$$0.203 - \frac{1.96}{\sqrt{37}} < \mu_z < 0.203 + \frac{1.96}{\sqrt{37}}$$

or

$$-0.119 < \mu_z < 0.525$$

Then, looking up in Table VII the values of r which come closest to $Z = 0.119$ and $Z = 0.525$, we get the 0.95 confidence interval

$$-0.12 < \rho < 0.48$$

for the population correlation coefficient.

EXERCISES 1 The following are the numbers of minutes it took 12 mechanics to assemble a piece of machinery in the morning, x, and in the late afternoon, y:

x	y
12	14
11	11
9	14
13	11
10	12
11	15
12	12
14	13
10	16
9	10
11	10
12	14

15.3 Correlation Analysis

Calculate r and test the null hypothesis of no correlation at the level of significance $\alpha = 0.05$.

2 The following table shows the percentages of the vote received by eight candidates for the U.S. Senate in the 1976 election, x, and the corresponding percentages predicted by a poll, y:

x	y
50	43
42	44
57	58
55	59
44	41
48	53
53	52
59	62

Calculate r and test the null hypothesis of no correlation at the level of significance $\alpha = 0.01$.

3 The following data were obtained in a study of the relationship between the resistance (ohms) and the failure time (minutes) of certain overloaded resistors:

Resistance	Failure time
48	45
28	25
33	39
40	45
36	36
39	35
46	36
40	45
30	34
42	39
44	51
48	41
39	38
34	32
47	45

(a) Calculate r and test the null hypothesis of no correlation at the level of significance $\alpha = 0.05$.

(b) What percentage of the variation of the failure times is accounted for by differences in resistance?

4 Calculate r for the data of Exercise 1 on page 368 and test the null hypothesis of no correlation at the level of significance $\alpha = 0.01$.

5 With reference to Exercise 3 on page 369:
   (a) calculate r;
   (b) test the null hypothesis of no correlation at the level of significance $\alpha = 0.05$;
   (c) determine the percentage of the variation in the demand for the product that is due to differences in price.

6 Calculate r for the data of Exercise 4 on page 369 and determine, at the level of significance $\alpha = 0.05$, whether there is a real relationship between rainfall and the yield of wheat.

7 Determine r separately for each of the following sets of data:

(a)	x	y
	12	5
	8	15

(b)	x	y
	6	9
	14	11

Is there a way to determine the results without making any calculation at all?

8 Since r does not depend on the scales of x and y, its calculation can often be simplified by adding a suitable positive or negative number to each x, each y, or both, or by multiplying each x, each y, or both by arbitrary positive constants.
   (a) Rework Exercise 1 after subtracting 9 from each x and 10 from each y.
   (b) Calculate r for the data of Exercise 2 on page 369 after dividing each x (number of hours) by 2, and multiplying each y (chlorine residual) by 10 and then subtracting 9.

9 With reference to Exercise 8 on page 383, calculate r for each of the following pairs of variables:
   (a) income and age;
   (b) income and years college;
   (c) age and years college.

10 State in each case whether you would expect a positive correlation, a negative correlation, or no correlation:
   (a) the ages of husbands and wives;
   (b) the amount of rubber on tires and the number of miles they have been driven;
   (c) the number of hours that golfers practice and their scores;
   (d) shoe size and IQ;
   (e) pollen count and the sale of anti-allergy drugs;
   (f) income and education;
   (g) the number of sunny days in August in Detroit and the attendance at the Detroit Zoo;
   (h) shirt size and sense of humor;
   (i) number of persons getting flu shots and number of persons catching the flu.

15.3 **Correlation Analysis**

403

11 Correlation methods are sometimes used to study the relationship between two (time) series of data which are recorded annually, monthly, weekly, daily, and so on. Suppose, for instance, that in the years 1963–1976 a large textile manufacturer spent 0.8, 0.5, 0.8, 1.0, 1.0, 0.9, 0.8, 1.2, 1.0, 0.9, 0.8, 1.0, 1.0, and 0.8 million dollars on research and development, and that in these years its share of the market was 20.4, 18.6, 19.1, 18.0, 18.2, 19.6, 20.0, 20.4, 19.2, 20.5, 20.8, 18.9, 19.0, and 19.8 percent. To see whether and how the company's share of the market in a given year may be related to its expenditures on research and development in prior years, let x_t denote the company's research and development expenditures and y_t its market share in the year t, and calculate

(a) the correlation coefficient for y_t and x_{t-1};
(b) the correlation coefficient for y_t and x_{t-2};
(c) the correlation coefficient for y_t and x_{t-3};
(d) the correlation coefficient for y_t and x_{t-4}.

For instance, in part (a) calculate r after pairing the 1964 percentage share of the market with the 1963 expenditures on research and development, the 1965 market share with the 1964 expenditures, and so on, . . . , and in part (d) calculate r after pairing the 1967 percentage share of the market with the 1963 expenditures on research and development, the 1968 market share with the 1964 expenditures, and so on. These time-lag correlations are called **cross correlations**. To continue,

(e) test, at the level of significance $\alpha = 0.05$, whether the correlation coefficients obtained in parts (a) through (d) are significant;
(f) discuss the apparent duration of the effect of expenditures on research and development on the company's share of the market.

12 In a study of the relationship between the annual production of cotton and citrus fruits in Arizona, data for $n = 16$ years yielded $r = -0.56$. At the level of significance $\alpha = 0.05$, test the null hypothesis $\rho = -0.40$ against the alternative hypothesis $\rho \neq -0.40$.

13 In a study of the relationship between the death rate from lung cancer and the per capita consumption of cigarettes twenty years earlier, data for $n = 9$ countries yielded $r = 0.73$. At the level of significance $\alpha = 0.05$, test the null hypothesis $\rho = 0.50$ against the alternative hypothesis $\rho > 0.50$.

14 In a study of the relationship between the available heat (per cord) of green wood and air-dried wood, data for $n = 13$ kinds of wood yielded $r = 0.94$. Use the level of significance $\alpha = 0.01$ to test the null hypothesis $\rho = 0.75$ against the alternative hypothesis $\rho \neq 0.75$.

15 If $n = 18$ and $r = -0.64$ for certain paired data, test the null hypothesis $\rho = -0.30$ against the alternative hypothesis $\rho < -0.30$ at the level of significance $\alpha = 0.05$.

16 Assuming that the conditions underlying normal correlation analysis are met, use the Fisher Z transformation to construct approximate 0.95 confidence intervals for ρ when

(a) $r = 0.80$ and $n = 15$;
(b) $r = -0.22$ and $n = 30$;
(c) $r = 0.64$ and $n = 100$.

17 Assuming that the conditions underlying normal correlation analysis are met, use the Fisher Z transformation to construct approximate 0.99 confidence

intervals for ρ when
(a) $r = -0.87$ and $n = 19$;
(b) $r = 0.39$ and $n = 24$;
(c) $r = 0.16$ and $n = 40$.

18 On page 393 we calculated for the proficiency scores the total sum of squares and the residual sum of squares by direct substitution into $\sum (y - \bar{y})^2$ and $\sum (y - y')^2$. However, it can be shown that

$$\sum (y - \bar{y})^2 = \sum y^2 - n \cdot \bar{y}^2$$

and

$$\sum (y - y')^2 = \sum y^2 - a(\sum y) - b(\sum xy)$$

where a and b are the y-intercept and the slope of the least-squares line, and in most problems these two formulas will greatly simplify the calculation of these sums of squares. Use these computing formulas to recalculate the two sums of squares for the proficiency scores. The answer for $\sum (y - \bar{y})^2$ should be the same as on page 393, but the answer for $\sum (y - y')^2$ will differ somewhat from that obtained before due to the rounding of a and b to one decimal.

15.4
Rank Correlation

Since the assumptions on which the significance test of the preceding section is based are rather stringent, it is sometimes preferable to use a nonparametric alternative which can be applied under much more general conditions. This test of the null hypothesis of no correlation is based on the **rank-correlation coefficient** (often called **Spearman's rank correlation coefficient**), which is essentially the coefficient of correlation of the ranks of the x's and the y's within the two samples. As we shall see, the rank-correlation coefficient has the added advantage that it is usually easier to determine than r when no calculator is available.

To calculate the rank-correlation coefficient, we first rank the x's among themselves, giving rank 1 to the largest (or smallest) value, rank 2 to the second largest (or smallest) value, and so on; then we rank the y's similarly among themselves, find the sum of the squares of the differences, d, between the ranks of the x's and y's, and substitute into the formula

Rank-correlation coefficient

$$r' = 1 - \frac{6(\sum d^2)}{n(n^2 - 1)}$$

where n is the number of pairs of observations. When there are ties in rank, we assign to each of the tied observations the mean of the ranks which they jointly occupy. For instance, if the third and fourth largest values of a variable are the same, we assign each the rank $\dfrac{3+4}{2} = 3.5$, and if the fifth, sixth, and seventh largest values of a variable are the same, we assign each the rank $\dfrac{5+6+7}{3} = 6$.

EXAMPLE The numbers of hours which ten students (in a random sample) studied for an examination and their grades in the examination are shown in the first two columns of the following table:

Number of hours studied x	Grade in examination y	Rank of x	Rank of y	d	d^2
8	56	6.5	7	−0.5	0.25
5	44	8.5	9	−0.5	0.25
11	79	4	3	1.0	1.00
13	72	3	4	−1.0	1.00
10	70	5	5	0.0	0.00
5	54	8.5	8	0.5	0.25
18	94	1	1	0.0	0.00
15	85	2	2	0.0	0.00
2	33	10	10	0.0	0.00
8	65	6.5	6	0.5	0.25
					3.00

To calculate r', we rank the x's and y's as shown, determine the d's and their squares, and substitute $\sum d^2 = 3$ into the formula. Since $n = 10$ in this example, we get

$$r' = 1 - \frac{6 \cdot 3}{10(10^2 - 1)} = 0.98$$

The correlation coefficient r calculated for the original x's and y's is 0.96, as can easily be verified, and the difference between $r' = 0.98$ and $r = 0.96$ is quite small.

When there are no ties, r' equals the correlation coefficient r calculated for the two sets of ranks; when ties exist, there may be a small (but usually negligible) difference. By using ranks we lose some information (as in the rank-sum tests of Chapter 13), but rank-correlation methods have the advantage that r' is usually easier to determine than r (at least, when no calculator is available), they can be used to measure relationships between ordinal data (for instance, when items cannot be measured but ranked), and tests of significance based on them are relatively unrestrictive.

When using r' to test the null hypothesis of no correlation between two variables x and y, we do not have to make any assumptions about the nature of the populations sampled. Under the null hypothesis of no correlation (indeed, the null hypothesis that the x's and y's are randomly matched), the sampling distribution of r' has the mean 0 and the standard deviation[†]

$$\sigma_{r'} = \frac{1}{\sqrt{n-1}}$$

Since this sampling distribution can be approximated with a normal distribution even for relatively small values of n, we base the test of the null hypothesis on the statistic

Statistic for testing significance of r'

$$z = \frac{r' - 0}{1/\sqrt{n-1}} = r'\sqrt{n-1}$$

which has approximately the standard normal distribution.

EXAMPLE In our example we had $r' = 0.98$ and $n = 10$, so that

$$z = 0.98\sqrt{10-1} = 2.94$$

Since this value exceeds $z_{0.005} = 2.58$, we reject the null hypothesis of no correlation and conclude at the level of significance $\alpha = 0.01$ that there is, in fact, a real (positive) relationship between grades and the amount of time studied for the population sampled.

15.5

Multiple and Partial Correlation ★

In the beginning of this chapter we defined the correlation coefficient as a measure of the goodness of the fit of a least-squares line to a set of paired data. If predictions are to be made with an equation of the form

$$y' = b_0 + b_1x_1 + b_2x_2 + \cdots + b_kx_k$$

[†] There exists a correction for $\sigma_{r'}$ that accounts for ties in rank, but it is seldom used unless the number of ties is large (see the Bibliography on page 411).

as in Section 14.4, we define the **multiple correlation coefficient** in the same way we originally defined r. We take the square root of the quantity

$$\frac{\sum (y' - \bar{y})^2}{\sum (y - \bar{y})^2} = 1 - \frac{\sum (y - y')^2}{\sum (y - \bar{y})^2}$$

which is the proportion of the total variation of the y's that can be attributed to the relationship with the x's. The only difference is that we now calculate y' by means of the multiple regression equation instead of the equation $y' = a + bx$.

EXAMPLE Referring to the problem on page 381, where we derived the equation $y' = 20,197 + 4,149x_1 + 731x_2$ to estimate the price of a house in terms of the number of bedrooms, x_1, and the number of baths, x_2, let us merely state the results that for the given data $\sum (y - y')^2 = 686,719$ and $\sum (y - \bar{y})^2 = 185,955,000$. So,

$$1 - \frac{686,719}{185,955,000} = 0.9963$$

and it follows that the multiple correlation coefficient is $\sqrt{0.9963} = 0.998$. Actually, this example serves to illustrate that adding more independent variables in a correlation study is not always sufficiently productive to justify including the additional variable (or variables). In this case, the correlation coefficient for y and x_1 alone is $r = 0.996$, so very little seems to be gained by considering also the number of baths. However, the situation is quite different in Exercise 7 on page 411, where two independent variables together account for a much higher percentage of the total variation in y than does either x_1 or x_2 alone.

When we discussed the problem of correlation and causation, we showed that a strong correlation between two variables may be due entirely to their dependence on a third variable. We illustrated this with the examples of chewing gum sales and the crime rate, and child births and the number of storks.

EXAMPLE To give another illustration, let us consider the variables x_1 and x_2, where x_1 is the weekly amount of hot chocolate sold by a refreshment stand at a summer resort, and x_2 is the weekly number of visitors to the resort. If, on the basis of suitable data, we get $r = -0.30$ for these variables, this should come as a surprise—after all, we would expect more sales of hot chocolate when there are more visitors and vice versa, and hence a positive correlation.

However, if we think for a moment, we may surmise that the negative correlation of -0.30 may well be due to the fact that the variables x_1 and x_2 are both related to a third variable x_3, the average weekly temperature at the resort. If the temperature is high, there will be more visitors, but they will prefer cold drinks to hot chocolate; if the temperature is low, there will be fewer visitors, but they will prefer hot chocolate to cold drinks. So, let us suppose that further data yield $r = -0.70$ for x_1 and x_3, and $r = 0.80$ for x_2 and x_3. These values seem reasonable since low sales of hot chocolate should go with high temperatures and vice versa, while the number of visitors should be high when the temperature is high, and low when the temperature is low.

In the preceding example, we should really have investigated the relationship between x_1 and x_2 (hot chocolate sales and the number of visitors to the resort) when all other factors, primarily temperature, are held fixed. As it is seldom possible to control matters to such an extent, it has been found that a statistic called the **partial correlation coefficient** does a fair job of eliminating the effects of other variables. If we write the ordinary correlation coefficients for x_1 and x_2, x_1 and x_3, and x_2 and x_3, as r_{12}, r_{13}, and r_{23}, the partial correlation coefficient for x_1 and x_2 with x_3 fixed is given by

Partial correlation coefficient

$$r_{12.3} = \frac{r_{12} - r_{13} \cdot r_{23}}{\sqrt{1 - r_{13}^2}\sqrt{1 - r_{23}^2}}$$

EXAMPLE If we substitute into this formula the values of our example, we get

$$r_{12.3} = \frac{(-0.30) - (-0.70)(0.80)}{\sqrt{1 - (-0.70)^2}\sqrt{1 - (0.80)^2}} = 0.61$$

and this shows that, as we expected, there is a positive relationship between the sales of hot chocolate and the number of visitors to the resort when the effect of differences in temperature is eliminated.

We have given this example primarily to illustrate what we mean by partial correlation, but it also served to show again that ordinary correlation coefficients can be very misleading unless they are interpreted with care.

EXERCISES

1 Calculate r' for the data of Exercise 4 on page 369 and test the null hypothesis of no correlation at the level of significance $\alpha = 0.05$.

2 Calculate r' for the data of Exercise 5 on page 370 and test the null hypothesis of no correlation at the level of significance $\alpha = 0.01$.

15.5 **Multiple and Partial Correlation**

3 Calculate r' for the following data representing the statistics grades, x, and sociology grades, y, of 18 students:

x	y	x	y
88	85	73	68
52	54	61	63
93	84	57	66
48	62	52	70
85	73	73	84
77	79	96	89
93	89	73	70
70	83	69	66
69	70	73	70

Also test for significance at $\alpha = 0.05$.

4 The following shows how a panel of nutrition experts and a panel of heads of household ranked 15 breakfast foods on their palatability:

Breakfast food	Nutrition experts	Heads of household
A	4	5
B	7	4
C	11	8
D	8	14
E	1	2
F	3	6
G	10	12
H	9	7
I	5	1
J	13	15
K	14	9
L	2	3
M	15	10
N	6	11
O	12	13

Calculate r' as a measure of the consistency of the two rankings.

5 Calculate r' for the data of Exercise 3 on page 402 and test the null hypothesis of no correlation at the level of significance $\alpha = 0.05$.

6 The following are the rankings which three judges gave to the works of ten artists:

Judge A	5	8	4	2	3	1	10	7	9	6
Judge B	3	10	1	4	2	5	6	7	8	9
Judge C	8	5	6	4	10	2	3	1	7	9

Calculate r' for each pair of rankings and decide

(a) which two judges are most alike in their opinions about these artists;

(b) which two judges differ the most in their opinions about these artists.

7 Use the least-squares equation obtained in Exercise 8 on page 383 to calculate y' for each of the five executives, determine the two sums of squares $\sum (y - y')^2$ and $\sum (y - \bar{y})^2$, and calculate the multiple correlation coefficient. Compare the value obtained for the multiple correlation coefficient with the values of r obtained in parts (a) and (b) of Exercise 9 on page 403.

8 Use the least-squares equation obtained in Exercise 9 on page 383 (or that obtained in Exercise 10 on page 383) to calculate y' for each of the six specimens of steel, determine the two sums of squares $\sum (y - y')^2$ and $\sum (y - \bar{y})^2$, and calculate the multiple correlation coefficient.

9 Using the results of Exercise 9 on page 403, find

(a) the partial correlation coefficient for income and age when "years college" is held fixed;

(b) The partial correlation coefficient for income and "years college" when age is held fixed.

10 With reference to the example on page 409, calculate the partial correlation coefficient for x_1 and x_3 (sales of hot chocolate and temperature) when x_2 (number of visitors) is held fixed. (*Hint*: Interchange the subscripts 2 and 3 everywhere in the formula for the partial correlation coefficient.)

BIBLIOGRAPHY   More detailed information about multiple and partial correlation may be found in

EZEKIEL, M., and FOX, K. A., *Methods of Correlation and Regression Analysis*, 3rd ed. New York: John Wiley & Sons, Inc., 1959.

and the correction in the formula for $\sigma_{r'}$ that is needed when there are many ties may be found in

SIEGEL, S., *Nonparametric Statistics for the Behavioral Sciences*. New York: McGraw-Hill Book Co., 1956.

An alternative way of testing the significance of r, based on a special table, is described in Exercise 5 on page 445; such a table may be found in

FREUND, J. E., and WILLIAMS, F. J., *Elementary Business Statistics: The Modern Approach*, 3rd ed. Englewood Cliffs, N.J.: Prentice-Hall, Inc., 1977.

Prior to the general availability of calculators, the calculation of r for mass data was often simplified by first grouping the data into a suitable two-way table, and then using a special formula for grouped data similar to the computing formula on page 395. Detailed descriptions of this method may be found in earlier editions of this book.

16

Analysis
of Variance

In this chapter we shall generalize the work of Sections 10.11 and 10.12 and consider the problem of deciding whether observed differences among more than two sample means can be attributed to chance, or whether there are real differences among the means of the populations sampled. For instance, we may want to decide on the basis of sample data whether there really is a difference in the effectiveness of three methods of teaching a foreign language, we may want to compare the average yields per acre of several varieties of wheat, we may want to see whether there really is a difference in the average mileage obtained with four kinds of gasoline, we may want to judge whether there really is a difference in the durability of five kinds of carpet, and so on. The method we shall introduce for this purpose is a powerful statistical tool called **analysis of variance,** ANOVA for short.

In the example of Section 10.12 we suggested that the reader would be well advised to read parts of this chapter before arriving at the conclusion that there *is* a difference in the heat-producing capacity of coal from the two mines. What we had in mind was the possibility that the significant difference between the two sample means may have been due to the data coming from different laboratories, or perhaps that the measurements were obtained by different techniques; then there is also the possibility that the specimens from either mine came from a particular vein, and hence their average heat-producing capacity may not be representative of the entire mine. So, we shall devote parts of this chapter to some of the things one has to watch in the **design of experiments** (in planning every detail), so that, with reasonable assurance, statistically significant results can be attributed to particular causes.

16.1

Differences Among k Means: An Example

EXAMPLE   Suppose we want to compare the cleansing action of three detergents on the basis of the following whiteness readings made on 15 swatches of white cloth, which were first soiled with India ink and then washed in an agitator-type machine with the respective detergents:

Detergent A: 77, 81, 71, 76, 80
Detergent B: 72, 58, 74, 66, 70
Detergent C: 76, 85, 82, 80, 77

The means of these three samples are 77, 68, and 80, and we want to know whether the differences among them are significant or whether they can be attributed to chance.

In general, in a problem like this, if $\mu_1, \mu_2, \ldots,$ and μ_k are the means of the k populations from which the samples are drawn, we want to test the null hypothesis $\mu_1 = \mu_2 = \cdots = \mu_k$ against the alternative hypothesis that these means are not all equal.[†] Evidently, this null hypothesis would be supported if the differences among the sample means are small, and the alternative hypothesis would be supported if at least some of the differences among the sample means are large. Thus, we need a precise measure of the discrepancies among the $\bar{x}$'s, and with it a rule which tells us when the discrepancies are so large that the null hypothesis should be rejected. Possible choices for such a measure are the standard deviation of the $\bar{x}$'s or their variance and it is the latter which we shall use here.

EXAMPLE   The mean of the three $\bar{x}$'s is $\dfrac{77 + 68 + 80}{3} = 75$; hence, their variance is

$$s_{\bar{x}}^2 = \frac{(77 - 75)^2 + (68 - 75)^2 + (80 - 75)^2}{3 - 1}$$

$$= 39$$

[†] In connection with work later in this chapter, it is desirable to write the means as $\mu_1 = \mu + \alpha_1, \mu_2 = \mu + \alpha_2, \ldots,$ and $\mu_k = \mu + \alpha_k$. Here

$$\mu = \frac{\mu_1 + \mu_2 + \cdots + \mu_k}{k}$$

is called the **grand mean** and the α's, whose sum is zero (see Exercise 11 on page 426) are called the **treatment effects**. In this notation, the null hypothesis $\mu_1 = \mu_2 = \cdots = \mu_k$ becomes $\alpha_1 = \alpha_2 = \cdots = \alpha_k = 0$, and the alternative hypothesis is that the α's are not all equal to zero.

where the subscript $\bar{x}$ is used to show that this is the variance of the sample means.

Let us now make two assumptions which are critical to the method of analysis we shall use: **First, it will be assumed that the populations we are sampling can be approximated closely with normal distributions; second, it will be assumed that these populations all have the same standard deviation** σ. With these assumptions, we note that, if the null hypothesis $\mu_1 = \mu_2 = \cdots = \mu_k$ is true, we can look upon the k samples as if they came from one and the same (normal) population and, hence, upon the variance of their means, $s_{\bar{x}}^2$, as an estimate of $\sigma_{\bar{x}}^2$, the square of the standard error of the mean. Now, since $\sigma_{\bar{x}} = \dfrac{\sigma}{\sqrt{n}}$ for samples from infinite populations, we can look upon $s_{\bar{x}}^2$ as an estimate of $\sigma_{\bar{x}}^2 = \left(\dfrac{\sigma}{\sqrt{n}}\right)^2 = \dfrac{\sigma^2}{n}$ and, therefore, upon $n \cdot s_{\bar{x}}^2$ as an estimate of σ^2.

EXAMPLE For our illustration, we have $n \cdot s_{\bar{x}}^2 = 5 \cdot 39 = 195$ as an estimate of σ^2, the common variance of the three populations.

If σ^2 were known, we could compare $n \cdot s_{\bar{x}}^2$ with σ^2 and reject the null hypothesis that the population means are all equal if this value is much larger than σ^2. However, in most practical problems σ^2 is not known and we have no choice but to estimate it on the basis of the sample data. Having assumed under the null hypothesis that the k samples do, in fact, come from identical populations, we could use any one of their variances, $s_1^2, s_2^2, \ldots,$ or s_k^2, as an estimate of σ^2, and we can also use their average.

EXAMPLE Averaging, or **pooling**, the three sample variances in our illustration, we get

$$\frac{s_1^2 + s_2^2 + s_3^2}{3}$$

$$= \frac{1}{3}\left[\frac{(77-77)^2 + (81-77)^2 + (71-77)^2 + (76-77)^2 + (80-77)^2}{5-1}\right.$$

$$+ \frac{(72-68)^2 + (58-68)^2 + (74-68)^2 + (66-68)^2 + (70-68)^2}{5-1}$$

$$\left. + \frac{(76-80)^2 + (85-80)^2 + (82-80)^2 + (80-80)^2 + (77-80)^2}{5-1}\right]$$

$$= 23$$

and we now have two estimates of σ^2,

$$n \cdot s_{\bar{x}}^2 = 195 \quad \text{and} \quad \frac{s_1^2 + s_2^2 + s_3^2}{3} = 23$$

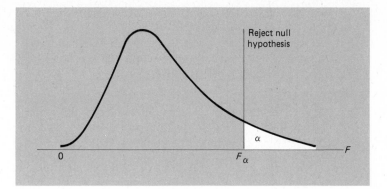

FIGURE 16.1

F distribution.

If the $\bar{x}$'s are far apart and the first of two such estimates of σ^2 (which is based on the variation among the sample means) is much larger than the second estimate (which is based on the variation within the samples and, hence, measures only variation that is due to chance), it stands to reason that the null hypothesis should be rejected. To put the comparison of the two estimates of σ^2 on a rigorous basis, we use the statistic

Statistic for test concerning differences among means

$$F = \frac{\text{estimate of } \sigma^2 \text{ based on the variation among the } \bar{x}\text{'s}}{\text{estimate of } \sigma^2 \text{ based on the variation within the samples}}$$

which is appropriately called a **variance ratio.**

If the null hypothesis is true and if the assumptions made are valid, the sampling distribution of this statistic is the *F* distribution, which we introduced in Chapter 11. Since the null hypothesis will be rejected only when *F* is large (that is, when the variation among the $\bar{x}$'s is too great to be attributed to chance), we base our decision on the criterion of Figure 16.1. For $\alpha = 0.05$ or 0.01, the values of F_α may be looked up in Table IV at the end of the book, and if we compare the means of *k* random samples of size *n*, we have $k - 1$ degrees of freedom for the numerator and $k(n - 1)$ degrees of freedom for the denominator.[†]

EXAMPLE   Returning to our illustration, we find that

$$F = \frac{195}{23} = 8.48$$

[†] In connection with the numerator degrees of freedom, the numerator of *F* is $n \cdot s_{\bar{x}}^2$ and $s_{\bar{x}}^2$ is the variance of *k* means and, hence, has $k - 1$ degrees of freedom in accordance with the terminology introduced in the footnote to page 252. As for the denominator degrees of freedom, $k(n - 1)$, the denominator of *F* is the mean of *k* sample variances each of which has $n - 1$ degrees of freedom.

16.1   Differences Among *k* Means: An Example

415

and since this exceeds 6.93, the value of $F_{0.01}$ for $k - 1 = 3 - 1 = 2$ and $k(n - 1) = 3(5 - 1) = 12$ degrees of freedom, the null hypothesis must be rejected at the level of significance $\alpha = 0.01$. In other words, we conclude that the differences among the $\bar{x}$'s are too large to be attributed entirely to chance.

The technique we have just described is the simplest form of an **analysis of variances**. Although we could go ahead and perform tests like the above for differences among k means without further discussion, it will be instructive to look at the problem from an analysis-of-variance point of view, and we shall do so in Section 16.3.

16.2

The Design of Experiments: Randomization

In the example of the preceding section, it may seem perfectly natural to conclude that the three detergents are not equally effective; and yet, a moment's reflection will show that this conclusion is not so "natural" at all. For all we know, the swatches cleaned with detergent B may have been more soiled than the others, the washing times may have been longer for detergent C, there may have been differences in water hardness or water temperature, and even the instruments used to make the whiteness readings may have gone out of adjustment after the readings for detergents A and C were made.

It is entirely possible, of course, that the differences among the three sample means, 77, 68, and 80, are due largely to differences in the quality of the three detergents, but we have just listed several other factors which could be responsible. It is important to remember that **a significance test may show that differences among sample means are too large to be attributed to chance, but it cannot say why the differences occurred.**

In general, if we want to show that one factor (among various others) can be considered the cause of an observed phenomenon, we must somehow make sure that none of the other factors can reasonably be held responsible. There are various ways in which this can be done; for instance, we can conduct a **rigorously controlled experiment** in which all variables except the one of concern are held fixed. To do this in the example dealing with the three detergents, we might soil the swatches with exactly equal amounts of India ink and oil, always use the same washing time, water of exactly the same temperature and hardness, and inspect the measuring instruments after each use. Under these rigid conditions, significant differences among the sample means cannot be due to differently soiled swatches, or differences in washing time, water temperature, water hardness, or measuring instruments. On the positive side, the differences show that the detergents are not all equally effective *if they are used in this narrowly restricted way*. Of course, we cannot say whether the same differences would exist if the washing time were longer or shorter, if the water had a different temperature or hardness, and so on.

In most cases, "overcontrolled" experiments like the one just described do not really provide us with the kind of information we want. Also, such experiments are rarely possible in actual practice; for example, it would have been difficult in our illustration to be sure that the instruments really were measuring identically on repeated washings or that some other factor, not thought of or properly controlled, was not responsible for the observed differences in whiteness. So, we look for alternatives. At the other extreme we can conduct an experiment in which none of the extraneous factors is controlled, but in which we protect ourselves against their effects by **randomization.** That is, we design, or plan, the experiment in such a way that the variations caused by these extraneous factors can all be combined under the general heading of "chance."

EXAMPLE In the illustration dealing with the three detergents, we could accomplish this by numbering from 1 to 15 the swatches (which need not be equally soiled), specifying the order in which they are to be washed and measured (at random, if this matters), and then selecting at random the five swatches to be washed with each of the three detergents.

When all the variations due to the uncontrolled extraneous factors can thus be included under the heading of chance variation, we refer to the design of the experiment as a **completely randomized design.**

As should be apparent, randomization protects against the effects of the extraneous factors only in a probabilistic sort of way. For instance, in our example it is possible, though very unlikely, that detergent A will be randomly assigned to the five swatches which happen to be the least soiled, or that the water happens to be coldest when we wash the five swatches with detergent B. It is partly for this reason that we often try to control some of the factors and randomize the others, and thus use designs that are somewhere between the two extremes which we have described.

16.3
One-Way Analysis of Variance

The basic idea of analysis of variance is to express a measure of the total variation in a set of data as a sum of terms, which can be attributed to specific sources, or causes, of variation; in its simplest form, it applies to experiments which are planned as completely randomized designs. With regard to the example of Section 16.1, two such sources of variation would be (1) actual differences in the cleansing action of the three detergents, and (2) chance, which in problems of this kind is usually called the **experimental error.** As a measure of the total variation in a set of data which consists of k samples, we shall use the **total sum of squares**[†]

$$SST = \sum_{i=1}^{k} \sum_{j=1}^{n} (x_{ij} - \bar{x}_{..})^2$$

where x_{ij} is the jth observation of the ith sample ($i = 1, 2, \ldots, k$ and $j = 1, 2, \ldots, n$), and $\bar{x}_{..}$ is the **grand mean,** the mean of all the kn measurements or observations. Note that if we divide the total sum of squares SST by $kn - 1$, we would get the variance of all the data.

If we let $\bar{x}_{i.}$ denote the mean of the ith sample (for $i = 1, 2, \ldots, k$), we can now write the following identity, which forms the basis of a **one-way analysis of variance**[‡]:

Identity for one-way analysis of variance

$$SST = n \cdot \sum_{i=1}^{k} (\bar{x}_{i.} - \bar{x}_{..})^2 + \sum_{i=1}^{k} \sum_{j=1}^{n} (x_{ij} - \bar{x}_{i.})^2$$

Looking closely at the two terms into which the total sum of squares SST has been partitioned, we find that the first term is a measure of the variation among the sample means; in fact, if we divide it by $k - 1$ we get the quantity which we denoted $n \cdot s_{\bar{x}}^2$ on page 414. Similarly, the second term is a measure of the variation within the individual samples, and if we divide this term by $k(n - 1)$, we get the quantity which we put into the denominator of F in Section 16.1.

It is customary to refer to the first term, the quantity which measures the variation among the sample means, as the **treatment sum of squares** $SS(Tr)$, and to the second term, which measures the variation within the samples, as the **error sum of squares** SSE. This terminology is explained by the fact that most analysis-of-variance techniques were originally developed

---

[†] The use of double subscripts and double summations is explained briefly in Section 3.10.
[‡] This identity may be derived by writing the total sum of squares as

$$SST = \sum_{i=1}^{k} \sum_{j=1}^{n} (x_{ij} - \bar{x}_{..})^2$$

$$= \sum_{i=1}^{k} \sum_{j=1}^{n} [(\bar{x}_{i.} - \bar{x}_{..}) + (x_{ij} - \bar{x}_{i.})]^2$$

and then expanding the squares $[(\bar{x}_{i.} - \bar{x}_{..}) + (x_{ij} - \bar{x}_{i.})]^2$ by means of the binomial theorem and simplifying algebraically.

in connection with agricultural experiments where different fertilizers, for example, were regarded as different **treatments** applied to the soil. So, we shall refer to the three detergents in our example as three different treatments, and in other problems we may refer to four nationalities as four different treatments, five kinds of advertising campaigns as five different treatments, and so on. The word "error" in "error sum of squares" pertains to the experimental error, or chance.

EXAMPLE Before we go any further, let us verify the identity $SST = SS(Tr) + SSE$ with reference to the numerical example of Section 16.1. Substituting into the formulas for the different sums of squares, we get

$$
\begin{aligned}
SST = {}& (77 - 75)^2 + (81 - 75)^2 + (71 - 75)^2 \\
& + (76 - 75)^2 + (80 - 75)^2 + (72 - 75)^2 \\
& + (58 - 75)^2 + (74 - 75)^2 + (66 - 75)^2 \\
& + (70 - 75)^2 + (76 - 75)^2 + (85 - 75)^2 \\
& + (82 - 75)^2 + (80 - 75)^2 + (77 - 75)^2 \\
= {}& 666 \\
SS(Tr) = {}& 5[(77 - 75)^2 + (68 - 75)^2 + (80 - 75)^2] \\
= {}& 390
\end{aligned}
$$

and

$$
\begin{aligned}
SSE = {}& (77 - 77)^2 + (81 - 77)^2 + (71 - 77)^2 + (76 - 77)^2 \\
& + (80 - 77)^2 + (72 - 68)^2 + (58 - 68)^2 + (74 - 68)^2 \\
& + (66 - 68)^2 + (70 - 68)^2 + (76 - 80)^2 + (85 - 80)^2 \\
& + (82 - 80)^2 + (80 - 80)^2 + (77 - 80)^2 \\
= {}& 276
\end{aligned}
$$

and it can be seen that

$$
SS(Tr) + SSE = 390 + 276 = 666 = SST
$$

To test the null hypothesis $\mu_1 = \mu_2 = \cdots = \mu_k$ (or $\alpha_1 = \alpha_2 = \cdots = \alpha_k = 0$ in the notation of the footnote to page 413) against the alternative hypothesis that the treatment means are not all equal (or equivalently that the treatment effects are not all zero), we now proceed as in Section 16.1 and compare $SS(Tr)$ with SSE by means of an F-statistic. In practice, we usually exhibit the necessary work in an **analysis-of-variance table** as follows:

Source of variation	Degrees of freedom	Sum of squares	Mean square	F
Treatments	$k-1$	$SS(Tr)$	$MS(Tr) = \dfrac{SS(Tr)}{k-1}$	$\dfrac{MS(Tr)}{MSE}$
Error	$k(n-1)$	SSE	$MSE = \dfrac{SSE}{k(n-1)}$	
Total	$kn-1$	SST		

Here the second column lists the degrees of freedom (the number of independent deviations from the mean on which the sums of squares are based), the fourth column lists the **mean squares** $MS(Tr)$ and MSE, which are obtained by dividing the corresponding sums of squares by their degrees of freedom, and the right-hand column gives the value of the F-statistic as the ratio of the two mean squares. These two mean squares are, in fact, the two estimates of σ^2 referred to on page 414; also, the numerator and denominator degrees of freedom for the F test, $k-1$ and $k(n-1)$, are shown opposite "Treatments" and "Error" in the "Degrees of freedom" column. The significance test is the same as before; we compare F with F_α for $k-1$ and $k(n-1)$ degrees of freedom.

EXAMPLE Referring again to our example dealing with the three detergents, we now construct the following analysis-of-variance table:

Source of variation	Degrees of freedom	Sum of squares	Mean square	F
Treatments	2	390	195	8.48
Error	12	276	23	
Total	14	666		

Since $F = \dfrac{195}{23} = 8.48$ exceeds 6.93, the value of $F_{0.01}$ for 2 and 12 degrees of freedom, we find (as before) that the null hypothesis must be rejected at the level of significance $\alpha = 0.01$.

The numbers which we used in our illustration were intentionally chosen so that the calculations would be relatively easy. In actual practice, the calculation of the sums of squares can be quite tedious unless we use the following computing formulas, in which $T_{i.}$ denotes the total of the observations for the ith treatment (that is, the sum of the values in the ith sample), and $T_{..}$ denotes the grand total of all the data:

Computing formulas for sums of squares

$$SST = \sum_{i=1}^{k} \sum_{j=1}^{n} x_{ij}^2 - \frac{1}{kn} \cdot T_{..}^2$$

$$SS(Tr) = \frac{1}{n} \cdot \sum_{i=1}^{k} T_{i.}^2 - \frac{1}{kn} \cdot T_{..}^2$$

and by subtraction

$$SSE = SST - SS(Tr)$$

EXAMPLE   Let us show how these computing formulas are used by recalculating SST, $SS(Tr)$, and SSE for the example dealing with the three detergents. Since $T_{1.} = 385$, $T_{2.} = 340$, $T_{3.} = 400$, $T_{..} = 1{,}125$, and $\sum\sum x^2 = 85{,}041$, substitution into the formulas yields

$$SST = 85{,}041 - \frac{1}{15}(1{,}125)^2$$

$$= 85{,}041 - 84{,}375$$

$$= 666$$

$$SS(Tr) = \frac{1}{5}\left[385^2 + 340^2 + 400^2\right] - 84{,}375$$

$$= 390$$

and

$$SSE = 666 - 390 = 276$$

Of course, these results are identical with those obtained before.

The method we have discussed here applies only when each sample has the same number of observations, but minor modifications make it applicable also to situations where the sample sizes are not all equal. If there are n_i observations for the ith treatment (in the ith sample), the computing formulas for the sums of squares become

Computing formulas for sums of squares (unequal sample sizes)

$$SST = \sum_{i=1}^{k} \sum_{j=1}^{n_i} x_{ij}^2 - \frac{1}{N} \cdot T_{..}^2$$

$$SS(Tr) = \sum_{i=1}^{k} \frac{T_{i.}^2}{n_i} - \frac{1}{N} \cdot T_{..}^2$$

$$SSE = SST - SS(Tr)$$

where $N = n_1 + n_2 + \cdots + n_k$. The only other change is that the total number of degrees of freedom is $N - 1$, and the degrees of freedom for treatments and error are $k - 1$ and $N - k$.

EXAMPLE The manager of a restaurant wants to determine whether the sales of chicken dinners depend on how this entree is described on the menu. He has three kinds of menus printed, listing chicken dinners among the other entrees, featuring them as "Chef's Special," and as "Gourmet's Delight," and he intends to use each kind of menu on six different Sundays. Actually, the manager collects only the following data, showing the number of chicken dinners sold on twelve Sundays:

Listed among other entrees	76, 94, 85, 77
Featured as Chef's Special	109, 117, 102, 92, 115
Featured as Gourmet's Delight	100, 83, 102

The means of these three samples are 83, 107, and 95, and this suggests that the different descriptions may well have an effect on the sales of the dinner. The samples are very small, however, and it remains to be seen whether the observed differences among the means are significant, say, at the level of significance $\alpha = 0.05$.

To perform an analysis of variance, we first determine the totals $T_{1.} = 332, T_{2.} = 535, T_{3.} = 285, T_{..} = 1,152,$ and $\sum\sum x^2 =$

112,722. Then we get

$$SST = 112{,}722 - \frac{1}{12}(1{,}152)^2 = 112{,}722 - 110{,}592$$

$$= 2{,}130$$

$$SS(Tr) = \frac{332^2}{4} + \frac{535^2}{5} + \frac{285^2}{3} - 110{,}592$$

$$= 1{,}284$$

and

$$SSE = 2{,}130 - 1{,}284 = 846$$

and this leads to the following analysis-of-variance table:

Source of variation	Degrees of freedom	Sum of squares	Mean square	F
Treatments	2	1,284	642	6.83
Error	9	846	94	
Total	11	2,130		

Since $F = 6.83$ exceeds 4.26, the value of $F_{0.05}$ for 2 and 9 degrees of freedom, we find that the null hypothesis (that the different descriptions do not affect the sales of the dinner) will have to be rejected. See, however, Exercise 10 on page 426.

EXERCISES

1 The following are the mileages which a test driver got with four gallons each of five brands of gasoline:

Brand A: 30, 25, 27, 26
Brand B: 29, 26, 29, 28
Brand C: 32, 32, 35, 37
Brand D: 29, 34, 32, 33
Brand E: 32, 26, 31, 27

(a) Use the method of Section 16.1 and the level of significance $\alpha = 0.01$ to test whether the differences among the five sample means can be attributed to chance.

(b) Perform an analysis of variance, using the computing formulas to calculate the required sums of squares, and compare the value of the F statistic with that obtained in part (a).

(c) What can we conclude, if the driver used the same car throughout the experiment, drove it over a track simulating freeway driving conditions, and randomized the order in which he used the twenty gallons of gasoline?

2 To compare three methods of teaching the programming of a certain digital computer, random samples of size 4 were taken from each of three groups of a company's trainees taught, respectively, by method X (straight teaching-machine instruction), method Y (personal instruction and some direct experience working with the computer), and method Z (personal instruction but no work with the computer itself), and the following are the scores obtained by these trainees on an appropriate achievement test:

Method X: 83, 77, 73, 79
Method Y: 92, 94, 88, 98
Method Z: 80, 85, 84, 87

(a) Use the method of Section 16.1 and the level of significance $\alpha = 0.05$ to test whether the differences among the three sample means can be attributed to chance.

(b) Perform an analysis of variance, using the computing formulas to calculate the required sums of squares, and compare the value of the F-statistic with that obtained in part (a).

(c) What can we conclude, if the company randomly assigns its trainees to the three methods of instruction? What can we conclude, if the company assigns only trainees with very high IQ's to method Y?

3 An experiment is performed to determine which of three different golf-ball designs, A, B, and C, will give the greatest distance when driven. Criticize the experiment if
(a) one golf pro hits all the design A balls, another all the design B balls, and a third all the design C balls;
(b) all the balls are hit with drivers of the same manufacture;
(c) all the design A balls are hit first, the design B balls next, and the design C balls last;
(d) all the design A balls are hit from a tee, while the design B and design C balls are hit without tees;
(e) all the design A balls are driven from the No. 1 tee, all the design B balls are driven from the No. 2 tee, and all the design C balls are driven from the No. 3 tee.

4 The following are eight consecutive weeks' earnings (in dollars) of three door-to-door cosmetics salespersons employed by a firm:

Salesperson 1:   176, 212, 188, 206, 200, 184, 193, 209
Salesperson 2:   187, 193, 184, 198, 210, 199, 180, 195
Salesperson 3:   164, 203, 180, 187, 223, 196, 189, 211

Test at the level of significance $\alpha = 0.05$ whether the differences among the average weekly earnings of the three salespersons are significant.

5   The following are the numbers of mistakes made in five successive days by four technicians working for a photographic laboratory:

Technician I: 8, 11, 7, 9, 10
Technician II: 9, 11, 6, 14, 10
Technician III: 8, 13, 11, 9, 13
Technician IV: 13, 5, 9, 10, 7

Test at the level of significance $\alpha = 0.01$ whether the differences among the four sample means can be attributed to chance.

6   With reference to Exercise 9 on page 344, perform an analysis of variance to test, at the level of significance $\alpha = 0.05$, whether the differences among the average scores obtained with the four bowling balls are significant.

7   The following data show the yields of soybeans (in bushels per acre) planted two inches apart on essentially similar plots with the rows 20, 24, 28, and 32 inches apart:

20 in	24 in	28 in	32 in
23.1	21.7	21.9	19.8
22.8	23.0	21.3	20.4
23.2	22.4	21.6	19.3
23.4	21.1	20.2	18.5
23.6	21.9	21.6	19.1
21.7	23.4	23.8	21.9

Test at the level of significance $\alpha = 0.05$ whether the differences among the four mean yields can be attributed to chance.

8   The following are the numbers of words per minute which a secretary typed on several occasions on four different typewriters:

Typewriter C: 71, 75, 69, 77, 61, 72, 71, 78
Typewriter D: 68, 71, 74, 66, 69, 67, 70, 62
Typewriter E: 75, 70, 81, 73, 78, 72
Typewriter F: 62, 59, 71, 68, 63, 65, 72, 60, 64

Use the level of significance $\alpha = 0.05$ to test whether the differences among the means of the four samples can be attributed to chance.

9   To study its performance, a newly designed motorboat was timed over a marked course under various wind and water conditions. Use the following data (in minutes) to test, at the level of significance $\alpha = 0.05$, the null hypothesis that the

boat's performance is not affected by the differences in wind and water conditions:

Calm conditions: 26, 19, 16, 22
Moderate conditions: 25, 27, 25, 20, 18, 23
Choppy conditions: 23, 25, 28, 31, 26

10 With reference to the example on page 422, suppose that the two largest values of the second sample, 117 and 115, were the numbers of chicken dinners sold on Mother's Day and Father's Day. Delete these two values and reanalyze the remaining data by means of an appropriate analysis of variance.

11 With reference to the notation introduced in the footnote to page 413, show that the sum of the α's, the treatment effects, is equal to zero.

16.4

The Design of Experiments: Blocking

To introduce another important concept in the design of experiments in addition to controlling and randomizing, let us consider the following illustration:

EXAMPLE A reading comprehension test is given to random samples of three 8th graders each from four schools, and the results are

School A: 87, 70, 92
School B: 43, 75, 56
School C: 70, 66, 50
School D: 67, 85, 79

The means of these four samples are 83, 58, 62, and 77, and since the differences among them are very large, it would seem reasonable to conclude that there are some real differences in the average reading comprehension of 8th graders in the four schools. This does not follow, however, from a one-way analysis of variance. We get

Source of variation	Degrees of freedom	Sum of squares	Mean square	F
Treatments	3	1,278	426	2.90
Error	8	1,176	147	
Total	11	2,454		

and since $F = 2.90$ does not exceed 4.07, the value of $F_{0.05}$ for 3 and 8 degrees of freedom, the null hypothesis (that the population means are all equal) cannot be rejected at the level of significance $\alpha = 0.05$.

The reason for this is that there are not only considerable differences among the four means, but also very large differences among the values within the samples. In the first sample they range from 70 to 92, in the second sample from 43 to 75, in the third sample from 50 to 70, and in the fourth sample from 67 to 85. Giving this some thought, it would seem reasonable to conclude that these differences within the samples may well be due to differences in intelligence, an extraneous factor (we might call it a "nuisance" factor) which was randomized by taking a random sample of 8th graders from each school. Thus, variations due to differences in intelligence were included in the experimental error; this "inflated" the error sum of squares which went into the denominator of the F-statistic, and the results were not significant.

To avoid this kind of situation, we could hold the extraneous factor fixed, but this will seldom give us the information we want. In our example, we could limit the study to 8th graders with an IQ of, say, 105, but then the results would apply only to 8th graders with IQ's of 105. Another possibility is to vary the known source of variability (the extraneous factor) deliberately over as wide a range as necessary, and to do it in such a way that the variability it causes can be measured and, hence, eliminated from the experimental error. This means that we should plan the experiment in such a way that we can perform a **two-way analysis of variance,** in which the total variability of the data is partitioned into three components attributed, respectively, to treatments (in our example, the four schools), the extraneous factor, and experimental error.

EXAMPLE   As we shall see later, this can be accomplished in our example by randomly selecting from each school one 8th grader with a low IQ, one 8th grader with an average IQ, and one 8th grader with a high IQ, where "low," "average," and "high" are presumably defined in a rigorous way. Suppose, then, we proceed in this way and get the results shown in the following table:

	Low IQ	Average IQ	High IQ
School A	71	92	89
School B	44	51	85
School C	50	64	72
School D	67	81	86

What we have done here is called **blocking,** and the three levels of intelligence are called **blocks.** In general, blocks are the levels at which we hold an extraneous factor fixed, so that we can measure its contribution to the total variability of the data by means of a two-way analysis of variance. The design which we used for our example is called a **complete block design;** it is "complete" in the sense that each treatment appears the same number of times in each block. In our example, there is one 8th grader from each school in each block.

16.5

Two-Way Analysis of Variance

To present the theory of a two-way analysis of variance, we shall use the terminology introduced in the preceding section and refer to the two variables (factors) under consideration as "treatments" and "blocks;" alternatively, we could refer to them as **factor** A and **factor** B, or as **rows** and **columns.**

Before we go into any details, let us point out that there are essentially two ways of analyzing such two-variable experiments, and they depend on whether the two variables are independent, or whether they **interact.**

EXAMPLE Suppose that a tire manufacturer is experimenting with different kinds of treads, and that he finds that one kind is especially good for use on dirt roads while another kind is especially good for use on hard pavement. If this is the case, we say that there is an interaction between road conditions and tread design. On the other hand, if each of the treads is affected equally by the different road conditions, we would say that there is no interaction and that the two variables (road conditions and tread design) are independent.

In this book we shall study only the case where there is no interaction.
To formulate the hypotheses to be tested in the two-variable case, let us write μ_{ij} for the population mean which corresponds to the ith treatment and the jth block (in our numerical example, the average reading comprehension score of 8th graders with the intelligence level j in the ith school) and express it as

$$\mu_{ij} = \mu + \alpha_i + \beta_j$$

As in the footnote to page 413, μ is the grand mean (the average of all the population means μ_{ij}) and the α_i are the treatment effects (whose sum is zero). Correspondingly, we refer to the β_j as the **block effects** (whose sum is also zero), and write the two null hypotheses we want to test as

$$\alpha_1 = \alpha_2 = \cdots = \alpha_k = 0$$

and

$$\beta_1 = \beta_2 = \cdots = \beta_n = 0$$

The alternative to the first null hypothesis (which in our illustration amounts to the hypothesis that the average reading comprehension of 8th graders is the same in all four schools) is that the treatment effects α_i are not all zero; the alternative to the second null hypothesis (which in our illustration amounts to the hypothesis that the average reading comprehension of 8th graders is the same for all three levels of intelligence) is that the block effects β_j are not all zero.

To test the second of the null hypotheses, we need a quantity, similar to the treatment sum of squares, which measures the variation among the block means (58, 72, and 83 in our example). So, if we let $T_{.j}$ denote the total of all the values in the jth block, substitute it for $T_{i.}$ in the computing formula for $SS(Tr)$ on page 421, sum on j instead of i, and interchange n and k, we obtain, analogous to $SS(Tr)$ the **block sum of squares**

Computing formula for block sum of squares

$$SSB = \frac{1}{k} \cdot \sum_{j=1}^{n} T_{.j}^2 - \frac{1}{kn} \cdot T_{..}^2$$

In a two-way analysis of variance (with no interaction) we compute SST and $SS(Tr)$ according to the formulas on page 421, SSB according to the formula of the preceding paragraph, and then get SSE by subtraction. Since $SST = SS(Tr) + SSB + SSE$, we have

Error sum of squares (Two-way analysis of variance)

$$SSE = SST - SS(Tr) - SSB$$

Observe that the error sum of squares for a two-way analysis of variance does not equal the error sum of squares for a one-way analysis of variance performed on the same data, even though we denote both with the symbol SSE. In fact, we are now partitioning the error sum of squares for the one-way analysis of variance into two terms: the block sum of squares, SSB, and the remainder which is the new error sum of squares, SSE.

We can now construct the following analysis-of-variance table for a two-way analysis of variance (with no interaction):

Source of variation	Degrees of freedom	Sum of squares	Mean square	F
Treatments	$k-1$	$SS(Tr)$	$MS(Tr) = \dfrac{SS(Tr)}{k-1}$	$\dfrac{MS(Tr)}{MSE}$
Blocks	$n-1$	SSB	$MSB = \dfrac{SSB}{n-1}$	$\dfrac{MSB}{MSE}$
Error	$(k-1)(n-1)$	SSE	$MSE = \dfrac{SSE}{(k-1)(n-1)}$	
Total	$kn-1$	SST		

The mean squares are again the sums of squares divided by their respective degrees of freedom, and the two F values are the mean squares for treatments and blocks divided by the mean square for error. Also, the degrees of freedom for blocks is $n-1$ (like that for treatments with n substituted for k), and the degrees of freedom for error is found by subtracting the degrees of freedom for treatments and blocks from $kn-1$, the total number of degrees of freedom; it is

$$(kn-1) - (k-1) - (n-1) = kn - k - n + 1$$

$$= (k-1)(n-1)$$

Thus, in the significance test for treatments the numerator and denominator degrees of freedom for F are $k-1$ and $(k-1)(n-1)$, and in the significance test for blocks the numerator and denominator degrees of freedom for F are $n-1$ and $(k-1)(n-1)$.

EXAMPLE Returning now to our illustration and the data on page 427, we find that $T_{1.} = 252$, $T_{2.} = 180$, $T_{3.} = 186$, $T_{4.} = 234$, $T_{.1} = 232$, $T_{.2} = 288$, $T_{.3} = 332$, $T_{..} = 852$, $k = 4$, $n = 3$, and $\sum\sum x^2 =$

63,414, so that

$$SST = 63,414 - \frac{1}{12}(852)^2$$

$$= 63,414 - 60,492$$

$$= 2,922$$

$$SS(Tr) = \frac{1}{3}[252^2 + 180^2 + 186^2 + 234^2] - 60,492$$

$$= 1,260$$

$$SSB = \frac{1}{4}[232^2 + 288^2 + 332^2] - 60,492$$

$$= 1,256$$

and the error sum of squares is

$$SSE = 2,922 - 1,260 - 1,256$$

$$= 406$$

This leads to the following analysis-of-variance table:

Source of variation	Degrees of freedom	Sum of squares	Mean square	F
Treatments	3	1,260	420	6.21
Blocks	2	1,256	628	9.28
Error	6	406	67.67	
Total	11	2,922		

Again using the level of significance $\alpha = 0.05$, we find that $F = 6.21$ exceeds 4.76, the value of $F_{0.05}$ for 3 and 6 degrees of freedom, and that $F = 9.28$ exceeds 5.14, the value of $F_{0.05}$ for 2 and 6 degrees of freedom. Thus, both null hypotheses must be rejected, and we conclude that the average reading comprehension of 8th graders is not the same in the four schools, and it is not the same for the three levels of intelligence. Observe that by blocking we were able to show that the differences among the means obtained for the four schools are significant, whereas without blocking, in the experiment described on page 426, the differences among the means were not significant.

EXERCISES

1 The following are the cholesterol contents (in milligrams per package) which four laboratories obtained for 6-ounce packages of three very similar diet foods:

	Laboratory 1	Laboratory 2	Laboratory 3	Laboratory 4
Diet food A	3.7	2.8	3.1	3.4
Diet food B	3.1	2.6	2.7	3.0
Diet food C	3.5	3.4	3.0	3.3

Perform a two-way analysis of variance and test, at the level of significance $\alpha = 0.05$, whether

   (a) the differences among the means obtained for the three diet foods are significant;
   (b) the differences among the means obtained by the four laboratories are significant.

2 Four different, although supposedly equivalent, forms of a standardized achievement test in science were given to each of five students, and the following are the scores which they obtained:

	Student C	Student D	Student E	Student F	Student G
Form 1	77	62	52	66	68
Form 2	85	63	49	65	76
Form 3	81	65	46	64	79
Form 4	88	72	55	60	66

Perform a two-way analysis of variance to test at the level of significance $\alpha = 0.01$ whether it is reasonable to treat the four forms as equivalent. If each student separately randomized the order in which he or she took the four tests (by using random numbers or some kind of gambling device), we refer to the design of this experiment as a **randomized block design.** The purpose of this randomization is to take care of such possible extraneous factors as fatigue or, perhaps, the experience gained from repeatedly taking the test.

3 An experiment performed to judge how the distance traveled by a missile is affected by four different fuels and two different types of launchers yielded the

following data (in nautical miles):

	Fuel I	Fuel II	Fuel III	Fuel IV
Launcher X	64.5	51.3	35.8	45.6
Launcher Y	42.4	41.7	49.4	61.8

Perform a two-way analysis of variance to test at the level of significance $\alpha = 0.05$ whether there are significant differences

(a) between the means obtained for the launchers;

(b) among the means obtained for the fuels.

4 Regarding the weeks as blocks, reanalyze the data of Exercise 4 on page 424 by means of a two-way analysis of variance. Use the level of significance $\alpha = 0.05$ for both tests of significance.

5 A laboratory technician measures the breaking strength of each of five kinds of linen threads by using four different measuring instruments, I_1, I_2, I_3, and I_4, and obtains the following results (in ounces):

	I_1	I_2	I_3	I_4
Thread 1	20.9	20.4	19.9	21.9
Thread 2	25.0	26.2	27.0	24.8
Thread 3	25.5	23.1	21.5	24.4
Thread 4	24.8	21.2	23.5	25.7
Thread 5	19.6	21.2	22.1	21.1

Perform a two-way analysis of variance, using $\alpha = 0.05$ for both tests of significance.

16.6

The Design of Experiments: Replication

In Section 16.4 we showed how we can increase the amount of information to be gained from an experiment by blocking, that is, by eliminating the effect of an extraneous factor. Another way to increase the amount of information to be gained from an experiment is to increase the volume of the data. For instance, in the example on page 426 we might increase the size of the samples and give the reading comprehension test to twenty 8th graders from each school instead of three. For more complicated designs, the same thing can be accomplished by executing the entire experiment more than once, and this is called **replication.** With reference to the example on page 427, we might conduct the experiment (select and test the twelve 8th graders) in one week, and then replicate (repeat) the entire experiment in the next week.

Conceptually, replication does not present any difficulties, but computationally it does, and that is why we shall not go into this any further. Indeed, if an experiment requiring a two-way analysis of variance is replicated, it will then require a three-way analysis of variance, since replication, itself, may be a source of variations in the data. For instance, this would be the case in our example if it got very hot and humid during the second week, making it difficult for the students to concentrate.

16.7

Latin Squares ★

We have seen how blocking can be used to eliminate the variability due to one extraneous factor from the experimental error, and, in principle, two or more extraneous sources of variation can be handled in the same way. The only real problem is that this may inflate the size of an experiment beyond practical bounds.

EXAMPLE   Suppose that in the example dealing with the reading comprehension of eighth graders we would also like to eliminate whatever variability there may be due to differences in age (12, 13, or 14) and in sex. Allowing for all possible combinations of intelligence, age, and sex, we will have to use $3 \cdot 3 \cdot 2 = 18$ different blocks, and if there is to be one eighth grader from each school in each block, we will have to select and test $18 \cdot 4 = 72$ eighth graders in all. If we also wanted to eliminate whatever variability there may be due to ethnic background, for which we might consider five categories, this would raise the required number of eighth graders to $72 \cdot 5 = 360$.

In this section we will show how problems like this can sometimes be resolved, at least in part, by using a **Latin square design**; at the same time, we hope to impress upon the reader that it is through proper design that experiments can be made to yield a wealth of information.

EXAMPLE   A market research organization wants to compare four ways of packaging a breakfast food, but it is concerned about possible regional differences in the popularity of the breakfast food, and also about the effects of promoting the breakfast food in different ways. So, it decides to test market the different kinds of packaging in the northeastern, southeastern, northwestern, and southwestern parts of the United States and to promote them with discounts, lotteries, coupons, and two-for-one sales. Thus, there are $4 \cdot 4 = 16$ blocks (combinations of regions and methods of promotion) and it would take $16 \cdot 4 = 64$ market areas (cities) to promote each kind of packaging once within each block. It is of interest to note, however, that with proper planning 16 market areas (cities) will suffice. To illustrate, let us consider the following arrangement, called a **Latin square,** in which the letters A, B, C, and D represent the four kinds of packaging:

	Discounts	Lotteries	Coupons	Two-for-one sales
Northeast	A	B	C	D
Southeast	B	C	D	A
Northwest	C	D	A	B
Southwest	D	A	B	C

In general, a Latin square is a square array of the letters A, B, C, $D, \ldots$, of the English (Latin) alphabet, which is such that each letter occurs once and only once in each row and in each column.

The above Latin square, looked upon as an experimental design, suggests that discounts be used with packaging A in a city in the Northeast, with packaging B in a city in the Southeast, with packaging C in a city in the Northwest, and with packaging D in a city in the Southwest; that lotteries be used with packaging B in a city in the Northeast, with packaging C in a city in the Southeast, with packaging D in a city in the Northwest, and with packaging A in a city in the Southwest; and so on. Note that each kind of promotion is used once in each region and once with each kind of packaging; each kind of packaging is used once in each region and once with each kind of promotion; and each region is used once with each kind of packaging and once with each kind of promotion. As we shall see, this will enable us to perform an analysis of variance leading to significance tests for all three variables.

The analysis of an $r \times r$ Latin square is very similar to the two-way analysis of variance of Section 16.5: The total sum of squares and the sums of squares for rows and columns are calculated in the same way in which we previously calculated SST, $SS(Tr)$, and SSB, but we must find an extra sum of squares which measures the variability due to the variable represented by the letters A, B, C, $D, \ldots$, namely, a new treatment sum of squares. The formula for this sum of squares is

Treatment sum of squares for Latin square

$$SS(Tr) = \frac{1}{r} \cdot (T_A^2 + T_B^2 + T_C^2 + \cdots) - \frac{1}{r^2} \cdot T_{..}^2$$

16.7 Latin Squares

where T_A is the total of the observations corresponding to treatment A, T_B is the total of the observations corresponding to treatment B, and so forth. Finally, the error sum of squares is again obtained by subtraction:

Error sum of squares for Latin square

$$SSE = SST - SSR - SSC - SS(Tr)$$

where SSR and SSC are the sums of squares for rows and columns.

We can now construct the following analysis-of-variance table for the analysis of an $r \times r$ Latin square:

Source of variation	Degrees of freedom	Sum of squares	Mean square	F
Rows	$r - 1$	SSR	$MSR = \dfrac{SSR}{r - 1}$	$\dfrac{MSR}{MSE}$
Columns	$r - 1$	SSC	$MSC = \dfrac{SSC}{r - 1}$	$\dfrac{MSC}{MSE}$
Treatments	$r - 1$	$SS(Tr)$	$MS(Tr) = \dfrac{SS(Tr)}{r - 1}$	$\dfrac{MS(Tr)}{MSE}$
Error	$(r - 1)(r - 2)$	SSE	$MSE = \dfrac{SSE}{(r - 1)(r - 2)}$	
Total	$r^2 - 1$	SST		

The mean squares are again the sums of squares divided by their respective degrees of freedom, and the three F-values are the mean squares for rows, columns, and treatments divided by the mean square for error. The degrees of freedom for rows, columns, and treatments are all $r - 1$, and, by subtraction, the degrees of freedom for error is

$$(r^2 - 1) - (r - 1) - (r - 1) - (r - 1) = r^2 - 3r + 2$$
$$= (r - 1)(r - 2)$$

Thus, for each of the three significance tests the numerator and denominator degrees of freedom for F are $r - 1$ and $(r - 1)(r - 2)$.

EXAMPLE   To illustrate the analysis of a Latin square, suppose that in the breakfast-food example the market research organization actually gets the data shown in the following table, where the figures are one week's sales in $10 thousand:

	Discounts	Lotteries	Coupons	Two-for-one sales
Northeast	A 48	B 38	C 42	D 53
Southeast	B 39	C 43	D 50	A 54
Northwest	C 42	D 50	A 47	B 44
Southwest	D 46	A 48	B 46	C 52

Since the row totals are 181, 186, 183, and 192, the column totals are 175, 179, 185, and 203, the treatment totals are 197, 167, 179, and 199, the grand total is 742, $r = 4$, and $\sum\sum x^2 = 34{,}756$, we get

$$SST = 34{,}756 - \frac{1}{16}(742)^2 = 34{,}756 - 34{,}410.25$$

$$= 345.75$$

$$SSR = \frac{1}{4}[181^2 + 186^2 + 183^2 + 192^2] - 34{,}410.25$$

$$= 17.25$$

$$SSC = \frac{1}{4}[175^2 + 179^2 + 185^2 + 203^2] - 34{,}410.25$$

$$= 114.75$$

$$SS(Tr) = \frac{1}{4}[197^2 + 167^2 + 179^2 + 199^2] - 34{,}410.25$$

$$= 174.75$$

and, hence,

$$SSE = 345.75 - 17.25 - 114.75 - 174.75$$

$$= 39.00$$

Thus, the analysis-of-variance table becomes

Source of variation	Degrees of freedom	Sum of squares	Mean square	F
Rows	3	17.25	$\dfrac{17.25}{3} = 5.75$	$\dfrac{5.75}{6.5} = 0.9$
Columns	3	114.75	$\dfrac{114.75}{3} = 38.25$	$\dfrac{38.25}{6.5} = 5.9$
Treatments	3	174.75	$\dfrac{174.75}{3} = 58.25$	$\dfrac{58.25}{6.5} = 9.0$
Error	6	39.00	$\dfrac{39.00}{6} = 6.5$	
Total	15	345.75		

Since the values of F for treatments and columns, but not for rows, exceed 4.76, the value of $F_{0.05}$ for $4 - 1 = 3$ and $(4 - 1)(4 - 2) = 6$ degrees of freedom, we find that the differences due to packaging and those due to promotions are significant at $\alpha = 0.05$, while those due to regions are not.

The construction of Latin squares (that is, square arrays in which each of the letters appears once and only once in each row and in each column) is a problem of pure mathematics; so far as applied work in statistics is concerned, Latin square patterns may be looked up in tables (for instance, the one listed in the Bibliography on page 443).

16.8

The Design of Experiments: Some Further Considerations ★

There are many other experimental designs besides the ones we have discussed in this chapter, and they serve a great variety of special purposes. Widely used, for example, are the **incomplete block designs,** which apply when it is impossible to have each treatment in each block.

EXAMPLE The need for such designs arises, say, when we want to compare 13 kinds of tires but cannot put them all on a test car at the same time, or when a taste tester has to compare 10 kinds of wine, but cannot compare more than three of the wines at a time. With regard to the first example, only four of the 13 tires can be tested at the same time, and we might use the following experimental design, where the tires are numbered 1–13:

Number of test run	Kinds of tires			
1	1	2	4	10
2	2	3	5	11
3	3	4	6	12
4	4	5	7	13
5	5	6	8	1
6	6	7	9	2
7	7	8	10	3
8	8	9	11	4
9	9	10	12	5
10	10	11	13	6
11	11	12	1	7
12	12	13	2	8
13	13	1	3	9

There are 13 test runs, or blocks, and since each kind of tire appears together with each other kind of tire once within the same block, the design is referred to as a **balanced incomplete block design**. The fact that each kind of tire appears together with each other kind of tire once within the same block is important; it facilitates the statistical analysis, for it assures that the same amount of information is available for comparing each pair of tires.

In the wine-tasting problem we cannot construct a design which is such that each wine is compared directly (by the same taste tester) with each other wine once and only once during the course of the experiment. Each wine has to be compared directly with nine other wines, in each comparison a wine is compared directly with two others, and 9 is not divisible by 2. To get around this difficulty, we may compare each wine with each other wine twice, as in the following balanced incomplete block design, where the wines are numbered 1–10:

Comparison	Wines	Comparison	Wines
1	1 2 3	16	6 8 9
2	2 5 8	17	3 7 10
3	3 4 7	18	1 8 10
4	1 4 6	19	2 5 9
5	5 7 8	20	6 7 10
6	4 6 9	21	1 3 5
7	1 7 9	22	2 6 7
8	2 8 10	23	3 8 9
9	3 9 10	24	2 4 10
10	5 6 10	25	3 5 6
11	1 2 4	26	1 6 8
12	2 3 6	27	2 7 9
13	3 4 8	28	4 7 8
14	4 5 9	29	1 9 10
15	1 5 7	30	4 5 10

The analysis of incomplete block designs is fairly complicated, and we shall not go into it here, as it has been our purpose only to demonstrate what can be accomplished by the careful design of an experiment.

1 Suppose we want to compare the number of defective pieces produced by four machine operators working on five different machine parts (1, 2, 3, 4, and 5) in three different shifts (I, II, and III).

    (a) If the machine operators are to be regarded as the "treatments," list the blocks (combinations of machine parts and shifts) that would be required if each part is to be produced in each shift.

    (b) How many observations would be required if each machine operator is to work once on each machine part during each shift?

2 An agronomist wants to compare the yield of 16 varieties of corn, and at the same time study the effect of five different fertilizers and two methods of irrigation. How many test plots must he plant if each variety of corn is to be grown once with each possible combination of fertilizers and methods of irrigation?

3 A manufacturer of pharmaceuticals wants to market a new cold remedy which is actually a combination of four medications, and he wants to experiment first with two dosages for each medication. If A_L and A_H denote the low and high dosage of medication A, B_L and B_H the low and high dosage of medication B, C_L and C_H the low and high dosage of medication C, and D_L and D_H the low and high dosage of medication D, list the 16 preparations he has to test if each dosage of each medication is to be used once in combination with each dosage of each of the other medications.

4 Making use of the fact that each of the letters must occur once and only once in each row and each column, complete the following Latin squares:

(a)

	A	
B		

(b)

A	D		
	B		
		C	B
			D

5 The sample data in the following 3×3 Latin square are the grades in an American history test obtained by nine college students of various ethnic backgrounds and of various professional interests, who were taught by instructors A, B, and C:

Ethnic backgrounds

	Mexican	German	Polish
Law	A 75	B 86	C 69
Medicine	B 95	C 79	A 86
Engineering	C 70	A 83	B 93

16.8 The Design of Experiments: Some Further Considerations

Use the level of significance $\alpha = 0.05$ to test the null hypothesis that

    (a) having a different instructor has no effect on the grades;

    (b) differences in ethnic background have no effect on the grades;

    (c) differences in professional interest have no effect on the grades.

In practice, we seldom use 3×3 Latin squares without replication, since there are only $(3 - 1)(3 - 2) = 2$ degrees of freedom for the experimental error, and this means that we have very little information about the possible size of chance variations.

6   To compare four different golf-ball designs, A, B, C, and D, each kind was driven by each of four golf pros, P_1, P_2, P_3, and P_4, using once each the four different drivers, D_1, D_2, D_3, and D_4. The distances from the tee to the points where the balls came to rest (in yards) were as shown in the following table:

	D_1	D_2	D_3	D_4
P_1	D 231	B 215	A 261	C 199
P_2	C 234	A 300	B 280	D 266
P_3	A 301	C 208	D 247	B 255
P_4	B 253	D 258	C 210	A 290

Analyze this Latin square, using the level of significance $\alpha = 0.05$ for the various tests of significance.

7   To test their ability to make decisions under pressure, the nine senior executives of a company are to be interviewed by each of four psychologists. As it takes a psychologist a full day to interview three of the executives, the schedule for the interviews is arranged as follows, where the nine executives are denoted A, B, C, D, E, F, G, H, and I:

Day	Psychologist	Executives		
March 2	I	B	C	?
March 3	I	E	F	G
March 4	I	H	I	A
March 5	II	C	?	H
March 6	II	B	F	A
March 9	II	D	E	?
March 10	III	D	G	A
March 11	III	C	F	?
March 12	III	B	E	H
March 13	IV	B	?	I
March 16	IV	C	?	A
March 17	IV	D	F	H

Replace the six question marks with the appropriate letters, given that each of the nine executives is to be interviewed together with each of the other executives once and only once on the same day. Note that this will make the arrangement a balanced incomplete block design, which may be important because each executive is tested together with each other executive once under identical conditions.

8　A newspaper regularly prints the columns of seven writers but has room for only three in each edition. Complete the following schedule, in which the writers are numbered 1–7, so that each writer's column appears three times per week, and a column of each writer appears together with a column of each other writer once per week:

	Writers		
Monday	1	2	3
Tuesday	4		
Wednesday	1	4	5
Thursday	2		
Friday	1	6	7
Saturday	5		
Sunday	2	4	6

BIBLIOGRAPHY　The following are some of the many books that have been written on the subject of analysis of variance:

GUENTHER, W. C., *Analysis of Variance*. Englewood Cliffs, N.J.: Prentice-Hall, Inc., 1964.

LI, C. C., *Introduction to Experimental Statistics*. New York: McGraw-Hill Book Co., 1964.

SNEDECOR, G. W., and COCHRAN, W. G., *Statistical Methods, 6th ed*. Ames, Iowa: Iowa University Press, 1973.

Problems relating to the design of experiments are also treated in the above books and in

ANDERSON, V. L., and McLEAN, R. A., *Design of Experiments: A Realistic Approach*. New York: Marcel Dekker, Inc., 1974.

COCHRAN, W. G., and COX, G. M., *Experimental Design, 2nd ed*. New York: John Wiley & Sons, Inc., 1957.

FINNEY, D. J., *An Introduction to the Theory of Experimental Design*. Chicago: The University of Chicago Press, 1960.

HICKS, C. R., *Fundamental Concepts in the Design of Experiments*. New York: Holt, Rinehart and Winston, Inc., 1964.

ROMANO, A., *Applied Statistics for Science and Industry*. Boston: Allyn and Bacon, Inc., 1977.

A table of Latin squares of size $r = 3, 4, 5, \ldots,$ and 12 may be found in the above-mentioned book by W. G. Cochran and G. M. Cox.

Further Exercises for Chapters

14, 15, and 16

1 The following are the weight losses of certain machine parts (in milligrams) due to friction, when three different lubricants were used under controlled conditions:

Lubricant A: 13, 20, 14, 12, 18, 14, 23, 20
Lubricant B: 11, 6, 14, 10, 16, 11, 14, 11
Lubricant C: 12, 12, 13, 11, 4, 5, 10, 8

Test at the level of significance $\alpha = 0.01$ whether the differences among the means of these three samples can be attributed to chance.

2 Suppose that we are given the following sample data to study the relationship between the grades students get in a certain examination, their IQ's, and the number of hours they studied for the test:

Number of hours studied x_1	IQ x_2	Grade in examination y
9	116	70
3	101	38
16	102	90
19	99	99
6	118	59
11	111	75
14	96	77
12	120	84
6	101	49
9	100	61

Fit an equation of the form $y' = b_0 + b_1 x_1 + b_2 x_2$ to the given data, and use it to predict the grade of a student who has an IQ of 107 and studies 13 hours for the examination.

3 With reference to the preceding exercise, calculate the correlation coefficient for the IQ's and the grades, and also the partial correlation coefficient for the IQ's and the grades when the number of hours studied is held fixed.

4 Verify the figures in the analysis-of-variance table on page 426.

5 To facilitate significance tests for r, some books give a table of values of $r_{\alpha/2}$, for which the probability is $\alpha/2$ that they will be exceeded by a sample value of r when the null hypothesis of no correlation is true. Thus, we reject the null hypothesis of no correlation at the level of significance α if the value of r obtained for a set of paired data exceeds $r_{\alpha/2}$ or is less than $-r_{\alpha/2}$. To find values of $r_{\alpha/2}$, which depend on n (the number of paired observations), we take the formula for t on page 398 with $t = t_{\alpha/2}$, namely,

$$t_{\alpha/2} = \frac{r\sqrt{n-2}}{\sqrt{1-r^2}}$$

and solve for r. The number of degrees of freedom for $t_{\alpha/2}$ is $n - 2$.

(a) Verify that the resulting formula for $r_{\alpha/2}$ is

$$r_{\alpha/2} = \frac{t_{\alpha/2}}{\sqrt{n - 2 + t_{\alpha/2}^2}}$$

(b) Find $r_{0.025}$ and $r_{0.005}$ for $n = 10$.
(c) Find $r_{0.025}$ and $r_{0.005}$ for $n = 20$.
(d) Find $r_{0.025}$ and $r_{0.005}$ for $n = 120$, given that $t_{0.025} = 1.980$ and $t_{0.005} = 2.617$ for 118 degrees of freedom.

6 A *bon vivant*, wishing to ascertain the cause of his frequent hangovers, conducted the following experiment: On the first night, he drank nothing but whiskey and water; on the second night he drank vodka and water; on the third night he drank gin and water; and on the fourth night he drank rum and water. On each of the following mornings he had a hangover, and he concludes that it was the common factor, the water, that made him ill.

(a) Explain why this conclusion is unwarranted.
(b) Give a less obvious example of an experiment having the same shortcoming.

7 With reference to Exercise 2 on page 369, find
(a) a 0.95 confidence interval for the mean chlorine residual after 5 hours;
(b) 0.95 limits of prediction for the chlorine residual after 5 hours.

8 Correlation methods are sometimes used to study the internal structure (or systematic patterns) in series of data (time series) which are recorded on an annual, monthly, weekly, . . . , basis. Consider, for example, the diagram of Figure E.1, which shows a line chart of a company's annual sales for the years 1958–1977. There is an obvious linear trend in the series as shown by the least-squares line, and to look for further patterns, we can study the deviations from the line,

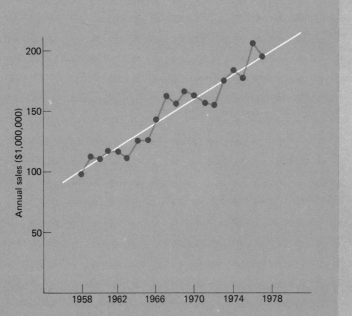

FIGURE E.1

Diagram for Exercise 8.

$y - y'$, which are -2, 6, 0, 3, -2, -13, -5, -10, 1, 18, 6, 10, 1, -8, -15, 0, 3, -7, 15, and 0 million dollars for the years 1958–1977. Letting y_t denote the deviation from the line in the year t, calculate

    (a) the correlation coefficient for y_t and y_{t-1};

    (b) the correlation coefficient for y_t and y_{t-2};

    (c) the correlation coefficient for y_t and y_{t-3};

    (d) the correlation coefficient for y_t and y_{t-4};

    (e) the correlation coefficient for y_t and y_{t-5}.

For instance, in part (a) calculate r after pairing the 1959 deviation from the line with the 1958 deviation from the line, the 1960 deviation from the line with the 1959 deviation from the line, and so on, ..., and in part (e) calculate r after pairing the 1963 deviation from the line with the 1958 deviation from the line, the 1964 deviation from the line with the 1959 deviation from the line, and, finally, the 1977 deviation from the line with the 1972 deviation from the line. These lag-time correlations are also called **autocorrelations.** To continue,

    (f) test at the level of significance $\alpha = 0\ 0.05$ whether the correlation coefficients obtained in parts (a) through (e) are significant;

    (g) discuss the possible existence of a cyclical (or repeating) pattern in the series of data.

9 In the years 1969–1973 the total value of New Zealand's exports was 1,194, 1,250, 1,380, 1,636, and 2,094 million SDR's (Special Drawing Rights, or "paper gold" issued by the International Monetary Fund for the use by its members). Use the coding of Exercise 7 on page 382 to fit a parabola to these data.

10 If $r = 0.25$ for the ages of a group of students and their knowledge of foreign affairs, what percentage of the variation of their knowledge of foreign affairs can be attributed to differences in age?

11 Suppose that a clothing manufacturer wants to compare three different kinds of sewing machines, A, B, and C, and that she wants to know, in particular, how they perform with three different kinds of needles, a, b, and c, and three kinds of threads I, II, and III.

    (a) Construct a 3×3 Latin square.

    (b) Use the Latin square of part (a) to design an experiment in which three of the nine possible combinations of needles and threads are assigned to each sewing machine, so that each sewing machine will be used once with each kind of needle and once with each kind of thread.

    (c) Suppose that the clothing manufacturer wants to use in the experiment three sewing machine operators, P, Q, and R. With reference to the Latin square design of part (b), indicate how she might assign three of the nine combination of needles and threads to each of these operators so that each operator works once with each machine, once with each kind of needle, and once with each kind of thread. (There exists a systematic way of planning this kind of experiment, but use trial and error here to make the assignment.)

12 For each of the following, construct a confidence interval for β at the indicated degree of confidence:

    (a) Exercise 2 on page 369, degree of confidence 0.99;

    (b) Exercise 4 on page 369, degree of confidence 0.95.

13 The figures in the following 5×5 Latin square are the numbers of minutes engines E_1, E_2, E_3, E_4, and E_5, tuned up by mechanics M_1, M_2, M_3, M_4, and M_5, ran with a gallon of fuel A, B, C, D, or E:

	E_1	E_2	E_3	E_4	E_5
M_1	A 31	B 24	C 20	D 20	E 18
M_2	B 21	C 27	D 23	E 25	A 31
M_3	C 21	D 27	E 25	A 29	B 21
M_4	D 21	E 25	A 33	B 25	C 22
M_5	E 21	A 37	B 24	C 24	D 20

Use the level of significance $\alpha = 0\ 0.01$ to test

(a) the null hypothesis that there is no difference in the performance of the five engines;

(b) the null hypothesis that the persons who tuned up these engines have no effect on their performance;

(c) the null hypothesis that the engines perform equally well with each of the fuels.

14 The following are the rankings which two sports writers gave to twelve tennis players:

	Writer I	Writer II
Player A	2	1
Player B	5	8
Player C	10	11
Player D	1	4
Player E	7	6
Player F	6	5
Player G	12	12
Player H	3	2
Player I	11	9
Player J	4	3
Player K	8	10
Player L	9	7

Calculate r' as a measure of the consistency of the two rankings.

15 The following data pertain to the cosmic ray doses measured at various altitudes:

Altitude (100 feet) x	Dose rate (mrem/year) y
0.5	28
4.5	30
7.8	32
12.0	36
48.0	58
53.0	69

Fit an exponential curve and use it to estimate the cosmic radiation at 6,000 feet.

16 Among the nine persons interviewed in a poll, three are Easterners, three are Southerners, and three are Westerners. By profession, three of them are teachers, three are lawyers, and three are doctors, and no two of the same profession come from the same part of the United States. Also, three are Democrats, three are Republicans, and three are Independents, and no two of the same political affiliation are of the same profession or come from the same part of the United States. If one of the teachers is an Easterner and an Independent, another teacher

is a Southerner and a Republican, and one of the lawyers is a Southerner and a Democrat, what is the political affiliation of the doctor who is a Westerner? (*Hint:* Construct a 3 × 3 Latin square; this exercise is a simplified version of a famous problem posed by R. A. Fisher in his classical work *The Design of Experiments.*)

17 The following data pertain to a study of the effects of environmental pollution on wildlife; in particular, the effect of DDT on the thickness of the eggshells of certain birds:

DDT residue in yolk lipids (*parts per million*)	*Thickness of eggshells* (*millimeters*)
117	0.49
65	0.52
303	0.37
98	0.53
122	0.49
150	0.42

Calculate r for these data and test its significance at the level of significance $\alpha = 0.05$.

18 With reference to Exercise 10 on page 344, perform a one-way analysis of variance at the level of significance $\alpha = 0.05$ to test the null hypothesis that the differences in dosage have no effect.

19 It can be shown that the t test of Section 10.12 is equivalent to a one-way analysis of variance performed on the data; in fact, it can be shown that the relationship between the respective test statistics is $F = t^2$.

   (a) Verify that $F_{0.05}$ for 1 and 10 degrees of freedom equals $t_{0.025}^2$ for 10 degrees of freedom.

   (b) Verify that $F_{0.01}$ for 1 and 15 degrees of freedom equals $t_{0.005}^2$ for 15 degrees of freedom.

   (c) Rework the example on page 278, the one dealing with the heat-producing capacity of coal from two mines, by performing a one-way analysis of variance.

20 Use the coding of Exercise 3 on page 382 to rework Exercise 2 on page 381.

21 Suppose that in connection with Exercise 1 on page 368 we want to predict, or estimate, how many weeks are required for a gain of 100 words per minute. We could substitute $y = 100$ into the least-squares equation obtained in part (a) of that exercise, but this would not yield a prediction in the least-squares sense. To make a prediction of x in terms of y as good as possible, we must minimize the sum of the squares of the errors in the x-direction (namely, the sum of the squares of the horizontal deviations from the points to the line.) If we write the equation of the line as $x = a + by$, the values of a and b may be obtained with the two formulas on page 367, with x substituted throughout for y, and vice versa. Find this kind of regression line for the data of Exercise 1 on page 368, and use it to predict the value of x when $y = 100$.

22 A school has seven department heads who are assigned to seven different committees as shown in the following table:

	Department heads
Textbooks	Dodge, Fleming, Griffith, Anderson
Athletics	Bowman, Evans, Griffith, Anderson
Band	Bowman, Carlson, Fleming, Anderson
Dramatics	Bowman, Carlson, Dodge, Griffith
Tenure	Carlson, Evans, Fleming, Griffith
Salaries	Bowman, Dodge, Evans, Fleming
Discipline	Carlson, Dodge, Evans, Anderson

(a) Verify that this arrangement is a balanced incomplete block design.
(b) If Dodge, Bowman, and Carlson are (in that order) appointed chairperson of the first three committees, how will the chairpersons of the other four committees have to be chosen so that each of the department heads is chairperson of one of the committees?

23 The following are the numbers of defectives produced by four workmen operating, in turn, three different machines:

	Machine 1	Machine 2	Machine 3
Workman 1	27	29	21
Workman 2	31	30	27
Workman 3	24	26	22
Workman 4	21	25	20

Perform a two-way analysis of variance and test, at the level of significance $\alpha = 0\ 0.05$, whether

(a) the differences among the means obtained for the four workmen are significant;
(b) the differences among the means obtained for the three machines are significant.

24 In fitting curves to business data, we sometimes describe the over-all pattern (trend, cycle, etc.) by means of a **moving average** rather than a specific kind of curve. A moving average is obtained by replacing each value in a series of equally spaced data (except for the first few and the last few) by the mean of itself and some of the values directly preceding it and directly following it. For instance, in a **three-year moving average** each annual figure is replaced by the mean of itself and the annual figures corresponding to the two adjacent years; in a **five-year moving** average each annual figure is replaced by the mean of itself and the annual figures corresponding to the two years which precede it and the two years which follow it.

(a) The following figures are the factory sales of motor vehicles (in millions) in the United States for the years 1945–1974: 0.7, 3.1, 4.8, 5.3, 6.3, 8.0,

6.8, 5.5, 7.3, 6.6, 9.2, 6.9, 7.2, 5.1, 6.7, 7.9, 6.7, 8.2, 9.1, 9.3, 11.1, 10.3, 9.0, 10.7, 10.1, 8.2, 10.6, 11.3, 12.6, and 10.0. Construct a three-year moving average and plot it on a chart showing also the original data. Connect successive points of the moving average.

(b) The acquisition of real estate investments by U.S. life insurance companies for the years 1947–1973 were 2.2, 2.8, 2.6, 2.6, 2.7, 3.4, 2.1, 3.9, 3.7, 3.6, 4.6, 4.6, 4.5, 3.0, 4.3, 3.6, 4.9, 4.9, 4.5, 5.1, 6.7, 8.2, 7.9, 8.6, 10.4, 9.8, and 11.8. Construct a five-year moving average and plot it on a chart showing also the original data. Connect successive points of the moving average.

25 Assuming that the conditions underlying normal correlation analysis are met, use the Fisher Z-transformation to construct

(a) an approximate 0.95 confidence interval for ρ when $r = 0.44$ and $n = 20$;

(b) an approximate 0.99 confidence interval for ρ when $r = -0.32$ and $n = 38$.

Statistical Tables

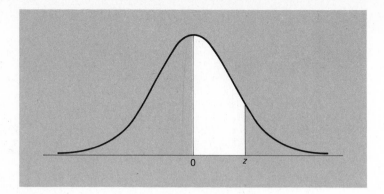

The entries in Table I are the probabilities that a random variable having the standard normal distribution takes on a value between 0 and z; they are given by the area of the white region under the curve in the figure shown above.

TABLE I   Normal-Curve Areas

z	.00	.01	.02	.03	.04	.05	.06	.07	.08	.09
0.0	.0000	.0040	.0080	.0120	.0160	.0199	.0239	.0279	.0319	.0359
0.1	.0398	.0438	.0478	.0517	.0557	.0596	.0636	.0675	.0714	.0753
0.2	.0793	.0832	.0871	.0910	.0948	.0987	.1026	.1064	.1103	.1141
0.3	.1179	.1217	.1255	.1293	.1331	.1368	.1406	.1443	.1480	.1517
0.4	.1554	.1591	.1628	.1664	.1700	.1736	.1772	.1808	.1844	.1879
0.5	.1915	.1950	.1985	.2019	.2054	.2088	.2123	.2157	.2190	.2224
0.6	.2257	.2291	.2324	.2357	.2389	.2422	.2454	.2486	.2517	.2549
0.7	.2580	.2611	.2642	.2673	.2704	.2734	.2764	.2794	.2823	.2852
0.8	.2881	.2910	.2939	.2967	.2995	.3023	.3051	.3078	.3106	.3133
0.9	.3159	.3186	.3212	.3238	.3264	.3289	.3315	.3340	.3365	.3389
1.0	.3413	.3438	.3461	.3485	.3508	.3531	.3554	.3577	.3599	.3621
1.1	.3643	.3665	.3686	.3708	.3729	.3749	.3770	.3790	.3810	.3830
1.2	.3849	.3869	.3888	.3907	.3925	.3944	.3962	.3980	.3997	.4015
1.3	.4032	.4049	.4066	.4082	.4099	.4115	.4131	.4147	.4162	.4177
1.4	.4192	.4207	.4222	.4236	.4251	.4265	.4279	.4292	.4306	.4319
1.5	.4332	.4345	.4357	.4370	.4382	.4394	.4406	.4418	.4429	.4441
1.6	.4452	.4463	.4474	.4484	.4495	.4505	.4515	.4525	.4535	.4545
1.7	.4554	.4564	.4573	.4582	.4591	.4599	.4608	.4616	.4625	.4633
1.8	.4641	.4649	.4656	.4664	.4671	.4678	.4686	.4693	.4699	.4706
1.9	.4713	.4719	.4726	.4732	.4738	.4744	.4750	.4756	.4761	.4767
2.0	.4772	.4778	.4783	.4788	.4793	.4798	.4803	.4808	.4812	.4817
2.1	.4821	.4826	.4830	.4834	.4838	.4842	.4846	.4850	.4854	.4857
2.2	.4861	.4864	.4868	.4871	.4875	.4878	.4881	.4884	.4887	.4890
2.3	.4893	.4896	.4898	.4901	.4904	.4906	.4909	.4911	.4913	.4916
2.4	.4918	.4920	.4922	.4925	.4927	.4929	.4931	.4932	.4934	.4936
2.5	.4938	.4940	.4941	.4943	.4945	.4946	.4948	.4949	.4951	.4952
2.6	.4953	.4955	.4956	.4957	.4959	.4960	.4961	.4962	.4963	.4964
2.7	.4965	.4966	.4967	.4968	.4969	.4970	.4971	.4972	.4973	.4974
2.8	.4974	.4975	.4976	.4977	.4977	.4978	.4979	.4979	.4980	.4981
2.9	.4981	.4982	.4982	.4983	.4984	.4984	.4985	.4985	.4986	.4986
3.0	.4987	.4987	.4987	.4988	.4988	.4989	.4989	.4989	.4990	.4990

Also, for $z = 4.0$, 5.0, and 6.0, the areas are 0.49997, 0.4999997, and 0.499999999.

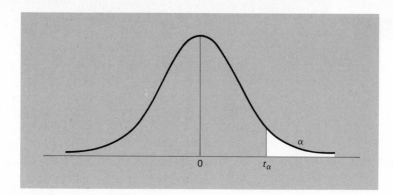

The entries in Table II are values for which the area to their right under the t distribution with given degrees of freedom (the white area in the figure shown above) is equal to α.

TABLE II Values of t[†]

d.f.	$t_{.100}$	$t_{.050}$	$t_{.025}$	$t_{.010}$	$t_{.005}$	d.f.
1	3.078	6.314	12.706	31.821	63.657	1
2	1.886	2.920	4.303	6.965	9.925	2
3	1.638	2.353	3.182	4.541	5.841	3
4	1.533	2.132	2.776	3.747	4.604	4
5	1.476	2.015	2.571	3.365	4.032	5
6	1.440	1.943	2.447	3.143	3.707	6
7	1.415	1.895	2.365	2.998	3.499	7
8	1.397	1.860	2.306	2.896	3.355	8
9	1.383	1.833	2.262	2.821	3.250	9
10	1.372	1.812	2.228	2.764	3.169	10
11	1.363	1.796	2.201	2.718	3.106	11
12	1.356	1.782	2.179	2.681	3.055	12
13	1.350	1.771	2.160	2.650	3.012	13
14	1.345	1.761	2.145	2.624	2.977	14
15	1.341	1.753	2.131	2.602	2.947	15
16	1.337	1.746	2.120	2.583	2.921	16
17	1.333	1.740	2.110	2.567	2.898	17
18	1.330	1.734	2.101	2.552	2.878	18
19	1.328	1.729	2.093	2.539	2.861	19
20	1.325	1.725	2.086	2.528	2.845	20
21	1.323	1.721	2.080	2.518	2.831	21
22	1.321	1.717	2.074	2.508	2.819	22
23	1.319	1.714	2.069	2.500	2.807	23
24	1.318	1.711	2.064	2.492	2.797	24
25	1.316	1.708	2.060	2.485	2.787	25
26	1.315	1.706	2.056	2.479	2.779	26
27	1.314	1.703	2.052	2.473	2.771	27
28	1.313	1.701	2.048	2.467	2.763	28
29	1.311	1.699	2.045	2.462	2.756	29
inf.	1.282	1.645	1.960	2.326	2.576	inf.

[†] Based on Table 12 of *Biometrika Tables for Statisticians, Volume I* (Cambridge: Cambridge University Press, 1954), by permission of the *Biometrika* trustees.

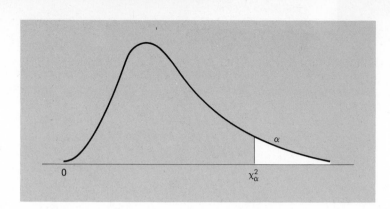

The entries in Table III are values for which the area to their right under the chi-square distribution with given degrees of freedom (the white area in the figure shown above) is equal to α.

TABLE III Values of χ^2†

d.f.	$\chi^2_{.995}$	$\chi^2_{.99}$	$\chi^2_{.975}$	$\chi^2_{.95}$	$\chi^2_{.05}$	$\chi^2_{.025}$	$\chi^2_{.01}$	$\chi^2_{.005}$	d.f.
1	.0000393	.000157	.000982	.00393	3.841	5.024	6.635	7.879	1
2	.0100	.0201	.0506	.103	5.991	7.378	9.210	10.597	2
3	.0717	.115	.216	.352	7.815	9.348	11.345	12.838	3
4	.207	.297	.484	.711	9.488	11.143	13.277	14.860	4
5	.412	.554	.831	1.145	11.070	12.832	15.086	16.750	5
6	.676	.872	1.237	1.635	12.592	14.449	16.812	18.548	6
7	.989	1.239	1.690	2.167	14.067	16.013	18.475	20.278	7
8	1.344	1.646	2.180	2.733	15.507	17.535	20.090	21.955	8
9	1.735	2.088	2.700	3.325	13.919	19.023	21.666	23.589	9
10	2.156	2.558	3.247	3.940	18.307	20.483	23.209	25.188	10
11	2.603	3.053	3.816	4.575	19.675	21.920	24.725	26.757	11
12	3.074	3.571	4.404	5.226	21.026	23.337	26.217	28.300	12
13	3.565	4.107	5.009	5.892	22.362	24.736	27.688	29.819	13
14	4.075	4.660	5.629	6.571	23.685	26.119	29.141	31.319	14
15	4.601	5.229	6.262	7.261	24.996	27.488	30.578	32.801	15
16	5.142	5.812	6.908	7.962	26.296	28.845	32.000	34.267	16
17	5.697	6.408	7.564	8.672	27.587	30.191	33.409	35.718	17
18	6.265	7.015	8.231	9.390	28.869	31.526	34.805	37.156	18
19	6.844	7.633	8.907	10.117	30.144	32.852	36.191	38.582	19
20	7.434	8.260	9.591	10.851	31.410	34.170	37.566	39.997	20
21	8.034	8.897	10.283	11.591	32.671	35.479	38.932	41.401	21
22	8.643	9.542	10.982	12.338	33.924	36.781	40.289	42.796	22
23	9.260	10.196	11.689	13.091	35.172	38.076	41.638	44.181	23
24	9.886	10.856	12.401	13.848	36.415	39.364	42.980	45.558	24
25	10.520	11.524	13.120	14.611	37.652	40.646	44.314	46.928	25
26	11.160	12.198	13.844	15.379	38.885	41.923	45.642	48.290	26
27	11.808	12.879	14.573	16.151	40.113	43.194	46.963	49.645	27
28	12.461	13.565	15.308	16.928	41.337	44.461	48.278	50.993	28
29	13.121	14.256	16.047	17.708	42.557	45.722	49.588	52.336	29
30	13.787	14.953	16.791	18.493	43.773	46.979	50.892	53.672	30

† Based on Table 8 of *Biometrika Tables for Statisticians, Volume I* (Cambridge University Press, 1954), by permission of the *Biometrika* trustees.

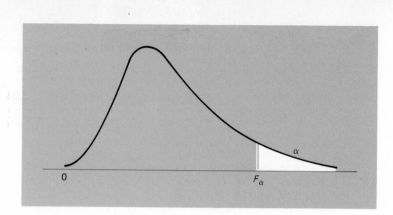

The entries in Tables IV(a) and IV(b) are values for which the area to their right under the F distribution with given degrees of freedom (the white area in the figure shown above) is equal to α.

TABLE IV(a) Values of $F_{0.05}$ [†]

Degrees of freedom for numerator

$v_2 \backslash v_1$	1	2	3	4	5	6	7	8	9	10	12	15	20	24	30	40	60	120	∞
1	161	200	216	225	230	234	237	239	241	242	244	246	248	249	250	251	252	253	254
2	18.5	19.0	19.2	19.2	19.3	19.3	19.4	19.4	19.4	19.4	19.4	19.4	19.4	19.5	19.5	19.5	19.5	19.5	19.5
3	10.1	9.55	9.28	9.12	9.01	8.94	8.89	8.85	8.81	8.79	8.74	8.70	8.66	8.64	8.62	8.59	8.57	8.55	8.53
4	7.71	6.94	6.59	6.39	6.26	6.16	6.09	6.04	6.00	5.96	5.91	5.86	5.80	5.77	5.75	5.72	5.69	5.66	5.63
5	6.61	5.79	5.41	5.19	5.05	4.95	4.88	4.82	4.77	4.74	4.68	4.62	4.56	4.53	4.50	4.46	4.43	4.40	4.37
6	5.99	5.14	4.76	4.53	4.39	4.28	4.21	4.15	4.10	4.06	4.00	3.94	3.87	3.84	3.81	3.77	3.74	3.70	3.67
7	5.59	4.74	4.35	4.12	3.97	3.87	3.79	3.73	3.68	3.64	3.57	3.51	3.44	3.41	3.38	3.34	3.30	3.27	3.23
8	5.32	4.46	4.07	3.84	3.69	3.58	3.50	3.44	3.39	3.35	3.28	3.22	3.15	3.12	3.08	3.04	3.01	2.97	2.93
9	5.12	4.26	3.86	3.63	3.48	3.37	3.29	3.23	3.18	3.14	3.07	3.01	2.94	2.90	2.86	2.83	2.79	2.75	2.71
10	4.96	4.10	3.71	3.48	3.33	3.22	3.14	3.07	3.02	2.98	2.91	2.85	2.77	2.74	2.70	2.66	2.62	2.58	2.54
11	4.84	3.98	3.59	3.36	3.20	3.09	3.01	2.95	2.90	2.85	2.79	2.72	2.65	2.61	2.57	2.53	2.49	2.45	2.40
12	4.75	3.89	3.49	3.26	3.11	3.00	2.91	2.85	2.80	2.75	2.69	2.62	2.54	2.51	2.47	2.43	2.38	2.34	2.30
13	4.67	3.81	3.41	3.18	3.03	2.92	2.83	2.77	2.71	2.67	2.60	2.53	2.46	2.42	2.38	2.34	2.30	2.25	2.21
14	4.60	3.74	3.34	3.11	2.96	2.85	2.76	2.70	2.65	2.60	2.53	2.46	2.39	2.35	2.31	2.27	2.22	2.18	2.13
15	4.54	3.68	3.29	3.06	2.90	2.79	2.71	2.64	2.59	2.54	2.48	2.40	2.33	2.29	2.25	2.20	2.16	2.11	2.07
16	4.49	3.63	3.24	3.01	2.85	2.74	2.66	2.59	2.54	2.49	2.42	2.35	2.28	2.24	2.19	2.15	2.11	2.06	2.01
17	4.45	3.59	3.20	2.96	2.81	2.70	2.61	2.55	2.49	2.45	2.38	2.31	2.23	2.19	2.15	2.10	2.06	2.01	1.96
18	4.41	3.55	3.16	2.93	2.77	2.66	2.58	2.51	2.46	2.41	2.34	2.27	2.19	2.15	2.11	2.06	2.02	1.97	1.92
19	4.38	3.52	3.13	2.90	2.74	2.63	2.54	2.48	2.42	2.38	2.31	2.23	2.16	2.11	2.07	2.03	1.98	1.93	1.88
20	4.35	3.49	3.10	2.87	2.71	2.60	2.51	2.45	2.39	2.35	2.28	2.20	2.12	2.08	2.04	1.99	1.95	1.90	1.84
21	4.32	3.47	3.07	2.84	2.68	2.57	2.49	2.42	2.37	2.32	2.25	2.18	2.10	2.05	2.01	1.96	1.92	1.87	1.81
22	4.30	3.44	3.05	2.82	2.66	2.55	2.46	2.40	2.34	2.30	2.23	2.15	2.07	2.03	1.98	1.94	1.89	1.84	1.78
23	4.28	3.42	3.03	2.80	2.64	2.53	2.44	2.37	2.32	2.27	2.20	2.13	2.05	2.01	1.96	1.91	1.86	1.81	1.76
24	4.26	3.40	3.01	2.78	2.62	2.51	2.42	2.36	2.30	2.25	2.18	2.11	2.03	1.98	1.94	1.89	1.84	1.79	1.73
25	4.24	3.39	2.99	2.76	2.60	2.49	2.40	2.34	2.28	2.24	2.16	2.09	2.01	1.96	1.92	1.87	1.82	1.77	1.71
30	4.17	3.32	2.92	2.69	2.53	2.42	2.33	2.27	2.21	2.16	2.09	2.01	1.93	1.89	1.84	1.79	1.74	1.68	1.62
40	4.08	3.23	2.84	2.61	2.45	2.34	2.25	2.18	2.12	2.08	2.00	1.92	1.84	1.79	1.74	1.69	1.64	1.58	1.51
60	4.00	3.15	2.76	2.53	2.37	2.25	2.17	2.10	2.04	1.99	1.92	1.84	1.75	1.70	1.65	1.59	1.53	1.47	1.39
120	3.92	3.07	2.68	2.45	2.29	2.18	2.09	2.02	1.96	1.91	1.83	1.75	1.66	1.61	1.55	1.50	1.43	1.35	1.25
∞	3.84	3.00	2.60	2.37	2.21	2.10	2.01	1.94	1.88	1.83	1.75	1.67	1.57	1.52	1.46	1.39	1.32	1.22	1.00

Degrees of freedom for denominator

[†] Reproduced from M. Merrington and C. M. Thompson. "Tables of percentage points of the inverted beta (F) distribution." *Biometrika*, vol. 33 (1943), by permission of the *Biometrika* trustees.

TABLE IV(b) Values of $F_{0.01}$[†]

Degrees of freedom for numerator

v_2	1	2	3	4	5	6	7	8	9	10	12	15	20	24	30	40	60	120	∞
1	4,052	5,000	5,403	5,625	5,764	5,859	5,928	5,982	6,023	6,056	6,106	6,157	6,209	6,235	6,261	6,287	6,313	6,339	6,366
2	98.5	99.0	99.2	99.2	99.3	99.3	99.4	99.4	99.4	99.4	99.4	99.4	99.4	99.5	99.5	99.5	99.5	99.5	99.5
3	34.1	30.8	29.5	28.7	28.2	27.9	27.7	27.5	27.3	27.2	27.1	26.9	26.7	26.6	26.5	26.4	26.3	26.2	26.1
4	21.2	18.0	16.7	16.0	15.5	15.2	15.0	14.8	14.7	14.5	14.4	14.2	14.0	13.9	13.8	13.7	13.7	13.6	13.5
5	16.3	13.3	12.1	11.4	11.0	10.7	10.5	10.3	10.2	10.1	9.89	9.72	9.55	9.47	9.38	9.29	9.20	9.11	9.02
6	13.7	10.9	9.78	9.15	8.75	8.47	8.26	8.10	7.98	7.87	7.72	7.56	7.40	7.31	7.23	7.14	7.06	6.97	6.88
7	12.2	9.55	8.45	7.85	7.46	7.19	6.99	6.84	6.72	6.62	6.47	6.31	6.16	6.07	5.99	5.91	5.82	5.74	5.65
8	11.3	8.65	7.59	7.01	6.63	6.37	6.18	6.03	5.91	5.81	5.67	5.52	5.36	5.28	5.20	5.12	5.03	4.95	4.86
9	10.6	8.02	6.99	6.42	6.06	5.80	5.61	5.47	5.35	5.26	5.11	4.96	4.81	4.73	4.65	4.57	4.48	4.40	4.31
10	10.0	7.56	6.55	5.99	5.64	5.39	5.20	5.06	4.94	4.85	4.71	4.56	4.41	4.33	4.25	4.17	4.08	4.00	3.91
11	9.65	7.21	6.22	5.67	5.32	5.07	4.89	4.74	4.63	4.54	4.40	4.25	4.10	4.02	3.94	3.86	3.78	3.69	3.60
12	9.33	6.93	5.95	5.41	5.06	4.82	4.64	4.50	4.39	4.30	4.16	4.01	3.86	3.78	3.70	3.62	3.54	3.45	3.36
13	9.07	6.70	5.74	5.21	4.86	4.62	4.44	4.30	4.19	4.10	3.96	3.82	3.66	3.59	3.51	3.43	3.34	3.25	3.17
14	8.86	6.51	5.56	5.04	4.70	4.46	4.28	4.14	4.03	3.94	3.80	3.66	3.51	3.43	3.35	3.27	3.18	3.09	3.00
15	8.68	6.36	5.42	4.89	4.56	4.32	4.14	4.00	3.89	3.80	3.67	3.52	3.37	3.29	3.21	3.13	3.05	2.96	2.87
16	8.53	6.23	5.29	4.77	4.44	4.20	4.03	3.89	3.78	3.69	3.55	3.41	3.26	3.18	3.10	3.02	2.93	2.84	2.75
17	8.40	6.11	5.19	4.67	4.34	4.10	3.93	3.79	3.68	3.59	3.46	3.31	3.16	3.08	3.00	2.92	2.83	2.75	2.65
18	8.29	6.01	5.09	4.58	4.25	4.01	3.84	3.71	3.60	3.51	3.37	3.23	3.08	3.00	2.92	2.84	2.75	2.66	2.57
19	8.19	5.93	5.01	4.50	4.17	3.94	3.77	3.63	3.52	3.43	3.30	3.15	3.00	2.92	2.84	2.76	2.67	2.58	2.49
20	8.10	5.85	4.94	4.43	4.10	3.87	3.70	3.56	3.46	3.37	3.23	3.09	2.94	2.86	2.78	2.69	2.61	2.52	2.42
21	8.02	5.78	4.87	4.37	4.04	3.81	3.64	3.51	3.40	3.31	3.17	3.03	2.88	2.80	2.72	2.64	2.55	2.46	2.36
22	7.95	5.72	4.82	4.31	3.99	3.76	3.59	3.45	3.35	3.26	3.12	2.98	2.83	2.75	2.67	2.58	2.50	2.40	2.31
23	7.88	5.66	4.76	4.26	3.94	3.71	3.54	3.41	3.30	3.21	3.07	2.93	2.78	2.70	2.62	2.54	2.45	2.35	2.26
24	7.82	5.61	4.72	4.22	3.90	3.67	3.50	3.36	3.26	3.17	3.03	2.89	2.74	2.66	2.58	2.49	2.40	2.31	2.21
25	7.77	5.57	4.68	4.18	3.86	3.63	3.46	3.32	3.22	3.13	2.99	2.85	2.70	2.62	2.53	2.45	2.36	2.27	2.17
30	7.56	5.39	4.51	4.02	3.70	3.47	3.30	3.17	3.07	2.98	2.84	2.70	2.55	2.47	2.39	2.30	2.21	2.11	2.01
40	7.31	5.18	4.31	3.83	3.51	3.29	3.12	2.99	2.89	2.80	2.66	2.52	2.37	2.29	2.20	2.11	2.02	1.92	1.80
60	7.08	4.98	4.13	3.65	3.34	3.12	2.95	2.82	2.72	2.63	2.50	2.35	2.20	2.12	2.03	1.94	1.84	1.73	1.60
120	6.85	4.79	3.95	3.48	3.17	2.96	2.79	2.66	2.56	2.47	2.34	2.19	2.03	1.95	1.86	1.76	1.66	1.53	1.38
∞	6.63	4.61	3.78	3.32	3.02	2.80	2.64	2.51	2.41	2.32	2.18	2.04	1.88	1.79	1.70	1.59	1.47	1.32	1.00

Degrees of freedom for denominator

[†] Reproduced from M. Merrington and C. M. Thompson, "Tables of percentage points of the inverted beta (F) distribution," *Biometrika*, vol. 33 (1943), by permission of the *Biometrika* trustees.

Statistical Tables

TABLE V(a) 0.95 Confidence Intervals for Proportions[†]

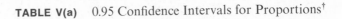

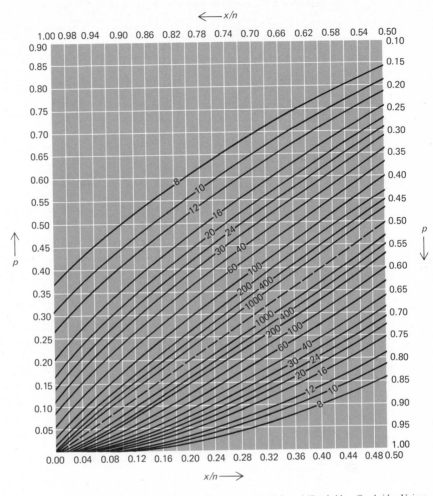

[†] Reproduced from Table 41 of *Biometrika Tables for Statisticians, Volume I* (Cambridge: Cambridge University Press, 1954), by permission of the *Biometrika* trustees.

TABLE V(b) 0.99 Confidence Intervals for Proportions[†]

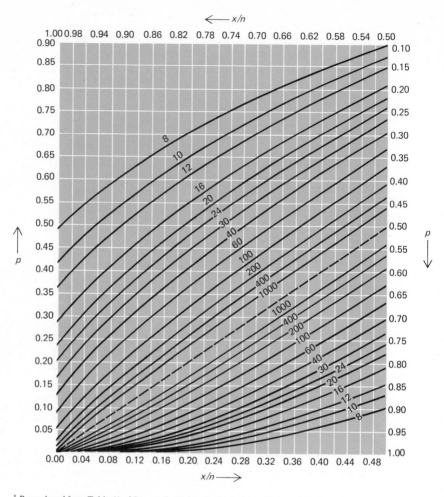
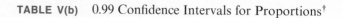

[†] Reproduced from Table 41 of *Biometrika Tables for Statisticians, Volume I* (Cambridge: Cambridge University Press, 1954), by permission of the *Biometrika* trustees.

TABLE VI   Binomial Probabilities

							p					
n	x	0.05	0.1	0.2	0.3	0.4	0.5	0.6	0.7	0.8	0.9	0.95
2	0	0.902	0.810	0.640	0.490	0.360	0.250	0.160	0.090	0.040	0.010	0.002
	1	0.095	0.180	0.320	0.420	0.480	0.500	0.480	0.420	0.320	0.180	0.095
	2	0.002	0.010	0.040	0.090	0.160	0.250	0.360	0.490	0.640	0.810	0.902
3	0	0.857	0.729	0.512	0.343	0.216	0.125	0.064	0.027	0.008	0.001	
	1	0.135	0.243	0.384	0.441	0.432	0.375	0.288	0.189	0.096	0.027	0.007
	2	0.007	0.027	0.096	0.189	0.288	0.375	0.432	0.441	0.384	0.243	0.135
	3		0.001	0.008	0.027	0.064	0.125	0.216	0.343	0.512	0.729	0.857
4	0	0.815	0.656	0.410	0.240	0.130	0.062	0.026	0.008	0.002		
	1	0.171	0.292	0.410	0.412	0.346	0.250	0.154	0.076	0.026	0.004	
	2	0.014	0.049	0.154	0.265	0.346	0.375	0.346	0.265	0.154	0.049	0.014
	3		0.004	0.026	0.076	0.154	0.250	0.346	0.412	0.410	0.292	0.171
	4			0.002	0.008	0.026	0.062	0.130	0.240	0.410	0.656	0.815
5	0	0.774	0.590	0.328	0.168	0.078	0.031	0.010	0.002			
	1	0.204	0.328	0.410	0.360	0.259	0.156	0.077	0.028	0.006		
	2	0.021	0.073	0.205	0.309	0.346	0.312	0.230	0.132	0.051	0.008	0.001
	3	0.001	0.008	0.051	0.132	0.230	0.312	0.346	0.309	0.205	0.073	0.021
	4			0.006	0.028	0.077	0.156	0.259	0.360	0.410	0.328	0.204
	5				0.002	0.010	0.031	0.078	0.168	0.328	0.590	0.774
6	0	0.735	0.531	0.262	0.118	0.047	0.016	0.004	0.001			
	1	0.232	0.354	0.393	0.303	0.187	0.094	0.037	0.010	0.002		
	2	0.031	0.098	0.246	0.324	0.311	0.234	0.138	0.060	0.015	0.001	
	3	0.002	0.015	0.082	0.185	0.276	0.312	0.276	0.185	0.082	0.015	0.002
	4		0.001	0.015	0.060	0.138	0.234	0.311	0.324	0.246	0.098	0.031
	5			0.002	0.010	0.037	0.094	0.187	0.303	0.393	0.354	0.232
	6				0.001	0.004	0.016	0.047	0.118	0.262	0.531	0.735
7	0	0.698	0.478	0.210	0.082	0.028	0.008	0.002				
	1	0.257	0.372	0.367	0.247	0.131	0.055	0.017	0.004			
	2	0.041	0.124	0.275	0.318	0.261	0.164	0.077	0.025	0.004		
	3	0.004	0.023	0.115	0.227	0.290	0.273	0.194	0.097	0.029	0.003	
	4		0.003	0.029	0.097	0.194	0.273	0.290	0.227	0.115	0.023	0.004
	5			0.004	0.025	0.077	0.164	0.261	0.318	0.275	0.124	0.041
	6				0.004	0.017	0.055	0.131	0.247	0.367	0.372	0.257
	7					0.002	0.008	0.028	0.082	0.210	0.478	0.698
8	0	0.663	0.430	0.168	0.058	0.017	0.004	0.001				
	1	0.279	0.383	0.336	0.198	0.090	0.031	0.008	0.001			
	2	0.051	0.149	0.294	0.296	0.209	0.109	0.041	0.010	0.001		
	3	0.005	0.033	0.147	0.254	0.279	0.219	0.124	0.047	0.009		
	4		0.005	0.046	0.136	0.232	0.273	0.232	0.136	0.046	0.005	
	5			0.009	0.047	0.124	0.219	0.279	0.254	0.147	0.033	0.005
	6			0.001	0.010	0.041	0.109	0.209	0.296	0.294	0.149	0.051
	7				0.001	0.008	0.031	0.090	0.198	0.336	0.383	0.279
	8					0.001	0.004	0.017	0.058	0.168	0.430	0.663

All values omitted in this table are 0.0005 or less

TABLE VI   Binomial Probabilities (*Continued*)

p

n	x	0.05	0.1	0.2	0.3	0.4	0.5	0.6	0.7	0.8	0.9	0.95
9	0	0.630	0.387	0.134	0.040	0.010	0.002					
	1	0.299	0.387	0.302	0.156	0.060	0.018	0.004				
	2	0.063	0.172	0.302	0.267	0.161	0.070	0.021	0.004			
	3	0.008	0.045	0.176	0.267	0.251	0.164	0.074	0.021	0.003		
	4	0.001	0.007	0.066	0.172	0.251	0.246	0.167	0.074	0.017	0.001	
	5		0.001	0.017	0.074	0.167	0.246	0.251	0.172	0.066	0.007	0.001
	6			0.003	0.021	0.074	0.164	0.251	0.267	0.176	0.045	0.008
	7				0.004	0.021	0.070	0.161	0.267	0.302	0.172	0.063
	8					0.004	0.018	0.060	0.156	0.302	0.387	0.299
	9						0.002	0.010	0.040	0.134	0.387	0.630
10	0	0.599	0.349	0.107	0.028	0.006	0.001					
	1	0.315	0.387	0.268	0.121	0.040	0.010	0.002				
	2	0.075	0.194	0.302	0.233	0.121	0.044	0.011	0.001			
	3	0.010	0.057	0.201	0.267	0.215	0.117	0.042	0.009	0.001		
	4	0.001	0.011	0.088	0.200	0.251	0.205	0.111	0.037	0.006		
	5		0.001	0.026	0.103	0.201	0.246	0.201	0.103	0.026	0.001	
	6			0.006	0.037	0.111	0.205	0.251	0.200	0.088	0.011	0.001
	7			0.001	0.009	0.042	0.117	0.215	0.267	0.201	0.057	0.010
	8				0.001	0.011	0.044	0.121	0.233	0.302	0.194	0.075
	9					0.002	0.010	0.040	0.121	0.268	0.387	0.315
	10						0.001	0.006	0.028	0.107	0.349	0.599
11	0	0.569	0.314	0.086	0.020	0.004						
	1	0.329	0.384	0.236	0.093	0.027	0.005	0.001				
	2	0.087	0.213	0.295	0.200	0.089	0.027	0.005	0.001			
	3	0.014	0.071	0.221	0.257	0.177	0.081	0.023	0.004			
	4	0.001	0.016	0.111	0.220	0.236	0.161	0.070	0.017	0.002		
	5		0.002	0.039	0.132	0.221	0.226	0.147	0.057	0.010		
	6			0.010	0.057	0.147	0.226	0.221	0.132	0.039	0.002	
	7			0.002	0.017	0.070	0.161	0.236	0.220	0.111	0.016	0.001
	8				0.004	0.023	0.081	0.177	0.257	0.221	0.071	0.014
	9				0.001	0.005	0.027	0.089	0.200	0.295	0.213	0.087
	10					0.001	0.005	0.027	0.093	0.236	0.384	0.329
	11							0.004	0.020	0.086	0.314	0.569
12	0	0.540	0.282	0.069	0.014	0.002						
	1	0.341	0.377	0.206	0.071	0.017	0.003					
	2	0.099	0.230	0.283	0.168	0.064	0.016	0.002				
	3	0.017	0.085	0.236	0.240	0.142	0.054	0.012	0.001			
	4	0.002	0.021	0.133	0.231	0.213	0.121	0.042	0.008	0.001		
	5		0.004	0.053	0.158	0.227	0.193	0.101	0.029	0.003		
	6			0.016	0.079	0.177	0.226	0.177	0.079	0.016		
	7			0.003	0.029	0.101	0.193	0.227	0.158	0.053	0.004	
	8			0.001	0.008	0.042	0.121	0.213	0.231	0.133	0.021	0.002
	9				0.001	0.012	0.054	0.142	0.240	0.236	0.085	0.017
	10					0.002	0.016	0.064	0.168	0.283	0.230	0.099
	11						0.003	0.017	0.071	0.206	0.377	0.341
	12							0.002	0.014	0.069	0.282	0.540

TABLE VI   Binomial Probabilities (*Continued*)

n	x	0.05	0.1	0.2	0.3	0.4	0.5	0.6	0.7	0.8	0.9	0.95
13	0	0.513	0.254	0.055	0.010	0.001						
	1	0.351	0.367	0.179	0.054	0.011	0.002					
	2	0.111	0.245	0.268	0.139	0.045	0.010	0.001				
	3	0.021	0.100	0.246	0.218	0.111	0.035	0.006	0.001			
	4	0.003	0.028	0.154	0.234	0.184	0.087	0.024	0.003			
	5		0.006	0.069	0.180	0.221	0.157	0.066	0.014	0.001		
	6		0.001	0.023	0.103	0.197	0.209	0.131	0.044	0.006		
	7			0.006	0.044	0.131	0.209	0.197	0.103	0.023	0.001	
	8			0.001	0.014	0.066	0.157	0.221	0.180	0.069	0.006	
	9				0.003	0.024	0.087	0.184	0.234	0.154	0.028	0.003
	10				0.001	0.006	0.035	0.111	0.218	0.246	0.100	0.021
	11					0.001	0.010	0.045	0.139	0.268	0.245	0.111
	12						0.002	0.011	0.054	0.179	0.367	0.351
	13							0.001	0.010	0.055	0.254	0.513
14	0	0.488	0.229	0.044	0.007	0.001						
	1	0.359	0.356	0.154	0.041	0.007	0.001					
	2	0.123	0.257	0.250	0.113	0.032	0.006	0.001				
	3	0.026	0.114	0.250	0.194	0.085	0.022	0.003				
	4	0.004	0.035	0.172	0.229	0.155	0.061	0.014	0.001			
	5		0.008	0.086	0.196	0.207	0.122	0.041	0.007			
	6		0.001	0.032	0.126	0.207	0.183	0.092	0.023	0.002		
	7			0.009	0.062	0.157	0.209	0.157	0.062	0.009		
	8			0.002	0.023	0.092	0.183	0.207	0.126	0.032	0.001	
	9				0.007	0.041	0.122	0.207	0.196	0.086	0.008	
	10				0.001	0.014	0.061	0.155	0.229	0.172	0.035	0.004
	11					0.003	0.022	0.085	0.194	0.250	0.114	0.026
	12					0.001	0.006	0.032	0.113	0.250	0.257	0.123
	13						0.001	0.007	0.041	0.154	0.356	0.359
	14							0.001	0.007	0.044	0.229	0.488
15	0	0.463	0.206	0.035	0.005							
	1	0.366	0.343	0.132	0.031	0.005						
	2	0.135	0.267	0.231	0.092	0.022	0.003					
	3	0.031	0.129	0.250	0.170	0.063	0.014	0.002				
	4	0.005	0.043	0.188	0.219	0.127	0.042	0.007	0.001			
	5	0.001	0.010	0.103	0.206	0.186	0.092	0.024	0.003			
	6		0.002	0.043	0.147	0.207	0.153	0.061	0.012	0.001		
	7			0.014	0.081	0.177	0.196	0.118	0.035	0.003		
	8			0.003	0.035	0.118	0.196	0.177	0.081	0.014		
	9			0.001	0.012	0.061	0.153	0.207	0.147	0.043	0.002	
	10				0.003	0.024	0.092	0.186	0.206	0.103	0.010	0.001
	11				0.001	0.007	0.042	0.127	0.219	0.188	0.043	0.005
	12					0.002	0.014	0.063	0.170	0.250	0.129	0.031
	13						0.003	0.022	0.092	0.231	0.267	0.135
	14							0.005	0.031	0.132	0.343	0.366
	15								0.005	0.035	0.206	0.463

Statistical Tables

TABLE VII   Values of $Z = \dfrac{1}{2} \cdot \ln \dfrac{1 + r}{1 - r}$

r	.00	.01	.02	.03	.04	.05	.06	.07	.08	.09
0.0	0.000	0.010	0.020	0.030	0.040	0.050	0.060	0.070	0.080	0.090
0.1	0.100	0.110	0.121	0.131	0.141	0.151	0.161	0.172	0.182	0.192
0.2	0.203	0.213	0.224	0.234	0.245	0.255	0.266	0.277	0.288	0.299
0.3	0.310	0.321	0.332	0.343	0.354	0.365	0.377	0.388	0.400	0.412
0.4	0.424	0.436	0.448	0.460	0.472	0.485	0.497	0.510	0.523	0.536
0.5	0.549	0.563	0.576	0.590	0.604	0.618	0.633	0.648	0.662	0.678
0.6	0.693	0.709	0.725	0.741	0.758	0.775	0.793	0.811	0.829	0.848
0.7	0.867	0.887	0.908	0.929	0.950	0.973	0.996	1.020	1.045	1.071
0.8	1.099	1.127	1.157	1.188	1.221	1.256	1.293	1.333	1.376	1.422
0.9	1.472	1.528	1.589	1.658	1.738	1.832	1.946	2.092	2.298	2.647

For negative values of r put a minus sign in front of the corresponding Z's, and vice versa.

TABLE VIII Binomial Coefficients

n	$\binom{n}{0}$	$\binom{n}{1}$	$\binom{n}{2}$	$\binom{n}{3}$	$\binom{n}{4}$	$\binom{n}{5}$	$\binom{n}{6}$	$\binom{n}{7}$	$\binom{n}{8}$	$\binom{n}{9}$	$\binom{n}{10}$
0	1										
1	1	1									
2	1	2	1								
3	1	3	3	1							
4	1	4	6	4	1						
5	1	5	10	10	5	1					
6	1	6	15	20	15	6	1				
7	1	7	21	35	35	21	7	1			
8	1	8	28	56	70	56	28	8	1		
9	1	9	36	84	126	126	84	36	9	1	
10	1	10	45	120	210	252	210	120	45	10	1
11	1	11	55	165	330	462	462	330	165	55	11
12	1	12	66	220	495	792	924	792	495	220	66
13	1	13	78	286	715	1287	1716	1716	1287	715	286
14	1	14	91	364	1001	2002	3003	3432	3003	2002	1001
15	1	15	105	455	1365	3003	5005	6435	6435	5005	3003
16	1	16	120	560	1820	4368	8008	11440	12870	11440	8008
17	1	17	136	680	2380	6188	12376	19448	24310	24310	19448
18	1	18	153	816	3060	8568	18564	31824	43758	48620	43758
19	1	19	171	969	3876	11628	27132	50388	75582	92378	92378
20	1	20	190	1140	4845	15504	38760	77520	125970	167960	184756

For $r > 10$ it may be necessary to make use of the identity $\binom{n}{r} = \binom{n}{n-r}$.

TABLE IX Random Numbers[†]

04433	80674	24520	18222	10610	05794	37515
60298	47829	72648	37414	75755	04717	29899
67884	59651	67533	68123	17730	95862	08034
89512	32155	51906	61662	64130	16688	37275
32653	01895	12506	88535	36553	23757	34209
95913	15405	13772	76638	48423	25018	99041
55864	21694	13122	44115	01601	50541	00147
35334	49810	91601	40617	72876	33967	73830
57729	32196	76487	11622	96297	24160	09903
86648	13697	63677	70119	94739	25875	38829
30574	47609	07967	32422	76791	39725	53711
81307	43694	83580	79974	45929	85113	72268
02410	54905	79007	54939	21410	86980	91772
18969	75274	52233	62319	08598	09066	95288
87863	82384	66860	62297	80198	19347	73234
68397	71708	15438	62311	72844	60203	46412
28529	54447	58729	10854	99058	18260	38765
44285	06372	15867	70418	57012	72122	36634
86299	83430	33571	23309	57040	29285	67870
84842	68668	90894	61658	15001	94055	36308
56970	83609	52098	04184	54967	72938	56834
83125	71257	60490	44369	66130	72936	69848
55503	52423	02464	26141	68779	66388	75242
47019	76273	33203	29608	54553	25971	69573
84828	32592	79526	29554	84580	37859	28504
68921	08141	79227	05748	51276	57143	31926
36458	96045	30424	98420	72925	40729	22337
95752	59445	36847	87729	81679	59126	59437
26768	47323	58454	56958	20575	76746	49878
42613	37056	43636	58085	06766	60227	96414
95457	30566	65482	25596	02678	54592	63607
95276	17894	63564	95958	39750	64379	46059
66954	52324	64776	92345	95110	59448	77249
17457	18481	14113	62462	02798	54977	48349
03704	36872	83214	59337	01695	60666	97410
21538	86497	33210	60337	27976	70661	08250
57178	67619	98310	70348	11317	71623	55510
31048	97558	94953	55866	96283	46620	52087
69799	55380	16498	80733	96422	58078	99643
90595	61867	59231	17772	67831	33317	00520
33570	04981	98939	78784	09977	29398	93896
15340	93460	57477	13898	48431	72936	78160
64079	42483	36512	56186	99098	48850	72527
63491	05546	67118	62063	74958	20946	28147
92003	63868	41034	28260	79708	00770	88643
52360	46658	66511	04172	73085	11795	52594
74622	12142	68355	65635	21828	39539	18988
04157	50079	61343	64315	70836	82857	35335
86003	60070	66241	32836	27573	11479	94114
41268	80187	20351	09636	84668	42486	71303

[†]Based on parts of *Table of 105,000 Random Decimal Digits*. Interstate Commerce Commission, Bureau of Transport Economics and Statistics, Washington, D.C.

TABLE IX   Random Numbers (*Continued*)

48611	62866	33963	14045	79451	04934	45576
78812	03509	78673	73181	29973	18664	04555
19472	63971	37271	31445	49019	49405	46925
51266	11569	08697	91120	64156	40365	74297
55806	96275	26130	47949	14877	69594	83041
77527	81360	18180	97421	55541	90275	18213
77680	58788	33016	61173	93049	04694	43534
15404	96554	88265	34537	38526	67924	40474
14045	22917	60718	66487	46346	30949	03173
68376	43918	77653	04127	69930	43283	35766
93385	13421	67957	20384	58731	53396	59723
09858	52104	32014	53115	03727	98624	84616
93307	34116	49516	42148	57740	31198	70336
04794	01534	92058	03157	91758	80611	45357
86265	49096	97021	92582	61422	75890	86442
65943	79232	45702	67055	39024	57383	44424
90038	94209	04055	27393	61517	23002	96560
97283	95943	78363	36498	40662	94188	18202
21913	72958	75637	99936	58715	07943	23748
41161	37341	81838	19389	80336	46346	91895
23777	98392	31417	98547	92058	02277	50315
59973	08144	61070	73094	27059	69181	55623
82690	74099	77885	23813	10054	11900	44653
83854	24715	48866	65745	31131	47636	45137
61980	34997	41825	11623	07320	15003	56774
99915	45821	97702	87125	44488	77613	56823
48293	86847	43186	42951	37804	85129	28993
33225	31280	41232	34750	91097	60752	69783
06846	32828	24425	30249	78801	26977	92074
32671	45587	79620	84831	38156	74211	82752
82096	21913	75544	55228	89796	05694	91552
51666	10433	10945	55306	78562	89630	41230
54044	67942	24145	42294	27427	84875	37022
66738	60184	75679	38120	17640	36242	99357
55064	17427	89180	74018	44865	53197	74810
69599	60264	84549	78007	88450	06488	72274
64756	87759	92354	78694	63638	80939	98644
80817	74533	68407	55862	32476	19326	95558
39847	96884	84657	33697	39578	90197	80532
90401	41700	95510	61166	33757	23279	85523
78227	90110	81378	96659	37008	04050	04228
87240	52716	87697	79433	16336	52862	69149
08486	10951	26832	39763	02485	71688	90936
39338	32169	03713	93510	61244	73774	01245
21188	01850	69689	49426	49128	14660	14143
13287	82531	04388	64693	11934	35051	68576
53609	04001	19648	14053	49623	10840	31915
87900	36194	31567	53506	34304	39910	79630
81641	00496	36058	75899	46620	70024	88753
19512	50277	71508	20116	79520	06269	74173

TABLE IX   Random Numbers (*Continued*)

24418	23508	91507	76455	54941	72711	39406
57404	73678	08272	62941	02349	71389	45605
77644	98489	86268	73652	98210	44546	27174
68366	65614	01443	07607	11826	91326	29664
64472	72294	95432	53555	96810	17100	35066
88205	37913	98633	81009	81060	33449	68055
98455	78685	71250	10329	56135	80647	51404
48977	36794	56054	59243	57361	65304	93258
93077	72941	92779	23581	24548	56415	61927
84533	26564	91583	83411	66504	02036	02922
11338	12903	14514	27585	45068	05520	56321
23853	68500	92274	87026	99717	01542	72990
94096	74920	25822	98026	05394	61840	83089
83160	82362	00350	98536	38155	42661	02363
97425	47335	69709	01386	74319	04318	99387
83951	11954	24317	20345	18134	90062	10761
93085	35203	05740	03206	92012	42710	34650
33762	83193	58045	89880	78101	44392	53767
49665	85397	85137	30496	23469	42846	94810
37541	82627	80051	72521	35342	56119	97190
22145	85304	35348	82854	55846	18076	12415
27153	08662	61078	52433	22184	33998	87436
00301	49425	66682	25442	83668	66236	79655
43815	43272	73778	63469	50083	70696	13558
14689	86482	74157	46012	97765	27552	49617
16680	55936	82453	19532	49988	13176	94219
86938	60429	01137	86168	78257	86249	46134
33944	29219	73161	46061	30946	22210	79302
16045	67736	18608	18198	19468	76358	69203
37044	52523	25627	63107	30806	80857	84383
61471	45322	35340	35132	42163	60332	98851
47422	21296	16785	66393	39249	51463	95963
24133	39719	14484	58613	88717	29289	77360
67253	67064	10748	16006	16767	57345	42285
62382	76941	01635	35829	77516	98468	51686
98011	16503	09201	03523	87192	66483	55649
37366	24386	20654	85117	74078	64120	04643
73587	83993	54176	05221	94119	20108	78101
33583	68291	50547	96085	62180	27453	18567
02878	33223	39109	49536	56199	05993	71201
91498	41673	17195	33175	04994	09879	70337
91127	19815	30219	55591	21725	43827	78862
12997	55013	18662	81724	24305	37661	18956
96098	13651	15393	69995	14762	69734	89150
97627	17837	10472	18983	28387	99781	52977
40064	47981	31484	76603	54088	91095	00010
16239	68743	71374	55863	22672	91609	51514
58354	24913	20435	30965	17453	65623	93058
52567	65085	60220	84641	18273	49604	47418
06236	29052	91392	07551	83532	68130	56970

TABLE IX   Random Numbers (*Continued*)

94620	27963	96478	21559	19246	88097	44926
60947	60775	73181	43264	56895	04232	59604
27499	53523	63110	57106	20865	91683	80688
01603	23156	89223	43429	95353	44662	59433
00815	01552	06392	31437	70385	45863	75971
83844	90942	74857	52419	68723	47830	63010
06626	10042	93629	37609	57215	08409	81906
56760	63348	24949	11859	29793	37457	59377
64416	29934	00755	09418	14230	62887	92683
63569	17906	38076	32135	19096	96970	75917
22693	35089	72994	04252	23791	60249	83010
43413	59744	01275	71326	91382	45114	20245
09224	78530	50566	49965	04851	18280	14039
67625	34683	03142	74733	63558	09665	22610
86874	12549	98699	54952	91579	26023	81076
54548	49505	62515	63903	13193	33905	66936
73236	66167	49728	03581	40699	10396	81827
15220	66319	13543	14071	59148	95154	72852
16151	08029	36954	03891	38313	34016	18671
43635	84249	88984	80993	55431	90793	62603
30193	42776	85611	57635	51362	79907	77364
37430	45246	11400	20986	43996	73122	88474
88312	93047	12088	86937	70794	01041	74867
98995	58159	04700	90443	13168	31553	67891
51734	20849	70198	67906	00880	82899	66065
88698	41755	56216	66852	17748	04963	54859
51865	09836	73966	65711	41699	11732	17173
40300	08852	27528	84648	79589	95295	72895
02760	28625	70476	76410	32988	10194	94917
78450	26245	91763	73117	33047	03577	62599
50252	56911	62693	73817	98693	18728	94741
07929	66728	47761	81472	44806	15592	71357
09030	39605	87507	85446	51257	89555	75520
56670	88445	85799	76200	21795	38894	58070
48140	13583	94911	13318	64741	64336	95103
36764	86132	12463	28385	94242	32063	45233
14351	71381	28133	68269	65145	28152	39087
81276	00835	63835	87174	42446	08882	27067
55524	86088	00069	59254	24654	77371	26409
78852	65889	32719	13758	23937	90740	16866
11861	69032	51915	23510	32050	52052	24004
67699	01009	07050	73324	06732	27510	33761
50064	39500	17450	18030	63124	48061	59412
93126	17700	94400	76075	08317	27324	72723
01657	92602	41043	05686	15650	29970	95877
13800	76690	75133	60456	28491	03845	11507
98135	42870	48578	29036	69876	86563	61729
08313	99293	00990	13595	77457	79969	11339
90974	83965	62732	85161	54330	22406	86253
33273	61993	88407	69399	17301	70975	99129

TABLE X   Random Normal Numbers[†]

1.801	0.459	1.102	−1.072	−0.336	0.942	−0.290	−0.716	1.396	−0.466
−0.175	−0.754	−0.134	1.231	1.483	−0.149	0.555	1.401	−1.142	0.205
−0.861	−1.460	0.526	0.239	−0.206	2.021	0.313	−0.253	−0.891	1.135
−0.577	0.335	−0.820	0.140	−0.333	0.426	0.209	−0.024	0.323	1.223
0.827	0.802	−0.457	0.560	0.643	−0.729	−0.249	0.338	−0.281	−1.804
−1.344	0.949	−1.459	−1.210	1.016	−0.148	−1.737	0.069	−1.185	0.040
1.476	1.262	−1.428	0.489	−0.523	−0.646	1.721	0.749	0.179	−0.922
0.527	−1.045	0.877	0.646	2.957	−0.972	−1.796	0.309	2.224	−0.070
−0.645	0.117	0.059	−0.080	−1.637	−0.746	1.256	2.520	−0.673	0.994
−0.514	−1.510	−0.714	−1.581	0.905	1.745	1.767	0.682	−0.648	−1.742
−0.656	−0.217	0.287	0.114	1.175	0.791	−0.263	−0.695	−1.348	1.239
−0.778	1.177	0.180	1.156	0.458	1.089	0.339	1.304	0.402	−0.831
0.352	−1.829	−0.645	0.236	0.641	0.920	−1.287	−0.187	−2.339	−0.237
1.352	−0.076	−1.962	0.827	0.252	1.621	0.770	1.324	0.488	−0.037
0.017	0.030	0.211	2.276	0.693	−1.733	0.773	0.652	−0.947	0.148
−0.218	−1.060	−0.553	1.043	2.305	0.380	−0.794	−1.498	1.088	−0.689
1.118	0.816	0.713	0.485	0.185	0.318	−1.050	0.110	0.563	1.177
−1.622	0.436	0.481	0.021	2.070	−0.845	−0.257	−0.680	−0.565	0.024
−1.103	−0.210	−1.088	−0.033	−1.022	0.366	−0.531	2.022	0.210	1.037
−0.677	−0.737	−0.950	−1.517	1.148	0.377	−0.397	−1.902	−0.748	−1.753
1.110	1.120	1.163	1.577	−1.172	−0.133	−0.213	0.154	−0.435	0.218
−0.278	0.569	0.586	1.523	−0.244	−0.170	−1.274	0.874	−1.020	−0.809
0.178	1.314	0.462	−0.253	−0.122	0.108	−1.256	−0.137	1.043	−0.135
0.312	−2.287	−0.655	−1.459	0.075	−0.457	−0.206	−0.326	0.489	−0.149
0.469	−2.066	−0.973	−1.009	−1.410	0.505	0.459	−0.572	−1.186	0.978
−0.730	1.650	0.760	−0.520	−0.671	−0.122	−0.324	−0.202	0.411	−2.103
0.834	0.280	0.744	0.598	0.122	−0.460	−1.310	−1.271	−0.917	0.650
−1.397	−1.053	0.412	1.286	−0.820	−0.371	0.826	−0.666	0.505	0.733
0.238	−0.668	1.861	0.051	0.460	0.079	1.008	−0.487	0.306	−0.061
0.102	−0.907	−0.833	1.103	−0.921	0.145	−0.904	−0.401	0.553	−1.422
−0.160	0.567	−0.638	0.355	0.427	−0.695	−0.846	0.359	1.500	−0.926
0.496	1.179	−0.776	0.511	−1.325	0.275	−0.130	−0.123	1.175	−0.102
0.307	−0.328	−2.474	−0.121	1.371	0.266	1.235	1.827	−0.296	−2.715
−0.559	0.523	1.264	−0.018	−2.791	0.139	1.515	1.976	0.173	−1.728
0.658	−0.261	0.004	−1.296	0.568	−1.215	0.104	0.178	1.126	1.134
−0.856	−2.278	−0.140	−0.164	1.416	−0.043	0.243	−1.399	−0.448	0.120
2.778	0.245	0.282	0.301	−1.506	1.805	1.798	1.078	1.629	−0.648
0.543	0.761	−2.038	−0.533	−0.594	1.742	0.487	1.432	−0.210	−0.358
−0.008	−0.445	−2.551	0.935	1.961	−0.270	−1.557	−1.318	−0.744	−0.860
−1.147	−1.151	−0.522	−2.118	−0.667	0.906	0.639	1.005	−0.480	−1.354
−0.851	0.585	0.672	0.481	−0.888	−0.480	0.041	0.345	−0.537	−0.589
0.023	0.609	0.623	0.356	0.279	−0.051	0.158	−0.353	0.776	0.102
−0.257	0.152	−1.413	0.175	0.149	−1.354	0.286	1.794	−0.571	−0.202
−0.421	−0.344	−0.803	0.832	0.256	−1.296	−1.390	0.379	0.955	0.366
−1.681	2.444	−1.025	1.178	−0.827	−0.200	0.727	0.778	0.169	−1.363
0.717	−1.666	1.071	−2.061	−1.367	−0.450	−0.038	−1.004	−1.240	0.901
−1.266	0.256	−1.312	−0.582	−0.351	−1.002	0.648	0.873	0.015	0.641
0.350	0.552	−1.549	−1.680	1.417	−0.769	−0.514	−1.900	1.017	−1.222
−0.186	0.006	0.148	0.560	−1.081	−0.637	−1.968	−0.623	0.009	−0.369
1.359	1.027	0.740	−2.067	0.543	1.099	0.543	0.064	0.589	−0.016

[†] Reproduced by permission from RAND CORPORATION, *A Million Random Digits with 100,000 Normal Deviates* (New York: The Free Press, 1955).

TABLE X   Random Normal Numbers (*Continued*)

0.048	1.040	−0.111	−0.120	1.396	−0.393	−0.220	0.422	0.233	0.197
−0.521	−0.563	−0.116	−0.512	−0.518	−2.194	2.261	0.461	−1.533	−1.836
−1.407	−0.213	0.948	−0.073	−1.474	−0.236	−0.649	1.555	1.285	−0.747
1.822	0.898	−0.691	0.972	−0.011	0.517	0.808	2.651	−0.650	0.592
1.346	−0.137	0.952	1.467	−0.352	0.309	0.578	−1.881	−0.488	−0.329
0.420	−1.085	−1.578	−0.125	1.337	0.169	0.551	−0.745	−0.588	1.810
−1.760	−1.868	0.677	0.545	1.465	0.572	−0.770	0.655	−0.574	1.262
−0.959	0.061	−1.260	−0.573	−0.646	−0.697	−0.026	−1.115	3.591	−0.519
0.561	−0.534	−1.730	−1.172	−0.261	−0.049	0.173	0.027	1.138	0.524
−0.717	0.254	0.421	−1.891	2.592	−1.443	−0.061	−2.520	−0.497	0.909
−2.097	−0.180	−1.298	−0.647	0.159	0.769	−0.735	−0.343	0.966	0.595
0.443	−0.191	0.705	0.420	−0.486	−1.038	−0.396	1.406	0.327	1.198
0.481	0.161	−0.044	−0.864	−0.587	−0.037	−1.304	−1.544	0.946	−0.344
−2.219	−0.123	−0.260	0.680	0.224	−1.217	0.052	0.174	0.692	−1.068
1.723	−0.215	−0.158	0.369	1.073	−2.442	−0.472	2.060	−3.246	−1.020
−0.937	1.253	0.321	−0.541	−0.648	0.265	1.487	−0.554	1.890	0.499
−0.568	−0.146	0.285	1.337	−0.840	0.361	−0.468	0.746	0.470	0.171
−1.717	−1.293	−0.556	−0.545	1.344	0.320	−0.087	0.418	1.076	1.669
−0.151	−0.266	0.920	−2.370	0.484	−1.915	−0.268	0.718	2.075	−0.975
2.278	−1.819	0.245	−0.163	0.980	−1.629	−0.094	−0.573	1.548	−0.896
−0.650	0.669	−0.761	0.154	0.872	0.914	−0.563	−1.434	−0.006	−0.975
−1.086	0.810	0.461	−0.528	2.130	−0.218	0.111	−0.412	−0.580	−1.487
−0.143	−1.196	−1.254	−0.133	0.937	−0.475	−2.348	0.618	−0.057	−0.710
−2.072	0.711	1.241	0.066	−0.341	0.356	1.220	0.431	0.263	−1.623
−0.394	−0.368	−2.108	0.605	0.485	2.068	0.687	−1.474	0.071	−1.196
0.174	−1.131	0.870	2.114	0.201	−0.373	−0.284	−0.234	−2.087	−1.304
0.020	0.102	−1.911	−1.132	1.267	0.420	0.791	1.548	−0.147	−0.453
0.297	0.449	−0.604	−0.858	−1.739	1.143	0.131	0.740	−1.596	0.165
1.160	0.253	0.716	−1.032	−0.595	−1.662	0.632	−0.315	−0.374	0.700
−0.351	−0.490	−0.632	−0.409	−0.116	−1.153	−0.266	−0.125	0.489	−0.366
−0.594	−0.214	−0.461	0.030	−0.595	−0.889	0.638	−0.488	0.418	−0.693
−1.882	1.890	−0.236	0.006	0.966	−0.723	0.229	−2.136	−1.017	−0.008
0.041	2.955	−1.526	2.114	−0.540	1.040	0.753	0.025	0.462	1.221
−0.403	1.237	−1.938	−1.704	−0.103	−0.346	1.214	0.826	0.336	−1.140
−0.068	0.599	0.192	1.503	−0.579	−1.485	−1.645	0.302	−1.348	0.553
−0.361	0.958	0.807	0.787	−0.547	−0.074	−1.378	−0.010	−1.096	0.789
−0.251	0.629	0.459	−0.165	0.016	0.489	−1.205	−0.260	−0.256	−0.399
−1.011	0.893	−0.741	−0.514	−0.576	−0.929	0.478	−0.374	1.950	−0.695
0.780	−2.464	−0.522	0.767	−1.657	−0.983	0.217	−0.529	−0.648	1.454
−0.712	−0.355	−0.564	1.052	−0.169	−0.410	1.543	−2.330	−0.008	−0.955
−0.612	−1.068	−0.644	−0.007	−0.835	0.623	0.093	0.105	−0.318	−0.228
−0.064	0.012	−0.676	0.349	0.303	1.539	0.792	−0.101	−0.344	−0.096
−0.379	1.504	2.375	0.498	−0.996	0.174	−1.268	−1.137	−0.618	0.173
1.145	−1.403	0.770	0.799	0.844	−1.361	−1.059	0.128	1.398	0.277
−0.117	0.585	−1.763	−0.632	0.239	−0.854	1.684	1.024	−0.067	−0.045
1.333	1.374	−0.515	−1.655	0.607	−0.885	−0.902	−1.010	−1.297	−0.139
−0.249	−0.747	1.044	−0.930	0.346	0.575	0.335	−1.159	−1.651	−1.642
−1.022	0.085	−1.441	−0.198	0.844	0.697	0.548	−0.080	0.656	0.443
−0.780	−0.534	−0.339	−0.642	−0.902	−0.827	0.071	−0.678	−0.359	−0.479
−0.687	−0.418	0.991	0.331	−1.003	0.061	−1.416	0.876	0.125	−2.246

TABLE XI   Logarithms

N	0	1	2	3	4	5	6	7	8	9
10	0000	0043	0086	0128	0170	0212	0253	0294	0334	0374
11	0414	0453	0492	0531	0569	0607	0645	0682	0719	0755
12	0792	0828	0864	0899	0934	0969	1004	1038	1072	1106
13	1139	1173	1206	1239	1271	1303	1335	1367	1399	1430
14	1461	1492	1523	1553	1584	1614	1644	1673	1703	1732
15	1761	1790	1818	1847	1875	1903	1931	1959	1987	2014
16	2041	2068	2095	2122	2148	2175	2201	2227	2253	2279
17	2304	2330	2355	2380	2405	2430	2455	2480	2504	2529
18	2553	2577	2601	2625	2648	2672	2695	2718	2742	2765
19	2788	2810	2833	2856	2878	2900	2923	2945	2967	2989
20	3010	3032	3054	3075	3096	3118	3139	3160	3181	3201
21	3222	3243	3263	3284	3304	3324	3345	3365	3385	3404
22	3424	3444	3464	3483	3502	3522	3541	3560	3579	3598
23	3617	3636	3655	3674	3692	3711	3729	3747	3766	3784
24	3802	3820	3838	3856	3874	3892	3909	3927	3945	3962
25	3979	3997	4014	4031	4048	4065	4082	4099	4116	4133
26	4150	4166	4183	4200	4216	4232	4249	4265	4281	4298
27	4314	4330	4346	4362	4378	4393	4409	4425	4440	4456
28	4472	4487	4502	4518	4533	4548	4564	4579	4594	4609
29	4624	4639	4654	4669	4683	4698	4713	4728	4742	4757
30	4771	4786	4800	4814	4829	4843	4857	4871	4886	4900
31	4914	4928	4942	4955	4969	4983	4997	5011	5024	5038
32	5051	5065	5079	5092	5105	5119	5132	5145	5159	5172
33	5185	5198	5211	5224	5237	5250	5263	5276	5289	5302
34	5315	5328	5340	5353	5366	5378	5391	5403	5416	5428
35	5441	5453	5465	5478	5490	5502	5514	5527	5539	5551
36	5563	5575	5587	5599	5611	5623	5635	5647	5658	5670
37	5682	5694	5705	5717	5729	5740	5752	5763	5775	5786
38	5798	5809	5821	5832	5843	5855	5866	5877	5888	5899
39	5911	5922	5933	5944	5955	5966	5977	5988	5999	6010
40	6021	6031	6042	6053	6064	6075	6085	6096	6107	6117
41	6128	6138	6149	6160	6170	6180	6191	6201	6212	6222
42	6232	6243	6253	6263	6274	6284	6294	6304	6314	6325
43	6335	6345	6355	6365	6375	6385	6395	6405	6415	6425
44	6435	6444	6454	6464	6474	6484	6493	6503	6513	6522
45	6532	6542	6551	6561	6571	6580	6590	6599	6609	6618
46	6628	6637	6646	6656	6665	6675	6684	6693	6702	6712
47	6721	6730	6739	6749	6758	6767	6776	6785	6794	6803
48	6812	6821	6830	6839	6848	6857	6866	6875	6884	6893
49	6902	6911	6920	6928	6937	6946	6955	6964	6972	6981
50	6990	6998	7007	7016	7024	7033	7042	7050	7059	7067
51	7076	7084	7093	7101	7110	7118	7126	7135	7143	7152
52	7160	7168	7177	7185	7193	7202	7210	7218	7226	7235
53	7243	7251	7259	7267	7275	7284	7292	7300	7308	7316
54	7324	7332	7340	7348	7356	7364	7372	7380	7388	7396

TABLE XI Logarithms (*Continued*)

N	0	1	2	3	4	5	6	7	8	9
55	7404	7412	7419	7427	7435	7443	7451	7459	7466	7474
56	7482	7490	7497	7505	7513	7520	7528	7536	7543	7551
57	7559	7566	7574	7582	7589	7597	7604	7612	7619	7627
58	7634	7642	7649	7657	7664	7672	7679	7686	7694	7701
59	7709	7716	7723	7731	7738	7745	7752	7760	7767	7774
60	7782	7789	7796	7803	7810	7818	7825	7832	7839	7846
61	7853	7860	7868	7875	7882	7889	7896	7903	7910	7917
62	7924	7931	7938	7945	7952	7959	7966	7973	7980	7987
63	7993	8000	8007	8014	8021	8028	8035	8041	8048	8055
64	8062	8069	8075	8082	8089	8096	8102	8109	8116	8122
65	8129	8136	8142	8149	8156	8162	8169	8176	8182	8189
66	8195	8202	8209	8215	8222	8228	8235	8241	8248	8254
67	8261	8267	8274	8280	8287	8293	8299	8306	8312	8319
68	8325	8331	8338	8344	8351	8357	8363	8370	8376	8382
69	8388	8395	8401	8407	8414	8420	8426	8432	8439	8445
70	8451	8457	8463	8470	8476	8482	8488	8494	8500	8506
71	8513	8519	8525	8531	8537	8543	8549	8555	8561	8567
72	8573	8579	8585	8591	8597	8603	8609	8615	8621	8627
73	8633	8639	8645	8651	8657	8663	8669	8675	8681	8686
74	8692	8698	8704	8710	8716	8722	8727	8733	8739	8745
75	8751	8756	8762	8768	8774	8779	8785	8791	8797	8802
76	8808	8814	8820	8825	8831	8837	8842	8848	8854	8859
77	8865	8871	8876	8882	8887	8893	8899	8904	8910	8915
78	8921	8927	8932	8938	8943	8949	8954	8960	8965	8971
79	8976	8982	8987	8993	8998	9004	9009	9015	9020	9025
80	9031	9036	9042	9047	9053	9058	9063	9069	9074	9079
81	9085	9090	9096	9101	9106	9112	9117	9122	9128	9133
82	9138	9143	9149	9154	9159	9165	9170	9175	9180	9186
83	9191	9196	9201	9206	9212	9217	9222	9227	9232	9238
84	9243	9248	9253	9258	9263	9269	9274	9279	9284	9289
85	9294	9299	9304	9309	9315	9320	9325	9330	9335	9340
86	9345	9350	9355	9360	9365	9370	9375	9380	9385	9390
87	9395	9400	9405	9410	9415	9420	9425	9430	9435	9440
88	9445	9450	9455	9460	9465	9469	9474	9479	9484	9489
89	9494	9499	9504	9509	9513	9518	9523	9528	9533	9538
90	9542	9547	9552	9557	9562	9566	9571	9576	9581	9586
91	9590	9595	9600	9605	9609	9614	9619	9624	9628	9633
92	9638	9643	9647	9652	9657	9661	9666	9671	9675	9680
93	9685	9689	9694	9699	9703	9708	9713	9717	9722	9727
94	9731	9736	9741	9745	9750	9754	9759	9763	9768	9773
95	9777	9782	9786	9791	9795	9800	9805	9809	9814	9818
96	9823	9827	9832	9836	9341	9845	9850	9854	9859	9863
97	9868	9872	9877	9881	9886	9890	9894	9899	9903	9908
98	9912	9917	9921	9926	9930	9934	9939	9943	9948	9952
99	9956	9961	9965	9969	9974	9978	9983	9987	9991	9996

TABLE XII   Values of e^{-x}

x	e^{-x}	x	e^{-x}	x	e^{-x}	x	e^{-x}
0.0	1.000	2.5	0.082	5.0	0.0067	7.5	0.00055
0.1	0.905	2.6	0.074	5.1	0.0061	7.6	0.00050
0.2	0.819	2.7	0.067	5.2	0.0055	7.7	0.00045
0.3	0.741	2.8	0.061	5.3	0.0050	7.8	0.00041
0.4	0.670	2.9	0.055	5.4	0.0045	7.9	0.00037
0.5	0.607	3.0	0.050	5.5	0.0041	8.0	0.00034
0.6	0.549	3.1	0.045	5.6	0.0037	8.1	0.00030
0.7	0.497	3.2	0.041	5.7	0.0033	8.2	0.00028
0.8	0.449	3.3	0.037	5.8	0.0030	8.3	0.00025
0.9	0.407	3.4	0.033	5.9	0.0027	8.4	0.00023
1.0	0.368	3.5	0.030	6.0	0.0025	8.5	0.00020
1.1	0.333	3.6	0.027	6.1	0.0022	8.6	0.00018
1.2	0.301	3.7	0.025	6.2	0.0020	8.7	0.00017
1.3	0.273	3.8	0.022	6.3	0.0018	8.8	0.00015
1.4	0.247	3.9	0.020	6.4	0.0017	8.9	0.00014
1.5	0.223	4.0	0.018	6.5	0.0015	9.0	0.00012
1.6	0.202	4.1	0.017	6.6	0.0014	9.1	0.00011
1.7	0.183	4.2	0.015	6.7	0.0012	9.2	0.00010
1.8	0.165	4.3	0.014	6.8	0.0011	9.3	0.00009
1.9	0.150	4.4	0.012	6.9	0.0010	9.4	0.00008
2.0	0.135	4.5	0.011	7.0	0.0009	9.5	0.00008
2.1	0.122	4.6	0.010	7.1	0.0008	9.6	0.00007
2.2	0.111	4.7	0.009	7.2	0.0007	9.7	0.00006
2.3	0.100	4.8	0.008	7.3	0.0007	9.8	0.00006
2.4	0.091	4.9	0.007	7.4	0.0006	9.9	0.00005

Table XIII contains the square roots of the numbers from 1.00 to 9.99, and also the square roots of these numbers multiplied by 10, spaced at intervals of 0.01. The square roots are all rounded to four decimals. To find the square root of any positive number rounded to three significant digits, we use the following rule in deciding whether to take the entry of the $\sqrt{n}$ or $\sqrt{10n}$ column:

> **Move the decimal point an even number of places to the right or to the left until a number greater than or equal to 1 but less than 100 is reached. If the resulting number is less than 10 go to the $\sqrt{n}$ column; if it is 10 or more go to the $\sqrt{10n}$ column.**

Thus, to find the square roots of 12,800, 379, and 0.0812, we go to the $\sqrt{n}$ column since the decimal point has to be moved, respectively, four places to the left, two places to the left, and two places to the right, to give 1.28, 3.79, and 8.12. Similarly, to find the square roots of 5,240, 0.281, and 0.0000259, we go to the $\sqrt{10n}$ column since the decimal point has to be moved, respectively, two places to the left, two places to the right, and six places to the right, to give 52.4, 28.1, and 25.9.

After we locate a square root in the appropriate column of Table XIII, we must be sure to get the decimal point in the right place. Here it will help to use the following rule:

> **Having previously moved the decimal point an even number of places to the left or right to get a number greater than or equal to 1 but less than 100, move the decimal point of the entry of the appropriate column in Table XIII half as many places in the opposite direction.**

For example, to determine the square root of 12,800, we first note that the decimal point has to be moved *four places to the left* to give 1.28. We then take the entry of the $\sqrt{n}$ column corresponding to 1.28, move its decimal point *two places to the right*, and get $\sqrt{12,800} = 113.14$. Similarly, to find the square root of 0.0000259, we note that the decimal point has to be moved *six places to the right* to give 25.9. We thus take the entry of the $\sqrt{10n}$ column corresponding to 2.59, move the decimal point *three places to the left*, and get $\sqrt{0.0000259} = 0.0050892$. In actual practice, if a number whose square root we want to find is rounded, the square root will have to be rounded to as many significant digits as the original number.

TABLE XIII Square Roots

n	$\sqrt{n}$	$\sqrt{10n}$	n	$\sqrt{n}$	$\sqrt{10n}$	n	$\sqrt{n}$	$\sqrt{10n}$
1.00	1.0000	3.1623	1.50	1.2247	3.8730	2.00	1.4142	4.4721
1.01	1.0050	3.1780	1.51	1.2288	3.8859	2.01	1.4177	4.4833
1.02	1.0100	3.1937	1.52	1.2329	3.8987	2.02	1.4213	4.4944
1.03	1.0149	3.2094	1.53	1.2369	3.9115	2.03	1.4248	4.5056
1.04	1.0198	3.2249	1.54	1.2410	3.9243	2.04	1.4283	4.5166
1.05	1.0247	3.2404	1.55	1.2450	3.9370	2.05	1.4318	4.5277
1.06	1.0296	3.2558	1.56	1.2490	3.9497	2.06	1.4353	4.5387
1.07	1.0344	3.2711	1.57	1.2530	3.9623	2.07	1.4387	4.5497
1.08	1.0392	3.2863	1.58	1.2570	3.9749	2.08	1.4422	4.5607
1.09	1.0440	3.3015	1.59	1.2610	3.9875	2.09	1.4457	4.5717
1.10	1.0488	3.3166	1.60	1.2649	4.0000	2.10	1.4491	4.5826
1.11	1.0536	3.3317	1.61	1.2689	4.0125	2.11	1.4526	4.5935
1.12	1.0583	3.3466	1.62	1.2728	4.0249	2.12	1.4560	4.6043
1.13	1.0630	3.3615	1.63	1.2767	4.0373	2.13	1.4595	4.6152
1.14	1.0677	3.3764	1.64	1.2806	4.0497	2.14	1.4629	4.6260
1.15	1.0724	3.3912	1.65	1.2845	4.0620	2.15	1.4663	4.6368
1.16	1.0770	3.4059	1.66	1.2884	4.0743	2.16	1.4697	4.6476
1.17	1.0817	3.4205	1.67	1.2923	4.0866	2.17	1.4731	4.6583
1.18	1.0863	3.4351	1.68	1.2961	4.0988	2.18	1.4765	4.6690
1.19	1.0909	3.4496	1.69	1.3000	4.1110	2.19	1.4799	4.6797
1.20	1.0954	3.4641	1.70	1.3038	4.1231	2.20	1.4832	4.6904
1.21	1.1000	3.4785	1.71	1.3077	4.1352	2.21	1.4866	4.7011
1.22	1.1045	3.4928	1.72	1.3115	4.1473	2.22	1.4900	4.7117
1.23	1.1091	3.5071	1.73	1.3153	4.1593	2.23	1.4933	4.7223
1.24	1.1136	3.5214	1.74	1.3191	4.1713	2.24	1.4967	4.7329
1.25	1.1180	3.5355	1.75	1.3229	4.1833	2.25	1.5000	4.7434
1.26	1.1225	3.5496	1.76	1.3266	4.1952	2.26	1.5033	4.7539
1.27	1.1269	3.5637	1.77	1.3304	4.2071	2.27	1.5067	4.7645
1.28	1.1314	3.5777	1.78	1.3342	4.2190	2.28	1.5100	4.7749
1.29	1.1358	3.5917	1.79	1.3379	4.2308	2.29	1.5133	4.7854
1.30	1.1402	3.6056	1.80	1.3416	4.2426	2.30	1.5166	4.7958
1.31	1.1446	3.6194	1.81	1.3454	4.2544	2.31	1.5199	4.8062
1.32	1.1489	3.6332	1.82	1.3491	4.2661	2.32	1.5232	4.8166
1.33	1.1533	3.6469	1.83	1.3528	4.2778	2.33	1.5264	4.8270
1.34	1.1576	3.6606	1.84	1.3565	4.2895	2.34	1.5297	4.8374
1.35	1.1619	3.6742	1.85	1.3601	4.3012	2.35	1.5330	4.8477
1.36	1.1662	3.6878	1.86	1.3638	4.3128	2.36	1.5362	4.8580
1.37	1.1705	3.7014	1.87	1.3675	4.3243	2.37	1.5395	4.8683
1.38	1.1747	3.7148	1.88	1.3711	4.3359	2.38	1.5427	4.8785
1.39	1.1790	3.7283	1.89	1.3748	4.3474	2.39	1.5460	4.8888
1.40	1.1832	3.7417	1.90	1.3784	4.3589	2.40	1.5492	4.8990
1.41	1.1874	3.7550	1.91	1.3820	4.3704	2.41	1.5524	4.9092
1.42	1.1916	3.7683	1.92	1.3856	4.3818	2.42	1.5556	4.9193
1.43	1.1958	3.7815	1.93	1.3892	4.3932	2.43	1.5588	4.9295
1.44	1.2000	3.7947	1.94	1.3928	4.4045	2.44	1.5620	4.9396
1.45	1.2042	3.8079	1.95	1.3964	4.4159	2.45	1.5652	4.9497
1.46	1.2083	3.8210	1.96	1.4000	4.4272	2.46	1.5684	4.9598
1.47	1.2124	3.8341	1.97	1.4036	4.4385	2.47	1.5716	4.9699
1.48	1.2166	3.8471	1.98	1.4071	4.4497	2.48	1.5748	4.9800
1.49	1.2207	3.8601	1.99	1.4107	4.4609	2.49	1.5780	4.9900

Statistical Tables

TABLE XIII   Square Roots (*Continued*)

n	$\sqrt{n}$	$\sqrt{10n}$	n	$\sqrt{n}$	$\sqrt{10n}$	n	$\sqrt{n}$	$\sqrt{10n}$
2.50	1.5811	5.0000	3.00	1.7321	5.4772	3.50	1.8708	5.9161
2.51	1.5843	5.0100	3.01	1.7349	5.4863	3.51	1.8735	5.9245
2.52	1.5875	5.0200	3.02	1.7378	5.4955	3.52	1.8762	5.9330
2.53	1.5906	5.0299	3.03	1.7407	5.5045	3.53	1.8788	5.9414
2.54	1.5937	5.0398	3.04	1.7436	5.5136	3.54	1.8815	5.9498
2.55	1.5969	5.0498	3.05	1.7464	5.5227	3.55	1.8841	5.9582
2.56	1.6000	5.0596	3.06	1.7493	5.5317	3.56	1.8868	5.9666
2.57	1.6031	5.0695	3.07	1.7521	5.5408	3.57	1.8894	5.9749
2.58	1.6062	5.0794	3.08	1.7550	5.5498	3.58	1.8921	5.9833
2.59	1.6093	5.0892	3.09	1.7578	5.5588	3.59	1.8947	5.9917
2.60	1.6125	5.0990	3.10	1.7607	5.5678	3.60	1.8974	6.0000
2.61	1.6155	5.1088	3.11	1.7635	5.5767	3.61	1.9000	6.0083
2.62	1.6186	5.1186	3.12	1.7664	5.5857	3.62	1.9026	6.0166
2.63	1.6217	5.1284	3.13	1.7692	5.5946	3.63	1.9053	6.0249
2.64	1.6248	5.1381	3.14	1.7720	5.6036	3.64	1.9079	6.0332
2.65	1.6279	5.1478	3.15	1.7748	5.6125	3.65	1.9105	6.0415
2.66	1.6310	5.1575	3.16	1.7776	5.6214	3.66	1.9131	6.0498
2.67	1.6340	5.1672	3.17	1.7804	5.6303	3.67	1.9157	6.0581
2.68	1.6371	5.1769	3.18	1.7833	5.6391	3.68	1.9183	6.0663
2.69	1.6401	5.1865	3.19	1.7861	5.6480	3.69	1.9209	6.0745
2.70	1.6432	5.1962	3.20	1.7889	5.6569	3.70	1.9235	6.0828
2.71	1.6462	5.2058	3.21	1.7916	5.6657	3.71	1.9261	6.0910
2.72	1.6492	5.2154	3.22	1.7944	5.6745	3.72	1.9287	6.0992
2.73	1.6523	5.2249	3.23	1.7972	5.6833	3.73	1.9313	6.1074
2.74	1.6553	5.2345	3.24	1.8000	5.6921	3.74	1.9339	6.1156
2.75	1.6583	5.2440	3.25	1.8028	5.7009	3.75	1.9365	6.1237
2.76	1.6613	5.2536	3.26	1.8055	5.7096	3.76	1.9391	6.1319
2.77	1.6643	5.2631	3.27	1.8083	5.7184	3.77	1.9416	6.1400
2.78	1.6673	5.2726	3.28	1.8111	5.7271	3.78	1.9442	6.1482
2.79	1.6703	5.2820	3.29	1.8138	5.7359	3.79	1.9468	6.1563
2.80	1.6733	5.2915	3.30	1.8166	5.7446	3.80	1.9494	6.1644
2.81	1.6763	5.3009	3.31	1.8193	5.7533	3.81	1.9519	6.1725
2.82	1.6793	5.3104	3.32	1.8221	5.7619	3.82	1.9545	6.1806
2.83	1.6823	5.3198	3.33	1.8248	5.7706	3.83	1.9570	6.1887
2.84	1.6852	5.3292	3.34	1.8276	5.7793	3.84	1.9596	6.1968
2.85	1.6882	5.3385	3.35	1.8303	5.7879	3.85	1.9621	6.2048
2.86	1.6912	5.3479	3.36	1.8330	5.7966	3.86	1.9647	6.2129
2.87	1.6941	5.3572	3.37	1.8358	5.8052	3.87	1.9672	6.2209
2.88	1.6971	5.3666	3.38	1.8385	5.8138	3.88	1.9698	6.2290
2.89	1.7000	5.3759	3.39	1.8412	5.8224	3.89	1.9723	6.2370
2.90	1.7029	5.3852	3.40	1.8439	5.8310	3.90	1.9748	6.2450
2.91	1.7059	5.3944	3.41	1.8466	5.8395	3.91	1.9774	6.2530
2.92	1.7088	5.4037	3.42	1.8493	5.8481	3.92	1.9799	6.2610
2.93	1.7117	5.4129	3.43	1.8520	5.8566	3.93	1.9824	6.2690
2.94	1.7146	5.4222	3.44	1.8547	5.8652	3.94	1.9849	6.2769
2.95	1.7176	5.4314	3.45	1.8574	5.8737	3.95	1.9875	6.2849
2.96	1.7205	5.4406	3.46	1.8601	5.8822	3.96	1.9900	6.2929
2.97	1.7234	5.4498	3.47	1.8628	5.8907	3.97	1.9925	6.3008
2.98	1.7263	5.4589	3.48	1.8655	5.8992	3.98	1.9950	6.3087
2.99	1.7292	5.4681	3.49	1.8682	5.9076	3.99	1.9975	6.3166

TABLE XIII Square Roots (*Continued*)

n	$\sqrt{n}$	$\sqrt{10n}$	n	$\sqrt{n}$	$\sqrt{10n}$	n	$\sqrt{n}$	$\sqrt{10n}$
4.00	2.0000	6.3246	4.50	2.1213	6.7082	5.00	2.2361	7.0711
4.01	2.0025	6.3325	4.51	2.1237	6.7157	5.01	2.2383	7.0781
4.02	2.0050	6.3403	4.52	2.1260	6.7231	5.02	2.2405	7.0852
4.03	2.0075	6.3482	4.53	2.1284	6.7305	5.03	2.2428	7.0922
4.04	2.0100	6.3561	4.54	2.1307	6.7380	5.04	2.2450	7.0993
4.05	2.0125	6.3640	4.55	2.1331	6.7454	5.05	2.2472	7.1063
4.06	2.0149	6.3718	4.56	2.1354	6.7528	5.06	2.2494	7.1134
4.07	2.0174	6.3797	4.57	2.1378	6.7602	5.07	2.2517	7.1204
4.08	2.0199	6.3875	4.58	2.1401	6.7676	5.08	2.2539	7.1274
4.09	2.0224	6.3953	4.59	2.1424	6.7750	5.09	2.2561	7.1344
4.10	2.0248	6.4031	4.60	2.1448	6.7823	5.10	2.2583	7.1414
4.11	2.0273	6.4109	4.61	2.1471	6.7897	5.11	2.2605	7.1484
4.12	2.0298	6.4187	4.62	2.1494	6.7971	5.12	2.2627	7.1554
4.13	2.0322	6.4265	4.63	2.1517	6.8044	5.13	2.2650	7.1624
4.14	2.0347	6.4343	4.64	2.1541	6.8118	5.14	2.2672	7.1694
4.15	2.0372	6.4420	4.65	2.1564	6.8191	5.15	2.2694	7.1764
4.16	2.0396	6.4498	4.66	2.1587	6.8264	5.16	2.2716	7.1833
4.17	2.0421	6.4576	4.67	2.1610	6.8337	5.17	2.2738	7.1903
4.18	2.0445	6.4653	4.68	2.1633	6.8411	5.18	2.2760	7.1972
4.19	2.0469	6.4730	4.69	2.1656	6.8484	5.19	2.2782	7.2042
4.20	2.0494	6.4807	4.70	2.1679	6.8557	5.20	2.2804	7.2111
4.21	2.0518	6.4885	4.71	2.1703	6.8629	5.21	2.2825	7.2180
4.22	2.0543	6.4962	4.72	2.1726	6.8702	5.22	2.2847	7.2250
4.23	2.0567	6.5038	4.73	2.1749	6.8775	5.23	2.2869	7.2319
4.24	2.0591	6.5115	4.74	2.1772	6.8848	5.24	2.2891	7.2388
4.25	2.0616	6.5192	4.75	2.1794	6.8920	5.25	2.2913	7.2457
4.26	2.0640	6.5269	4.76	2.1817	6.8993	5.26	2.2935	7.2526
4.27	2.0664	6.5345	4.77	2.1840	6.9065	5.27	2.2956	7.2595
4.28	2.0688	6.5422	4.78	2.1863	6.9138	5.28	2.2978	7.2664
4.29	2.0712	6.5498	4.79	2.1886	6.9210	5.29	2.3000	7.2732
4.30	2.0736	6.5574	4.80	2.1909	6.9282	5.30	2.3022	7.2801
4.31	2.0761	6.5651	4.81	2.1932	6.9354	5.31	2.3043	7.2870
4.32	2.0785	6.5727	4.82	2.1954	6.9426	5.32	2.3065	7.2938
4.33	2.0809	6.5803	4.83	2.1977	6.9498	5.33	2.3087	7.3007
4.34	2.0833	6.5879	4.84	2.2000	6.9570	5.34	2.3108	7.3075
4.35	2.0857	6.5955	4.85	2.2023	6.9642	5.35	2.3130	7.3144
4.36	2.0881	6.6030	4.86	2.2045	6.9714	5.36	2.3152	7.3212
4.37	2.0905	6.6106	4.87	2.2068	6.9785	5.37	2.3173	7.3280
4.38	2.0928	6.6182	4.88	2.2091	6.9857	5.38	2.3195	7.3348
4.39	2.0952	6.6257	4.89	2.2113	6.9929	5.39	2.3216	7.3417
4.40	2.0976	6.6332	4.90	2.2136	7.0000	5.40	2.3238	7.3485
4.41	2.1000	6.6408	4.91	2.2159	7.0071	5.41	2.3259	7.3553
4.42	2.1024	6.6483	4.92	2.2181	7.0143	5.42	2.3281	7.3621
4.43	2.1048	6.6558	4.93	2.2204	7.0214	5.43	2.3302	7.3689
4.44	2.1071	6.6633	4.94	2.2226	7.0285	5.44	2.3324	7.3756
4.45	2.1095	6.6708	4.95	2.2249	7.0356	5.45	2.3345	7.3824
4.46	2.1119	6.6783	4.96	2.2271	7.0427	5.46	2.3367	7.3892
4.47	2.1142	6.6858	4.97	2.2293	7.0498	5.47	2.3388	7.3959
4.48	2.1166	6.6933	4.98	2.2316	7.0569	5.48	2.3409	7.4027
4.49	2.1190	6.7007	4.99	2.2338	7.0640	5.49	2.3431	7.4095

TABLE XIII   Square Roots (*Continued*)

n	$\sqrt{n}$	$\sqrt{10n}$	n	$\sqrt{n}$	$\sqrt{10n}$	n	$\sqrt{n}$	$\sqrt{10n}$
5.50	2.3452	7.4162	6.00	2.4495	7.7460	6.50	2.5495	8.0623
5.51	2.3473	7.4229	6.01	2.4515	7.7524	6.51	2.5515	8.0685
5.52	2.3495	7.4297	6.02	2.4536	7.7589	6.52	2.5534	8.0747
5.53	2.3516	7.4364	6.03	2.4556	7.7653	6.53	2.5554	8.0808
5.54	2.3537	7.4431	6.04	2.4576	7.7717	6.54	2.5573	8.0870
5.55	2.3558	7.4498	6.05	2.4597	7.7782	6.55	2.5593	8.0932
5.56	2.3580	7.4565	6.06	2.4617	7.7846	6.56	2.5612	8.0994
5.57	2.3601	7.4632	6.07	2.4637	7.7910	6.57	2.5632	8.1056
5.58	2.3622	7.4699	6.08	2.4658	7.7974	6.58	2.5652	8.1117
5.59	2.3643	7.4766	6.09	2.4678	7.8038	6.59	2.5671	8.1179
5.60	2.3664	7.4833	6.10	2.4698	7.8102	6.60	2.5690	8.1240
5.61	2.3685	7.4900	6.11	2.4718	7.8166	6.61	2.5710	8.1302
5.62	2.3707	7.4967	6.12	2.4739	7.8230	6.62	2.5729	8.1363
5.63	2.3728	7.5033	6.13	2.4759	7.8294	6.63	2.5749	8.1425
5.64	2.3749	7.5100	6.14	2.4779	7.8358	6.64	2.5768	8.1486
5.65	2.3770	7.5166	6.15	2.4799	7.8422	6.65	2.5788	8.1548
5.66	2.3791	7.5233	6.16	2.4819	7.8486	6.66	2.5807	8.1609
5.67	2.3812	7.5299	6.17	2.4839	7.8549	6.67	2.5826	8.1670
5.68	2.3833	7.5366	6.18	2.4860	7.8613	6.68	2.5846	8.1731
5.69	2.3854	7.5432	6.19	2.4880	7.8677	6.69	2.5865	8.1792
5.70	2.3875	7.5498	6.20	2.4900	7.8740	6.70	2.5884	8.1854
5.71	2.3896	7.5565	6.21	2.4920	7.8804	6.71	2.5904	8.1915
5.72	2.3917	7.5631	6.22	2.4940	7.8867	6.72	2.5923	8.1976
5.73	2.3937	7.5697	6.23	2.4960	7.8930	6.73	2.5942	8.2037
5.74	2.3958	7.5763	6.24	2.4980	7.8994	6.74	2.5962	8.2098
5.75	2.3979	7.5829	6.25	2.5000	7.9057	6.75	2.5981	8.2158
5.76	2.4000	7.5895	6.26	2.5020	7.9120	6.76	2.6000	8.2219
5.77	2.4021	7.5961	6.27	2.5040	7.9183	6.77	2.6019	8.2280
5.78	2.4042	7.6026	6.28	2.5060	7.9246	6.78	2.6038	8.2341
5.79	2.4062	7.6092	6.29	2.5080	7.9310	6.79	2.6058	8.2401
5.80	2.4083	7.6158	6.30	2.5100	7.9373	6.80	2.6077	8.2462
5.81	2.4104	7.6223	6.31	2.5120	7.9436	6.81	2.6096	8.2523
5.82	2.4125	7.6289	6.32	2.5140	7.9498	6.82	2.6115	8.2583
5.83	2.4145	7.6354	6.33	2.5159	7.9561	6.83	2.6134	8.2644
5.84	2.4166	7.6420	6.34	2.5179	7.9624	6.84	2.6153	8.2704
5.85	2.4187	7.6485	6.35	2.5199	7.9687	6.85	2.6173	8.2765
5.86	2.4207	7.6551	6.36	2.5219	7.9750	6.86	2.6192	8.2825
5.87	2.4228	7.6616	6.37	2.5239	7.9812	6.87	2.6211	8.2885
5.88	2.4249	7.6681	6.38	2.5259	7.9875	6.88	2.6230	8.2946
5.89	2.4269	7.6746	6.39	2.5278	7.9937	6.89	2.6249	8.3006
5.90	2.4290	7.6811	6.40	2.5298	8.0000	6.90	2.6268	8.3066
5.91	2.4310	7.6877	6.41	2.5318	8.0062	6.91	2.6287	8.3126
5.92	2.4331	7.6942	6.42	2.5338	8.0125	6.92	2.6306	8.3187
5.93	2.4352	7.7006	6.43	2.5357	8.0187	6.93	2.6325	8.3247
5.94	2.4372	7.7071	6.44	2.5377	8.0250	6.94	2.6344	8.3307
5.95	2.4393	7.7136	6.45	2.5397	8.0312	6.95	2.6363	8.3367
5.96	2.4413	7.7201	6.46	2.5417	8.0374	6.96	2.6382	8.3427
5.97	2.4434	7.7266	6.47	2.5436	8.0436	6.97	2.6401	8.3487
5.98	2.4454	7.7330	6.48	2.5456	8.0498	6.98	2.6420	8.3546
5.99	2.4474	7.7395	6.49	2.5475	8.0561	6.99	2.6439	8.3606

TABLE XIII   Square Roots (*Continued*)

n	$\sqrt{n}$	$\sqrt{10n}$	n	$\sqrt{n}$	$\sqrt{10n}$	n	$\sqrt{n}$	$\sqrt{10n}$
7.00	2.6458	8.3666	7.50	2.7386	8.6603	8.00	2.8284	8.9443
7.01	2.6476	8.3726	7.51	2.7404	8.6660	8.01	2.8302	8.9499
7.02	2.6495	8.3785	7.52	2.7423	8.6718	8.02	2.8320	8.9554
7.03	2.6514	8.3845	7.53	2.7441	8.6776	8.03	2.8337	8.9610
7.04	2.6533	8.3905	7.54	2.7459	8.6833	8.04	2.8355	8.9666
7.05	2.6552	8.3964	7.55	2.7477	8.6891	8.05	2.8373	8.9722
7.06	2.6571	8.4024	7.56	2.7495	8.6948	8.06	2.8390	8.9778
7.07	2.6589	8.4083	7.57	2.7514	8.7006	8.07	2.8408	8.9833
7.08	2.6608	8.4143	7.58	2.7532	8.7063	8.08	2.8425	8.9889
7.09	2.6627	8.4202	7.59	2.7550	8.7121	8.09	2.8443	8.9944
7.10	2.6646	8.4261	7.60	2.7568	8.7178	8.10	2.8460	9.0000
7.11	2.6665	8.4321	7.61	2.7586	8.7235	8.11	2.8478	9.0056
7.12	2.6683	8.4380	7.62	2.7604	8.7293	8.12	2.8496	9.0111
7.13	2.6702	8.4439	7.63	2.7622	8.7350	8.13	2.8513	9.0167
7.14	2.6721	8.4499	7.64	2.7641	8.7407	8.14	2.8531	9.0222
7.15	2.6739	8.4558	7.65	2.7659	8.7464	8.15	2.8548	9.0277
7.16	2.6758	8.4617	7.66	2.7677	8.7521	8.16	2.8566	9.0333
7.17	2.6777	8.4676	7.67	2.7695	8.7579	8.17	2.8583	9.0388
7.18	2.6796	8.4735	7.68	2.7713	8.7636	8.18	2.8601	9.0443
7.19	2.6814	8.4794	7.69	2.7731	8.7693	8.19	2.8618	9.0499
7.20	2.6833	8.4853	7.70	2.7749	8.7750	8.20	2.8636	9.0554
7.21	2.6851	8.4912	7.71	2.7767	8.7807	8.21	2.8653	9.0609
7.22	2.6870	8.4971	7.72	2.7785	8.7864	8.22	2.8671	9.0664
7.23	2.6889	8.5029	7.73	2.7803	8.7920	8.23	2.8688	9.0719
7.24	2.6907	8.5088	7.74	2.7821	8.7977	8.24	2.8705	9.0774
7.25	2.6926	8.5147	7.75	2.7839	8.8034	8.25	2.8723	9.0830
7.26	2.6944	8.5206	7.76	2.7857	8.8091	8.26	2.8740	9.0885
7.27	2.6963	8.5264	7.77	2.7875	8.8148	8.27	2.8758	9.0940
7.28	2.6981	8.5323	7.78	2.7893	8.8204	8.28	2.8775	9.0995
7.29	2.7000	8.5381	7.79	2.7911	8.8261	8.29	2.8792	9.1049
7.30	2.7019	8.5440	7.80	2.7928	8.8318	8.30	2.8810	9.1104
7.31	2.7037	8.5499	7.81	2.7946	8.8374	8.31	2.8827	9.1159
7.32	2.7055	8.5557	7.82	2.7964	8.8431	8.32	2.8844	9.1214
7.33	2.7074	8.5615	7.83	2.7982	8.8487	8.33	2.8862	9.1269
7.34	2.7092	8.5674	7.84	2.8000	8.8544	8.34	2.8879	9.1324
7.35	2.7111	8.5732	7.85	2.8018	8.8600	8.35	2.8896	9.1378
7.36	2.7129	8.5790	7.86	2.8036	8.8657	8.36	2.8914	9.1433
7.37	2.7148	8.5849	7.87	2.8054	8.8713	8.37	2.8931	9.1488
7.38	2.7166	8.5907	7.88	2.8071	8.8769	8.38	2.8948	9.1542
7.39	2.7185	8.5965	7.89	2.8089	8.8826	8.39	2.8965	9.1597
7.40	2.7203	8.6023	7.90	2.8107	8.8882	8.40	2.8983	9.1652
7.41	2.7221	8.6081	7.91	2.8125	8.8938	8.41	2.9000	9.1706
7.42	2.7240	8.6139	7.92	2.8142	8.8994	8.42	2.9017	9.1761
7.43	2.7258	8.6197	7.93	2.8160	8.9051	8.43	2.9034	9.1815
7.44	2.7276	8.6255	7.94	2.8178	8.9107	8.44	2.9052	9.1869
7.45	2.7295	8.6313	7.95	2.8196	8.9163	8.45	2.9069	9.1924
7.46	2.7313	8.6371	7.96	2.8213	8.9219	8.46	2.9086	9.1978
7.47	2.7331	8.6429	7.97	2.8231	8.9275	8.47	2.9103	9.2033
7.48	2.7350	8.6487	7.98	2.8249	8.9331	8.48	2.9120	9.2087
7.49	2.7368	8.6545	7.99	2.8267	8.9387	8.49	2.9138	9.2141

TABLE XIII Square Roots (*Continued*)

n	$\sqrt{n}$	$\sqrt{10n}$	n	$\sqrt{n}$	$\sqrt{10n}$	n	$\sqrt{n}$	$\sqrt{10n}$
8.50	2.9155	9.2195	9.00	3.0000	9.4868	9.50	3.0822	9.7468
8.51	2.9172	9.2250	9.01	3.0017	9.4921	9.51	3.0838	9.7519
8.52	2.9189	9.2304	9.02	3.0033	9.4974	9.52	3.0854	9.7570
8.53	2.9206	9.2358	9.03	3.0050	9.5026	9.53	3.0871	9.7622
8.54	2.9223	9.2412	9.04	3.0067	9.5079	9.54	3.0887	9.7673
8.55	2.9240	9.2466	9.05	3.0083	9.5131	9.55	3.0903	9.7724
8.56	2.9257	9.2520	9.06	3.0100	9.5184	9.56	3.0919	9.7775
8.57	2.9275	9.2574	9.07	3.0116	9.5237	9.57	3.0935	9.7826
8.58	2.9292	9.2628	9.08	3.0133	9.5289	9.58	3.0952	9.7877
8.59	2.9309	9.2682	9.09	3.0150	9.5341	9.59	3.0968	9.7929
8.60	2.9326	9.2736	9.10	3.0166	9.5394	9.60	3.0984	9.7980
8.61	2.9343	9.2790	9.11	3.0183	9.5446	9.61	3.1000	9.8031
8.62	2.9360	9.2844	9.12	3.0199	9.5499	9.62	3.1016	9.8082
8.63	2.9377	9.2898	9.13	3.0216	9.5551	9.63	3.1032	9.8133
8.64	2.9394	9.2952	9.14	3.0232	9.5603	9.64	3.1048	9.8184
8.65	2.9411	9.3005	9.15	3.0249	9.5656	9.65	3.1064	9.8234
8.66	2.9428	9.3059	9.16	3.0265	9.5708	9.66	3.1081	9.8285
8.67	2.9445	9.3113	9.17	3.0282	9.5760	9.67	3.1097	9.8336
8.68	2.9462	9.3167	9.18	3.0299	9.5812	9.68	3.1113	9.8387
8.69	2.9479	9.3220	9.19	3.0315	9.5864	9.69	3.1129	9.8438
8.70	2.9496	9.3274	9.20	3.0332	9.5917	9.70	3.1145	9.8489
8.71	2.9513	9.3327	9.21	3.0348	9.5969	9.71	3.1161	9.8539
8.72	2.9530	9.3381	9.22	3.0364	9.6021	9.72	3.1177	9.8590
8.73	2.9547	9.3434	9.23	3.0381	9.6073	9.73	3.1193	9.8641
8.74	2.9563	9.3488	9.24	3.0397	9.6125	9.74	3.1209	9.8691
8.75	2.9580	9.3541	9.25	3.0414	9.6177	9.75	3.1225	9.8742
8.76	2.9597	9.3595	9.26	3.0430	9.6229	9.76	3.1241	9.8793
8.77	2.9614	9.3648	9.27	3.0447	9.6281	9.77	3.1257	9.8843
8.78	2.9631	9.3702	9.28	3.0463	9.6333	9.78	3.1273	9.8894
8.79	2.9648	9.3755	9.29	3.0480	9.6385	9.79	3.1289	9.8944
8.80	2.9665	9.3808	9.30	3.0496	9.6437	9.80	3.1305	9.8995
8.81	2.9682	9.3862	9.31	3.0512	9.6488	9.81	3.1321	9.9045
8.82	2.9698	9.3915	9.32	3.0529	9.6540	9.82	3.1337	9.9096
8.83	2.9715	9.3968	9.33	3.0545	9.6592	9.83	3.1353	9.9146
8.84	2.9732	9.4021	9.34	3.0561	9.6644	9.84	3.1369	9.9197
8.85	2.9749	9.4074	9.35	3.0578	9.6695	9.85	3.1385	9.9247
8.86	2.9766	9.4128	9.36	3.0594	9.6747	9.86	3.1401	9.9298
8.87	2.9783	9.4181	9.37	3.0610	9.6799	9.87	3.1417	9.9348
8.88	2.9799	9.4234	9.38	3.0627	9.6850	9.88	3.1432	9.9398
8.89	2.9816	9.4287	9.39	3.0643	9.6902	9.89	3.1448	9.9448
8.90	2.9833	9.4340	9.40	3.0659	9.6954	9.90	3.1464	9.9499
8.91	2.9850	9.4393	9.41	3.0676	9.7005	9.91	3.1480	9.9549
8.92	2.9866	9.4446	9.42	3.0692	9.7057	9.92	3.1496	9.9599
8.93	2.9883	9.4499	9.43	3.0708	9.7108	9.93	3.1512	9.9649
8.94	2.9900	9.4552	9.44	3.0725	9.7160	9.94	3.1528	9.9700
8.95	2.9917	9.4604	9.45	3.0741	9.7211	9.95	3.1544	9.9750
8.96	2.9933	9.4657	9.46	3.0757	9.7263	9.96	3.1559	9.9800
8.97	2.9950	9.4710	9.47	3.0773	9.7314	9.97	3.1575	9.9850
8.98	2.9967	9.4763	9.45	3.0790	9.7365	9.98	3.1591	9.9900
8.99	2.9983	9.4816	9.49	3.0806	9.7417	9.99	3.1607	9.9950

In exercises involving extensive calculations, the reader may well get answers differing somewhat from those given here due to rounding at various intermediate stages.

Answers to Odd-Numbered Exercises

3 (a) Description; (b) description; (c) generalization; (d) generalization; (e) description.

5 (a) Nominal; (b) ratio; (c) interval; (d) ordinal; (e) ratio.

7 It is meaningless to divide the data; time measurements in minutes are ratio data.

1 (a) Yes; (b) no; (c) yes; (d) no; (e) yes; (f) yes; (g) no; (h) no.

3 (a) Cannot be found; (b) 68; (c) none; (d) cannot be found; (e) 57; (f) cannot be found; (g) cannot be found; (h) cannot be found.

5 Possible answer is 140.0–149.9, 150.0–159.9, 160.0–169.9, 170.0–179.9, 180.0–189.9, 190.0–199.9, 200.0–209.9, 210.0–219.9, 220.0–229.9.

7 1–5, 6–10, 11–15, 16–20, 21–25, 26–30, 31–35.

9 (a) The class frequencies are 2, 4, 14, 25, 22, 9, 4; (b) the percentages are 2.5, 5, 17.5, 31.25, 27.5, 11.25, 5; (d) the cumulative frequencies are 0, 2, 6, 20, 45, 67, 76, 80.

11 (a) −0.5, 19.5, 39.5, 59.5, 79.5, 99.5; (b) 9.5, 29.5, 49.5, 69.5, 89.5; (f) the cumulative frequencies are 0, 18, 69, 135, 167, 180.

15

Maneuverability	Number of cars
Excellent	3
Very good	9
Good	20
Fair	6
Poor	1
Very poor	1
Total	40

1 (a) The data would constitute a sample if we are interested, say, in the sales tax collected by all drug stores in that state; (b) the data would constitute a population if we are interested only in the sales tax collected in the given month by the drug stores in the given city.

3 (a) 8; (b) 9; (c) 10.

5 (a) 58.1; (b) 57.

7 70.

9 At most 8.

11 (a) The means are 33.4, 33.2, 32.6, so that car A performed best; (b) the medians are 33.6, 34.1, 32.1, so that car B performed best.

13 Independent is the modal choice.

17 (a) 77.12; (b) $13,180; (c) 4.9.

19 (b) $40.00; (c) 360 mph.

PAGES 48–50	1	Median is \$153.75 and range is \$142.60.
	3	(a) 1.56; (b) 1.56.
	5	(a) 3.775; (b) 3.775; if the same constant is subtracted from each value, the standard deviation remains unchanged.
	7	8.56.
	9	The chicken dinner is relatively most overpriced.
	11	The first person.
	13	(a) 0.23 and 0.29; (b) 0.22 and 0.30.

PAGES 55–57	1	(a) $\bar{x} = 7.8$ and $s = 5.7$; (b) same results.
	3	$\bar{x} = 46.75$ and $s^2 = 744.52$.
	5	(a) 117.5; (b) 117.5; (c) 170.6.
	7	(a) 14.95 and 22.83; (b) 2.97, 6.83, and 12.00; (c) 26.40, 43.09, and 62.01.
	9	(a) 13.81, 23.77, 9.35, and 28.06; (b) 25.40, 82.86, 11.14, 12.78, 95.46 and 97.48.

PAGE 60	1	0.53.
	3	0.

PAGES 61–62	1	(a) $x_1 + x_2 + x_3 + x_4 + x_5 + x_6$; (b) $y_1 + y_2 + y_3 + y_4 + y_5$; (c) $x_1y_1 + x_2y_2 + x_3y_3$; (d) $x_1f_1 + x_2f_2 + x_3f_3 + x_4f_4 + x_5f_5 + x_6f_6 + x_7f_7 + x_8f_8$; (e) $x_3^2 + x_4^2 + x_5^2 + x_6^2 + x_7^2$; (f) $(x_1 + y_1) + (x_2 + y_2) + (x_3 + y_3) + (x_4 + y_4)$.
	3	(a) 10; (b) 40.
	5	(a) 1; (b) 3; (c) 33; (d) 39; (e) 29.
	7	(a) 7, 4, -2, and 10; (b) 4, 8, and 7.

PAGES 64–68	1	(a) 1.8 and 1.5; (b) 1.58 and 5.
	3	Differences between successive grades are of equal importance; for instance, difference between A and B equals difference between C and D.
	5	\$247.29.
	7	(b) 13.4; (c) 13.8; (d) 40.7.
	9	(a) Ratio; (b) ordinal; (c) ratio; (d) nominal.
	11	(a) $\frac{13}{16}$; (b) $\frac{79}{125}$.
	13	(a) Ordinal; (b) interval.
	15	Ordinal data should not be added; their total earnings (prize money) provide a better means of comparison.
	17	96 percent.
	19	(a) If the dean evaluates the grading during that particular year; (b) if the dean compares faculty members' grading in general.
	21	(a) 146; (b) 146 and 149; (c) mode does not exist.

1 There are three possibilities.

5 There are six possibilities.

7 24.

9 (a) 48; (b) 12; (c) 24.

11 336.

13 (a) 3,024; (b) 126.

15 720.

17 (a) 60; (b) 20; (c) 120; (d) if we label the r objects $o_1, o_2, \ldots,$ and o_r, there are $n!$ permutations, but sets of $r!$ of these lead to the same permutation without subscripts.

19 66.

21 (a) 12,870; (b) 8,008; (c) 560.

23 (a) 364; (b) 91.

27 (a) 1, 6, 15, 20, 15, 6, 1; 1, 7, 21, 35, 35, 21, 7, 1; 1, 8, 28, 56, 70, 56, 28, 8, 1.

1 (a) $\frac{1}{26}$; (b) $\frac{3}{13}$; (c) $\frac{1}{2}$; (d) $\frac{4}{13}$.

3 (a) $\frac{11}{50}$; (b) $\frac{7}{10}$; (c) $\frac{2}{25}$; (d) $\frac{18}{25}$.

5 (a) $\frac{1}{3}$; (b) $\frac{1}{11}$; (c) $\frac{2}{3}$.

7 0.25.

9 (a) $\frac{1}{3}$; (b) 2 to 1; (c) the friend.

11 (a) The odds are 11 to 2 that they will not all be $1 bills; (b) the odds against rolling "7 or 11" are 7 to 2; (c) the odds are 3 to 1 that Tom will not be chosen.

13 The probability is $\frac{1}{6}$.

15 (a) The probability of drawing two black balls is $\frac{33}{58}$; (b) the probability is $\frac{13}{15}$ that she will make a mistake.

17 The probability is greater than $\frac{5}{6}$.

21 (a) Even bet; (b) even bet.

1 (a) Each salesperson sells at least one car; (b) second salesperson sells at least two cars; (c) first salesperson sells more cars than the second salesperson.

3 (b) K is the event that the number of graduate assistants exceeds the number of professors by one: L is the event that altogether four persons are supervising the laboratory; M is the event that all five graduate assistants are there; (c) (1, 2), (1, 3), (2, 2), (2, 3); at most three of the graduate assistants are there; (d) K and L are mutually exclusive; K and M are mutually exclusive; L and M are mutually exclusive.

5 (a) $\{A, D\}$; (b) $\{C, E\}$; (c) $\{B\}$.

7 (0, 0, 0), (1, 0, 0), (0, 1, 0), (0, 0, 1), (1, 1, 0), (1, 0, 1), (0, 1, 1), (1, 1, 1).

9 (a) Not mutually exclusive; (b) mutually exclusive; (c) not mutually exclusive; (d) not mutually exclusive; (e) mutually exclusive; (f) not mutually exclusive; (g) mutually exclusive.

11 (a) Gas consumption will be low, maintenance cost will be low, and the engine can be sold at a profit; (b) gas consumption will be low, maintenance cost will be low, but the engine cannot be sold at a profit; (c) gas consumption will not be low, maintenance cost will not be low, but the engine can be sold at a profit; (d) maintenance cost will be low and the engine can be sold at a profit; (e) gas consumption will be low, but the engine cannot be sold at a profit; (f) maintenance cost will not be low.

13 (a) 4; (b) 5; (c) 2 and 7; (d) 7 and 8.

PAGES 110–114

1 (a) The probability that the juvenile delinquent has not dropped out of school; (b) the probability that the juvenile delinquent's parents are not on welfare; (c) the probability that the juvenile delinquent has dropped out of school and his/her parents are on welfare; (d) the probability that the juvenile delinquent has dropped out of school or his/her parents are not on welfare; (e) the probability that the juvenile delinquent has not dropped out of school and his/her parents are not on welfare; (f) the probability that the juvenile delinquent has not dropped out of school or his/her parents are on welfare.

3 (a) Sum of the two probabilities does not equal 1; (b) one of the probabilities is negative; (c) $0.77 + 0.08$ does not equal 0.95; (d) $0.73 + 0.47$ exceeds 1; (e) $P(R \cup S)$ cannot be less than $P(R)$; (f) probability for more than 12 calls cannot exceed probability for more than 10 calls; (g) $P(E \cap F)$ cannot exceed $P(E)$; (h) $0.48 + 0.36 + 0.12$ is less than 1.

5 (a) 0.9998; (b) 0.0007; (c) 0.9993.

7 The probabilities are consistent.

9 (a) 0.29; (b) 0.79; (c) 0.63; (d) 0.71.

11 (a) 0.43; (b) 0.67; (c) 0.11; (d) 0.59.

13 $P(T) = \frac{1}{3}, P(U) = \frac{2}{5}, P(V) = \frac{3}{5}$, and $P(W) = \frac{1}{5}$.

15 (a) 0.54; (b) 0.65; (c) 0.60; (d) 0.14; (e) 0.75; (f) 0.40.

17 (a) 0.32; (b) 0.68.

19 First step is based on Postulate 3, second step is based on Postulate 2, and fourth step is based on Postulate 1.

PAGES 120–123

1 (a) The probability that she will find a job with a good future given that she will find a job with a high starting salary; (b) the probability that she will not find a job with a high starting salary given that she will find a job with a good future; (c) the probability that she will find a job with a high starting salary given that she will not find a job with a good future; (d) the probability that she will not find a job with a good future given that she will not find a job with a high starting salary.

3 (a) $\frac{3}{5}$; (b) $\frac{7}{10}$; (c) $\frac{1}{5}$; (d) $\frac{3}{10}$; (e) $\frac{1}{3}$; (f) $\frac{3}{7}$.

7 (a) 0.90; (b) 0.15.

9 (a) $\frac{8}{19}$; (b) $\frac{8}{23}$.

11 (a) $\frac{1}{7}$; (b) $\frac{12}{35}$; (c) $\frac{18}{35}$.

13 (a) $\frac{1}{64}$; (b) $\frac{1}{64}$; (c) 625/1,296.

1 0.758.

3 (a) 0.760;   (b) 0.7184.

5 0.805.

7 (a) 0.290;   (b) 0.387.

9 Purposeful action.

1 $0.75.

3 (a) $24,000 and $24,000;   (b) $26,000 and $22,000.

5 $5\frac{13}{16}$.

7 $1,720.

9 The probability is less than $\frac{9}{38}$.

11 The utility is greater than 4.

1 (b) Hotel II;   (c) hotel I.

3 Does not matter.

5 Stock 2.

7 (a) Discontinue;   (b) hotel II;   (c) site which is 18 miles from yard.

9 (a) $35.10 and $1.80; worthwhile;   (b) $1,035,000 and $945,000; worthwhile.

1 (a) 18;   (b) 19;   (c) 20.

3 $1,695.

1 4,200.

3 (a) 0.729;   (b) 0.125;   (c) 0.860.

5 0.57.

7 (b) Reject;   (c) no.

9 362,880.

11 0.16.

13 5 to 1.

15 (a) $\frac{18}{25}$;   (b) $\frac{2}{3}$;   (c) $\frac{4}{5}$;   (d) $\frac{20}{27}$;   (e) $\frac{11}{25}$;   (f) $\frac{10}{21}$.

17 (a) True;   (b) false;   (c) true;   (d) true.

19 0.66.

21 (a) $168,000;   (b) 0;   (c) delay.

25 (a) 120;   (b) 10;   (c) 90.

27 $26.

29 (a) Not take the raincoat;   (b) take the raincoat;   (c) does not matter.

31 0.63.

33 $p = \frac{5}{8}$; if p is the probability that player who starts with a dollars wins his friend's b dollars before he loses his own a dollars, then $p = \dfrac{a}{a + b}$.

35 (a) 0.55; (b) 0.95; (c) 0.50.

37 (a) $\frac{6}{11}$; (b) $\frac{1}{22}$; (c) $\frac{9}{22}$.

39 390,625.

PAGES 167–171

1 (a) No; sum of the probabilities exceeds 1; (b) yes; (c) yes;
(d) no; negative value; (e) no; sum of the probabilities is less than 1.

3 (a) $\frac{160}{729}$; (b) $\frac{64}{729}$; (c) $\frac{13}{729}$.

5 (a) 0.2128; (b) 0.213.

7 (a) 0.27648; (b) 0.276.

9 (a) 0.155; (b) 0.943 or 0.941, depending on whether result is obtained by addition or subtraction from 1; (c) 0.722 or 0.720, depending on whether result is obtained by addition or subtraction from 1; (d) 0.859.

11 0.000, 0.002, 0.011, 0.042, 0.111, 0.201, 0.251, 0.215, 0.121, 0.040, 0.006.

13 (a) $\frac{1}{7}$; (b) $\frac{18}{35}$; (c) $\frac{12}{35}$.

15 0.317.

17 (a) $\frac{5}{21}$; (b) $\frac{1}{84}$.

19 (a) 0.113; (b) 0.111.

21 0.938.

23 0.165, 0.297, and 0.267.

25 (a) 0.041; (b) 0.131; (c) 0.210; (d) 0.618.

PAGES 178–180

1 $\mu = 1$ and $\sigma^2 = 1$.

3 (a) 1; (b) 1.37.

5 $\mu = 6$ and $\sigma = 1.55$.

9 (b) $\mu = 2.5$ and $\sigma = 0.96$.

11 $\mu = 3.101$ and $\sigma^2 = 3.087$.

13 The probability is at least 0.96.

PAGE 184

3 (a) 00–13, 14–40, 41–67, 68–85, 86–94, 95–98, 99.

5 000–117, 118–419, 420–743, 744–928, 929–988, 989–998, 999.

7 (a) 0000–2465, 2466–5917, 5918–8334, 8335–9462, 9463–9857, 9858–9968, 9969–9994, 9995–9999.

PAGES 197–200

1 (a) $\frac{1}{4}$; (b) $\frac{5}{8}$; yes; (c) 0.7625.

3 (a) 0.3078; (b) 0.4515; (c) 0.3156: (d) 0.6064; (e) 0.9032; (f) 0.2148; (g) 0.1598; (h) 0.1068; (i) 0.6444.

5 (a) 1.92; (b) 2.22; (c) -0.74; (d) -0.50; (e) 1.12; (f) 1.44 or -1.44; (g) 2.17 or -2.17.

7 (a) 1.28; (b) 1.64 or 1.65; (c) 1.96; (d) 2.05; (e) 2.33; (f) 2.57 or 2.58.

9 20.

15 (a) 0.259, 0.121, and 0.223; (b) 0.593 and 0.053; (c) 0.670.

1 (a) 0.0749;   (b) 0.2615.
3 (a) 0.0030;   (b) 0.1587;   (c) 0.8664.
5 (a) 0.0336;   (b) 0.0099;   (c) 0.7745;   (d) 0.1127.
7 (a) 0.7190;   (b) 0.7074.
9 (a) 0.0649;   (b) 0.9306;   (c) 0.6664;   (d) 0.2872.
11 (a) 0.0823;   (b) 0.2033.
13 0.1841.
15 (a) 0.2358;   (b) 0.4908;   (c) 0.9556.

1 (a) 15;   (b) 45;   (c) 300.
3 (a) $\frac{1}{495}$;   (b) 1/26,334.
5 (a) $\frac{1}{15}$;   (b) $\frac{2}{3}$;   (c) $\frac{2}{5}$.
7 320, 495, 457, 040, 314, 488, 418, 431.

1 56, 4, 40, 19, and 10; 3, 7, 87, 100, and 68; 18, 110, 45, 75, and 134; 44, 79, 42, 10, and 10; 161, 43, 50, 113, and 6; 30, 31, 10, 37, and 62; 21, 67, 17, 21, and 42; 5, 48, 11, 119, and 27; 97, 15, 14, 7, and 79; 71, 35, 27, 43, and 1.

3 (a) 115 and 125, 115 and 135, 115 and 185, 115 and 195, 115 and 205, 125 and 135, 125 and 185, 125 and 195, 125 and 205, 135 and 185, 135 and 195, 135 and 205, 185 and 195, 185 and 205, 195 and 205; the means are 120, 125, 150, 155, 160, 130, 155, 160, 165, 160, 165, 170, 190, 195, 200; probability is $\frac{8}{15}$.   (b) 115 and 185, 115 and 195, 115 and 205, 125 and 185, 125 and 195, 125 and 205, 135 and 185, 135 and 195, 135 and 205; the means are 150, 155, 160, 155, 160, 165, 160, 165, 170; probability is $\frac{2}{9}$.

7 115 and 125, 115 and 135, 125 and 135, 185 and 195, 185 and 205, 195 and 205; the means are 120, 125, 130, 190, 195, 200; the probability is 1.

9 (a) 81 and 510; 216.47;   (b) 216.47.

1 (b) 5 and 6, 5 and 7, 5 and 8, 5 and 9, 5 and 10, 6 and 7, 6 and 8, 6 and 9, 6 and 10, 7 and 8, 7 and 9, 7 and 10, 8 and 9, 8 and 10, 9 and 10; the means are 5.5, 6, 6.5, 7 7.5, 6.5, 7, 7.5, 8, 7.5, 8, 8.5, 8.5, 9, 9.5;   (c) 5.5, 6, 9 and 9.5 have the probability $\frac{1}{15}$, 6.5, 7, 8, and 8.5 have the probability $\frac{2}{15}$, 7.5 has the probability $\frac{3}{15}$;   (d) 7.5 and 1.08.

3 (a) 5 and 5, 5 and 6, 5 and 7, 5 and 8, 5 and 9, 5 and 10, 6 and 5, 6 and 6, 6 and 7, 6 and 8, 6 and 9, 6 and 10, 7 and 5, 7 and 6, 7 and 7, 7 and 8, 7 and 9, 7 and 10, 8 and 5, 8 and 6, 8 and 7, 8 and 8, 8 and 9, 8 and 10, 9 and 5, 9 and 6, 9 and 7, 9 and 8, 9 and 9, 9 and 10, 10 and 5, 10 and 6, 10 and 7, 10 and 8, 10 and 9, 10 and 10; the means are 5, 5.5, 6, 6.5, 7, 7.5, 5.5, 6, 6.5, 7, 7.5, 8, 6, 6.5, 7, 7.5, 8, 8.5, 6.5, 7, 7.5, 8, 8.5, 9, 7, 7.5, 8, 8.5, 9, 9.5, 7.5, 8, 8.5, 9, 9.5, 10;   (b) 5 and 10 have the probability $\frac{1}{36}$, 5.5 and 9.5 have the probability $\frac{2}{36}$, 6 and 9 have the probability $\frac{3}{36}$, 6.5 and 8.5 have the probability $\frac{4}{36}$, 7 and 8 have the probability $\frac{5}{36}$, 7.5 has the probability $\frac{6}{36}$;   (c) $\mu_{\bar{x}} = 7.5$ and $\sigma_{\bar{x}} = 1.21$.

5 (a) Divided by 2;   (b) divided by 1.5;   (c) divided by 3;   (d) multiplied by 4.

7 (a) 1.74;   (b) 0.56.

9 (a) 21.0; (b) 7.1.

11 The frequencies corresponding to 11, 12, . . . , and 21 are, respectively, 1, 1, 4, 4, 12, 12, 5, 8, 1, 1, and 1; the standard deviation is 1.97.

13 (a) 3.64 and 1.41; (b) 1.265.

15 (a) At least 0.84; (b) 0.9876.

19 0.0918.

1 0.9332.

3 0.2109.

5 (b) 1 and 3, 1 and 5, 1 and 7, 3 and 5, 3 and 7, 5 and 7; the means are 2, 3, 4, 4, 5, 6; 2, 3, 5, and 6 have the probability $\frac{1}{6}$, and 4 has the probability $\frac{2}{6}$; (c) $\mu_{\bar{x}} = 4$ and $\sigma_{\bar{x}} = \sqrt{5/3}$.

7 (a) $n_1 = 40$, $n_2 = 120$, $n_3 = 200$, $n_4 = 80$; (b) $n_1 = 50$, $n_2 = 120$, $n_3 = 150$, $n_4 = 120$.

9 Supplier A.

11 (a) At least $\frac{5}{9}$; (b) 0.8664.

17 (a) 0.326; (b) 0.325.

19 515, 034, 275, 209, 041, 147, 268, 288, 234, 412, 308, 242, 573, 504, 337.

21 0.0867.

23 The probabilities that the two are right are now 0.403 and 0.597, respectively.

25 (a) 000–119, 120–326, 327–684, 685–846, 847–942, 943–985, 986–999.

27 (a) 0.9722; (b) 0.5762.

29 $\mu = 1.02$ and $\sigma^2 = 0.8596$.

31 (a) Many persons will not give honest answers about personal habits; (b) the location introduces a bias; (c) there may be a defect in the machine that affects every 20th can, every 10th can, etc.; (d) longer logs have greater chance of being selected.

33 $n_1 = 27$, $n_2 = 36$, $n_3 = 18$, $n_4 = 9$.

1 With probability 0.95, error is less than 0.45 minutes.

3 43.5 mm $< \mu <$ 46.1 mm.

5 (a) $3,881 $< \mu <$ $4,005; (b) with probability 0.95, error is less than $47.

7 0.9030.

9 (a) 8.02 $< \mu <$ 8.58 days; (b) 76.7 $< \mu <$ 79.5.

11 246.

13 61.96 $< \mu <$ 65.72 gallons.

15 (a) With probability 0.99, error is less than 0.69 micrograms; (b) 1.67 $< \mu <$ 2.85 micrograms.

17 With probability 0.99, error is less than 0.075 gram.

19 $127.90.

21 (a) 0.8985; (b) $\mu_1 = 70.0$; (c) it reduces the probability from 0.8985 to 0.0150.

1 Type I error, Type II error.

3 The antipollution device for cars is not effective.

7 (a) 0.16;

(b)

Value of μ	Probability of Type II error
37.5	0.000
38.5	0.01
39.5	0.08
40.5	0.32
41.5	0.67
43.5	0.67
44.5	0.32
45.5	0.08
46.5	0.01
47.5	0.000

9 (a) 0.02; (b) 0.32; (c) 0.08.

11 Null hypothesis $\mu = 15$ and alternative hypothesis $\mu < 15$; make modification only if null hypothesis is rejected.

1 $z = -1.25$; cannot reject the null hypothesis.

3 $z = 3.02$; reject the null hypothesis.

5 $z = 2.35$; (a) reject the null hypothesis $\mu = 2.41$ ounces; (b) cannot reject the null hypothesis $\mu = 2.41$ ounces.

7 $z = 1.57$; cannot reject the claim.

9 $t = -1.47$; cannot reject the null hypothesis.

11 $t = 0.56$; cannot reject the null hypothesis.

13 $t = -2.58$; no significant effect.

15 $t = 5.66$; reject the claim.

17 $z = 1.79$; difference is not significant.

19 $z = -2.12$; difference is not significant.

21 $t = 3.02$; difference is significant.

23 $t = 1.11$; difference is not significant.

25 (a) $z = 4.88$; reject the null hypothesis that the true average increase in height is 0.5 inch; (b) $t = 4.17$; substantiates the claim.

27 $t = 2.20$; difference is not significant.

1 $0.0027 < \sigma < 0.0077$.

3 $1.12 < \sigma < 4.39$.

5 $3.93 < \sigma < 5.85$.

7 $8.37 < \sigma < 10.98$.

Answers to Odd-Numbered Exercises

1 $\chi^2 = 12.00$; cannot reject the null hypothesis.

3 $\chi^2 = 92.45$; reject the null hypothesis.

5 $z = 2.53$; reject the null hypothesis.

7 $z = -2$; reject the null hypothesis.

9 $F = 1.40$; cannot reject the null hypothesis.

11 $F = 1.66$; cannot reject the null hypothesis.

1 $0.47 < p < 0.61$.

3 $0.20 < p < 0.52$.

5 $0.361 < p < 0.539$.

7 (a) $0.315 < p < 0.45$; (b) $0.313 < p < 0.447$.

9 The probability is 0.95 that the error is less than 0.109.

11 The probability is 0.95 that the error is less than 0.039.

13 2401.

15 (a) 1,068; (b) 897.

19 (a) 0.50, 0.45, and 0.05; (b) 0.53, 0.44, and 0.03.

21 0.145, 0.053, 0.039, 0.105, and 0.658.

1 0 or 9 or more.

3 Reject the dean's claim.

5 Cannot reject the claim.

7 $z = -1.30$; cannot reject the claim.

9 $z = -2.00$; (a) reject the null hypothesis; (b) cannot reject the null hypothesis.

11 $z = 1.39$; cannot reject the claim.

1 $\chi^2 = 41.6$; difference is significant.

3 $\chi^2 = 9.0$; reject the null hypothesis.

5 (a) $z = 2.95$; reject the null hypothesis; (b) $z = -3.22$; the two solicitations are not equally effective.

7 (a) $z = 1.81$; reject the null hypothesis, that is, the test item discriminates between well-prepared and poorly-prepared students; (b) $z = -2.71$; reject the null hypothesis.

9 $\chi^2 = 14.92$; differences are significant.

11 $\chi^2 = 5.88$; differences are not significant.

1 $\chi^2 = 5.37$; no significant differences among the three groups.

3 $\chi^2 = 20.13$; the therapy is effective.

5 $\chi^2 = 52.8$; there is a relationship.

7 $\chi^2 = 4.01$; no significant relationship.

11 $\chi^2 = 1.69$; null hypothesis cannot be rejected, that is, data support the claim.

13 $\chi^2 = 21.4$; reject the claim.

1 Cannot reject the claim.

3 Cannot reject the null hypothesis.

5 Reject the null hypothesis.

7 Cannot reject the null hypothesis.

9 $z = 2.55$; reject the null hypothesis.

1 $z = 0.024$: cannot reject the null hypothesis.

3 $z = 3.32$; there is a difference.

5 $z = 1.30$; cannot reject the claim.

7 (a) $z = 0.11$; cannot reject the null hypothesis; (b) $z = 2.83$; reject the null hypothesis, that is, the populations have unequal dispersions.

9 $H = 4.51$; cannot reject the null hypothesis.

1 $z = 0.51$; cannot reject hypothesis of randomness.

5 $z = -0.11$; cannot reject hypothesis of randomness.

7 $z = 3.05$; reject hypothesis of randomness.

1 (b) 0.5557, 0.0869, and 0.0021.

3 $t = 4.06$; reject the null hypothesis.

5 $\chi^2 = 11.77$; difference is significant.

7 $0.0083 < \sigma < 0.0165$.

9 (a) 4.80; (b) 5.14.

11 $z = -3.28$; difference is significant.

13 0.8926.

15 $\chi^2 = 11.6$; differences among the proportions are significant.

17 753.

19 (a) 0.6543; (b) 4,221; (c) 0.9736.

21 $t = 0.73$; difference is not significant.

23 (a) $\chi^2 = 3.63$; differences are not significant; (b) $\chi^2 = 8.67$; they are not all equally effective.

25 $H = 4.86$; null hypothesis cannot be rejected.

27 $z = -3.5$; the commission has grounds to proceed against the manufacturer.

29 $F = 3.04$; cannot reject the null hypothesis.

31 $\chi^2 = 8.37$; it is evidence that the manuscript is not the work of the historian.

33 (a) $p < 0.008$, and introduce the more threatening kind of letter only if the null hypothesis can be rejected; (b) $p > 0.008$, and introduce the more threatening kind of letter unless the null hypothesis is rejected.

35 $\chi^2 = 7.02$; difference is significant.

37 (a) $4.62 < \sigma < 7.80$; (b) $4.63 < \sigma < 7.76$.

39 Same set.

Answers to Odd-Numbered Exercises

41 $17.66 < \mu < 26.34$.

43 (a) With probability 0.95, error is less than 0.47; (b) $23.92 < \mu < 25.48$.

45 (a) $0.29 < \mu_1 - \mu_2 < 1.51$; (b) $-58.8 < \mu_1 - \mu_2 < 918.8$;
(c) $-16.4 < \mu_1 - \mu_2 < 1.6$; (d) $-0.19 < \mu_1 - \mu_2 < 1.99$.

47 (a) $-0.372 < \delta < -0.204$; (b) $-0.186 < \delta < 0.066$; (c) $-0.116 < \delta < -0.032$.

PAGES 368–371

1 (a) $y' = -7.12 + 26.64x$; (b) 46.16.

3 (a) $y' = 257.11 - 14.15x$; (b) 44.86.

5 (a) $y' = 0.49 + 0.27x$; (b) 10.75.

7 $y' = 1.90 - 0.086x$.

9 (a) $y' = 25.14x$; (b) $y' = 4.44x$.

PAGES 381–383

1 $y' = 101(0.96)^x$; 34.3.

3 (a) $x = 9$; \$1,980,000; (b) $y' = 300(1.47)^x$; 6,410.

5 $y' = 6{,}115x^{-1.49}$; 108.

7 (b) $y' = 4.22 - 0.11x + 0.05x^2$; $y' = 4.58$ for $x = 4$;
(c) $y' = 5{,}943.7 + 2{,}967.5x + 755.9x^2$.

9 $y' = 197.7 + 37.2x_1 - 0.12x_2$.

PAGES 388–390

1 $t = 1.70$; cannot reject the null hypothesis.

3 (a) $t = -1.91$; cannot reject the null hypothesis; (b) $t = 2.84$; reject the null hypothesis; (c) $t = -4.0$; reject the null hypothesis.

5 (a) $y' = 1.26 + 1.48x$; 3.04 percent; (b) $t = 2.33$; reject the null hypothesis that when no money is spent on advertising, average operating profits are 0.8 percent of total sales.

7 (a) $1.64 < \alpha < 2.16$; (b) $0.34 < \alpha < 2.18$.

9 (a) 48.4–97.2; (b) 33.53–64.17.

PAGES 401–405

1 $r = -0.01$; not significant.

3 (a) $r = 0.70$; significant; (b) 49 percent.

5 (a) $r = -0.98$; (b) $t = -9.85$; significant; (c) 96 percent.

7 (a) $r = -1$; (b) $r = 1$; yes.

9 (a) 0.13; (b) 0.52; (c) -0.78.

11 (a) 0.13; (b) 0.81; (c) 0.61; (d) -0.18; (e) only those of parts (b) and (c) are significant; (f) the effect is felt most strongly after two years, after that it diminishes.

13 $z = 0.93$; cannot reject the null hypothesis.

15 $z = -1.735$; reject the null hypothesis.

17 (a) $-0.96 < \rho < -0.60$; (b) $-0.15 < \rho < 0.75$; (c) $-0.26 < \rho < 0.53$.

1 $r' = 0.976$; significant.

3 $r' = 0.859$; significant.

5 $r' = 0.716$; significant.

7 0.999

9 (a) 1;   (b) 1.

1 (a) $F = 6.54$; reject the null hypothesis;

(b)

Source of variation	Degrees of freedom	Sum of squares	Mean square	F
Treatments	4	136	34	6.54
Error	15	78	5.2	
Total	19	214		

(c) There is a difference in the mileage yield of the gasolines when it is used in the given car under freeway driving conditions.

3 (a) We cannot differentiate between possible differences due to the designs and possible differences due to the different golf pros; in the language of experimental design, these two possible sources of variation are **confounded**; (b) the results, whatever they may be, apply only to the particular driver;   (c) observed differences among designs may be due rather to fatigue, wear of equipment, or perhaps changes in the weather;   (d) observed differences among the results obtained for the different designs may be due instead to the use of tees;   (e) observed differences among designs may be due rather to differences in playing conditions at the different tees.

5 $F = 0.68$; not significant.

7 $F = 10.79$; significant.

9 $F = 3.18$; cannot reject the null hypothesis.

1 (a) For the diet foods $F = 6.62$, which is significant;   (b) for the laboratories $F = 4.86$, which is significant.

3 (a) For the launchers $F = 0.003$, which is not significant;   (b) for the fuels $F = 0.35$, which is not significant.

5 For the threads $F = 8.31$, which is significant; for the technicians $F = 0.58$, which is not significant.

1 (a) 1 and I, 2 and I, 3 and I, 4 and I, 5 and I, 1 and II, 2 and II, 3 and II, 4 and II, 5 and II, 1 and III, 2 and III, 3 and III, 4 and III, 5 and III;   (b) 60.

3 $A_L B_L C_L D_L$, $A_L B_L C_L D_H$, $A_L B_L C_H D_L$, $A_L B_L C_H D_H$, $A_L B_H C_L D_L$, $A_L B_H C_L D_H$, $A_L B_H C_H D_L$, $A_L B_H C_H D_H$, $A_H B_L C_L D_L$, $A_H B_L C_L D_H$, $A_H B_L C_H D_L$, $A_H B_L C_H D_H$, $A_H B_H C_L D_L$, $A_H B_H C_L D_H$, $A_H B_H C_H D_L$, $A_H B_H C_H D_H$.

5   (a) For instructors $F = 94.17$, which is significant;   (b) for ethnic background $F = 2.56$, which is not significant;   (c) for professional interests $F = 27.02$, which is significant.

7   D on March 2, G on March 5, I on March 9, I on March 11, G on March 13, E on March 16.

PAGES 444–451

1   $F = 9.26$; significant.

3   0.99.

5   (b) 0.632 and 0.765;   (c) 0.444 and 0.561;   (d) 0.179 and 0.234.

7   (a) 1.34–1.60;   (b) 1.16–1.78.

9   $y' = 1{,}377.9 + 218.6x + 66.43x^2$.

11   (a)

A	B	C
C	A	B
B	C	A

(b) IaA, IbB, IcC, IIaC, IIbA, IIcB, IIIaB, IIIbC, and IIIcA;   (c) assign IaA, IIcB, and IIIbC to P; assign IbB, IIaC, and IIIcA to Q; assign IcC, IIbA, and IIIaB to R.

13   (a) $F = 8.24$; significant;   (b) $F = 2.31$; not significant;   (c) $F = 31.28$; significant.

15   $y' = 28.3(1.02)^x$; 74.2.

17   $r = -0.92$; significant.

19   (c) $F = 8.71$, which is significant.

21   $x' = 0.329 + 0.037y$; 4.029.

23   (a) $F = 12.57$; significant;   (b) $F = 11.17$; significant.

25   (a) $0 < \rho < 0.74$;   (b) $-0.65 < \rho < 0.10$.

Index

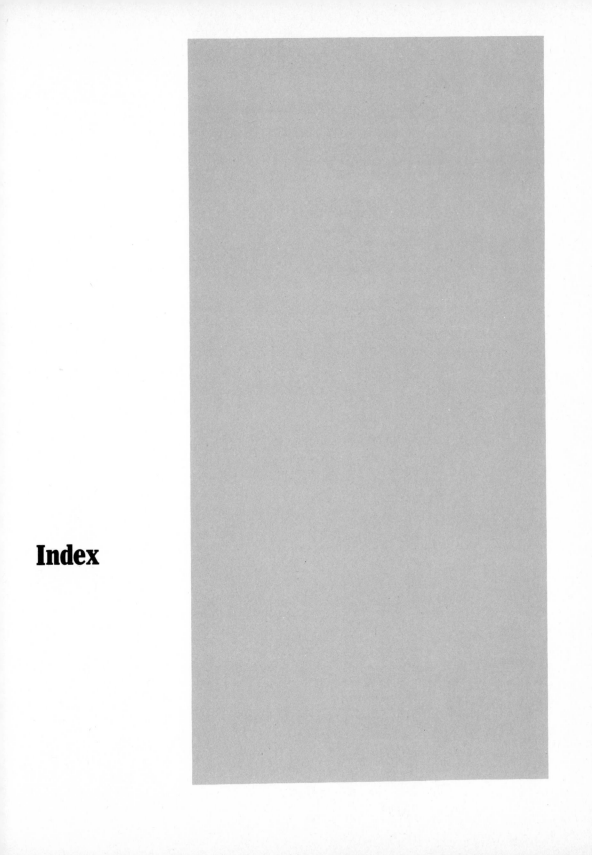

509

PROBLEMS OF ESTIMATION[†]

[†] Page numbers in italics refer to exercises.